Norbert **Wolnomiejski**
Zbigniew **Witek**

The Szczecin Lagoon Ecosystem:
The Biotic Community of the Great Lagoon and its Food Web Model

VERSITA

**Versita Discipline:
Life Sciences**

Published by Versita, Versita Ltd, 78 York Street, London W1H 1DP, Great Britain.

ISBN (paperback): 978-83-7656-048-9

ISBN (hardcover): 978-83-7656-049-6

ISBN (for electronic copy): 978-83-7656-050-2

Managing Editor: Elżbieta Jagielska
Katarzyna Michalczyk

Language Editor: Gaynor Ann Smith

www.versita.com

Cover illustration: © Adam Woźniczka

Published by [illegible] 2013 [illegible]

ISBN (paperback): 978-83 7656-0[illegible]

ISBN (hardcover): 978-83-[illegible]

[illegible]

[illegible]

[illegible]

Cover illustration © Adam [illegible]

Contents

Part 2. A MODEL OF THE GREAT LAGOON FOOD WEB
Zbigniew Witek and Norbert Wolnomiejski

List of Figures

List of Tables

Acknowledgements

We express our thanks to all co-workers for excellent cooperation during the sampling and laboratory analyses. Dr. Teresa Radziejewska (University of Szczecin) is warmly acknowledged for her contribution at the final stage of the manuscript preparation. This study was financed from the funds provided for statutory activities of the National Marine Fisheries Research Institute, Gdynia and the Pomeranian University, Słupsk.

Part 1. THE GREAT LAGOON BIOTA

Norbert Wolnomiejski and Zbigniew Witek

1.1 Introduction

The River Odra mouth area is a complex, multi-component physiographic system, shared by Poland and Germany, which provides a number of ecological, societal, and economic services (Tomaszewski, 1990). The Szczecin Lagoon, a major component of the system, is the main and final retention basin for the Odra water before it is discharged into the Baltic Sea. The Polish part of the Lagoon, called the Great Lagoon (Polish: *Zalew Wielki*), is of a great economic and ecological importance. In addition to supporting highly efficient commercial fisheries (50-60 kg fish ha^{-1} yr^{-1}), the Lagoon is a popular recreation and tourist area, and includes an important navigation route to Szczecin. In the 1960s-1990s, the area was regarded as having succumbed to advanced (and sometimes considered catastrophic) environmental degradation due to hypereutrophication and pollution. Later studies, including those of Wolnomiejski (1994) and Pastuszak et al. (2005) took the edge off those opinions and pointed to the persistently high ecological efficiency of the Lagoon's ecosystem and to its substantial homeostatic potential.

In 1998-2002, The Great Lagoon was a site of comprehensive hydrobiological research supported by the Polish Committee for Scientific Research project No. 6 PO4G 070 14 and by statutory funds of the Sea Fisheries Institute. The research supplied a wealth of data on the Lagoon's chemistry and major communities. The studies were carried out by researchers employed by the Sea Fisheries Institute's Research Station in Świnoujście, assisted by experts from the Nicolaus Copernicus University in Toruń and the University of Szczecin. The results obtained provided groundwork for a comprehensive assessment of the Great Lagoon's environment and biota, and for a multi-component analysis of the Lagoon's ecosystem functioning.

Part 1 of this work is aimed at summarising and discussing the knowledge obtained (its quantitative aspects in particular) on a number of communities in the Lagoon, including the assessment of the Lagoon's biological production. An important aspect of this study is that samples were collected each year at identical pre-determined time points (dates), thus providing 5- or 3-yr-long time series of data on major components of the producer and consumer communities in the Lagoon. Combined with auxiliary, literature data-based estimates of quantitative characteristics of certain other community components, the data collected may be treated as a basis on which to construct a food web model of the Polish part of the Szczecin Lagoon. Due to the paramount importance of the Lagoon for the functioning of the entire Odra-Pomeranian Bay system, such a model is urgently needed. When developed and tested, the model should substantially facilitate assessment of the overall biological production potential, biogeochemical macrofluxes, and ecological efficiency as well as matter and energy budgets and flows in the Lagoon's ecosystem. From the socio-economic standpoint, such a model should enhance the chance of appropriate use of natural resources of the area, particularly fish, and should come in aid of rational actions aimed at water quality improvement, in line with ecological, societal, and economic interests of the entire River Odra mouth area. Those are extremely important problems, as good water quality translates directly to a possibility of sustaining the richness of biological diversity and resources of the ecosystem as well as to the welfare of people using the waters in this German/Polish border region.

1.2 Area of Study - Habitat Characteristics

1.2.1 The Odra River Mouth System

The Odra River mouth, of which the Szczecin Lagoon is a major component, represents a complex system, composed of several elements. On the Polish side of the border it consists of branches of the downstream Odra reaches (the Lower Odra) with the deltaic Lake Dąbie, the Great Lagoon, the Świna strait with its main flow and side branches (including the storm delta), the Piastowski Canal, the Dziwna strait, and the Kamieński Lagoon. On the German side of the border, the system components include the Small Lagoon (German: *Kleines Haff*) and the Peene strait (*Peenestrom*) with two larger embayments. The Great Lagoon and the Small Lagoon are connected by a 7.5 km wide strait and together they constitute the main basin of the Szczecin Lagoon. The state border between Germany and Poland approximately matches the connection between the Great and Small Lagoons. Overall, according to Majewski (1980), the components listed above (but without Lake Dąbie and Lower Odra branches) cover 910 km^2 (Fig. 1.1). In the hydrographic sense, the *sensu lato* Odra 'estuarine' system also includes the southern part of the Pomeranian Bay.

The Szczecin Lagoon catchment covers almost 130 thou. km^2, 91.5% of which (i.e., almost 120 thou. km^2) being accounted for the River Odra catchment. The Odra contributes 97% to the total riverine input to the lagoon (Robakiewicz, 1993). In 1951-1995, the Odra flow rate averaged 574 m^3 s^{-1} (Fal et al., 2000), the average flow rate in 1998-2002 being very similar (Table 1.1).

Figure 1.1 The Odra River mouth area.

From the geological point of view, the Odra River mouth system can be classified as a flow-through coastal lagoon (Osadczuk et al., 2007). The hydrographic specificity of the system stems from a restricted connection with the Baltic Sea via three long straits (the Peene, the Świna, and the Dziwna). In the late 19th century, the Świna mouth has been additionally connected with the Lagoon through an artificial, navigable Piastowski Canal. The three straits

Table 1.1 Oder River water inflow in 1998-2002 (after Anon., 1998-2002).

Parameter \ Year	1998	1999	2000	2001	2002	Average	Coeff. of variation
Mean flow [$m^3 s^{-1}$]	599	602	463	617	631	**582**	**0.117**
Annual water discharge [km^3]	18.94	18.98	14.60	19.46	19.90	**18.38**	**0.117**

and the Canal serve as conduits for water exchange in both directions, from the Lagoon into the sea and back. Based on a series of measurements, Majewski (1980) found the Świna, the Peene, and the Dziwna to account for 78, 14, and 8% of the Odra discharge into the Baltic Sea, respectively. Almost identical results were produced by model calculations of Mohrholz and Lass (1998, in Tejakusuma, 2004).

As described by Mutko et al. (1994), the entire Odra mouth area is a physical entity functioning like a set of interconnected vessels. Similarly to other types of river mouth systems, the basic hydrographic regime and physico-chemical properties of water are controlled by riverine discharges, seawater intrusions, and mixing of the two types of water masses. Superimposed on those processes are the modulating effects of high water dynamics driven by wind- and ship-induced wave action, wind-driven circulation, and strong drift currents produced by storm surges following seawater intrusions and changes in the baric system. Seawater intrusions and their hydrological characteristics were analysed in detail by Chlubek (1975), Buchholz (1991), and Poleszczuk et al. (1993). The intrusions are a significant factor involved in environmental changes in the Lagoon and affect the biota via direct effects of increased salinity, water level rise, enhanced stagnation (up to total cessation of the riverine discharge over large areas), and intensified sediment metabolism (related to a higher ionic strength of a more saline water). Thus, increased salinity substantially modifies Lagoon's chemistry by intensifying recycling of certain sediment components and by depleting the oxygen content in the near-bottom layer when the flow stops.

The structure of flow fields in the Odra mouth area is highly complex and therefore its understanding requires mathematical modelling. Meteorological underpinnings of the variability in the local hydrography are included in, i.a., a three-dimensional hydrodynamic model (Kowalewska-Kalkowska et al., 2005) which combines a model of the Odra flow with a hydrographic model of the Baltic Sea. The resultant model allows to forecast, i.a., effects of sea water intrusions, and to reproduce, with a high degree of accuracy, individual phases of storm surges in the Lagoon.

1.2.2 The Great Lagoon

1.2.2.1 Morphology

As already mentioned, the key component of the Odra mouth system is the Szczecin Lagoon consisting of two major interconnected areas: the Small Lagoon (Kleines Haff) on the German side of the border, covering 277 km^2 and holding 1.02 km^3 water, and the Polish part, the Great Lagoon, with a surface area of 410 km^2 and water volume of 1.56 km^3 (Fig. 1.2). The above mentioned magnitudes of the surface area and capacity do not include other, minor parts of the mouth system, i.e., the Lower Odra, Lake Dąbie, the Świna straits, the Piastowski Canal, the Peenestrom with embayments, the Dziwna strait and the Kamieński Lagoon.

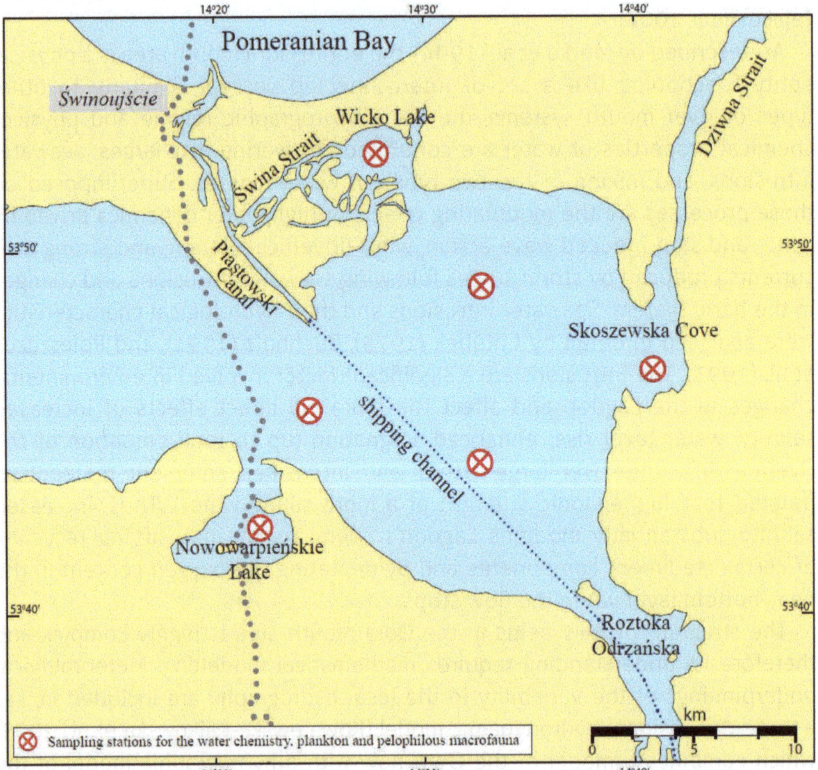

Figure 1.2 The Great Lagoon – the study area. Bottom depths at sampling stations were 5.5 – 6 m in the main basin, 4 m in Skoszewska Cove, 2.5 m in Lake Wicko, and 2 m in Lake Nowowarpieńskie.

The Great Lagoon features a number of more or less well defined sub-areas. The major part is the main basin (the open lagoon) covering 347 km^2, the remaining sub-areas being coves and embayments, i.e., the Roztoka Odrzańska (Odra mouth), Lake Nowowarpieńskie, the Skoszewska Cove, and Lake Wicko (Fig. 1.2). Longitudinally (from the Odra discharge into the Roztoka Odrzańska to the northern shore of Lake Wicko), the Great Lagoon extends along 33.5 km; the shoreline is 243 km long. The main basin is flanked by well-defined shallows (shoals). In the eastern part of the Lagoon, the shoals along the eastern shore are separated from the shore by a submerged ridge jutting up in places to 0.5 m below the water surface. The Lagoon and the Roztoka Odrzańska are intersected by a dredged, artificial shipping channel, its depth oscillating around 10 m.

The bottom of the main basin of the Great Lagoon is flat and featureless. The maximum natural depth is 8.5 m (close to the passage into the Kleines Haff), the average depth amounting to 3.8 m. Almost 96% of the Lagoon is shallower than 6 m, the area between the shore and the 2 m depth contour accounting for about 25% of the surface of the bottom (Majewski, 1980; Tejakusuma, 2004). The coves and embayments are very shallow (maximum depth of 2.0-4 m).

The main features of the Szczecin Lagoon have been described in detail by, i.a., Majewski (1980), Robakiewicz (1993) and by Radziejewska and Schernewski (2008).

> Data on sedimentary cover, water dynamics, physico-chemical parameters, and pollution presented further on pertain primarily to the **main basin** of the Great Lagoon which accounts for its major part (85% of the area and 92% of the water volume). Information pertaining to coves and embayments is only complementary or is used as an additional illustration of certain processes. This approach was necessary because the data published on the Lagoon, by the present authors and by others, concern almost exclusively the Lagoon's open part (the main basin).

1.2.2.2 The Bottom and the Sedimentary Cover

The bottom sediments in the Great Lagoon were described by, i.a., Wypych (1980), Osadczuk (2004), and Borówka et al. (2005). Silts cover the bottom in the central part of both the main basin and the coves. At a distance closer to the shores, the mud grades into sandy silts. The muddy sediment patches, up to the slopes of the shallows, are surrounded by very silty sands, in numerous places overlain by layers of bivalve shells (primarily those of *Dreissena polymorpha*). The slopes of the shallows are sandy. Near the shore, the bottom is sandy and/or shows the presence of deposits associated with those prevailing locally on the shores. Where the bottom is overgrown with macrophytes, it is covered by a layer of their remains. Maps of the Great Lagoon bottom sediments, along with data on grain size, organic matter and carbonate contents can be found in Wypych (1980), Osadczuk et al. (1996), Musielak et al. (1998), and Osadczuk (2004). Silts and sandy silts cover about 55% of the Great Lagoon bottom's surface (Wolnomiejski, 1994; Osadczuk, 2004; Tejakusuma, 2004). Those sediment types extend over 210 km^2 in the central basin and 15 km^2 in the embayments (totalling 225 km^2 of silt- and sandy silt-covered bottom).

The sediment-forming processes and bottom erosion remain in a specific equilibrium. The bathymetry of the Lagoon has been for a long time in a "climax"-like situation. The present isobath pattern in the Lagoon is very similar to that depicted in the German 1912 bathymetric chart of the area, no effects of shallowing being generally observed. Borówka et al. (2005) found it difficult to determine the contemporary sedimentation rate in the lagoonar sediment series due to bioturbation and sediment resuspension. In some areas, sedimentation may be altogether absent. The sediment-forming agents include autochthonous organic and inorganic matter derived from dead aquatic organisms, allochthonous particulates (bedload transport) brought in with the riverine runoff, materials originating from shore erosion, and dust supplied by atmospheric precipitation. As shown by, i.a., Falkowska (1985), Pęcherzewski (1991), and Woroń (1995), the coastal areas of the Baltic Sea are characterised by a particularly high atmospheric dust deposition rate. However, the importance of that deposition for the sediment formation has not been quantified. The budget developed by Nowak (1980) showed about 30 thou. t (dry weight) of river-borne (Odra runoff) suspended particulates that are deposited in the Great Lagoon annually, accounting for 10% of the total suspended particulate load transported by the Odra. According to Wiktor (1969), the entire zebra mussel (*Dreissena polymorpha*) population of the Great Lagoon "produces" 53 t of agglutinated deposit consisting of pseudofaeces and faeces per hour, and is thus an extremely important, additional sediment-forming agent. The zebra mussel plays also a great part in calcium carbonate deposition (about 79 thou. t yr^{-1}). It should be mentioned here that the present resources of the zebra mussel in the Lagoon (Woźniczka and Wolnomiejski, 2004) are very similar to those recorded by Wiktor (1969) in the 1950s-1960s.

Bottom erosion is a consequence of the very intensive hydrodynamics of the Lagoon, resulting in resuspension of the soft sediment and dampening resedimentation (mainly by wind-and ship movement-driven wave action), and in transporting the sediment in the form of suspended particulates (by wind-driven currents and eddies, drift currents, seawater incursions). Substantial amounts of sediments are deposited in the shipping channel and reduce its depth. According to Protasowicki et al. (1992 quoted by Poleszczuk, 1997), the maintenance of the channel requires removal of 1 million m^3 sediment a year.

As shown by Tejakusuma (2004), the critical wind speed required for producing waves which resuspend muddy sediment is 5-6 m s^{-1}. In addition, there is the ship-driven wave action triggered by ships moving along the channel (Nowak, 1980). Particularly prone to resuspension are the so-called liquid sediments, strongly hydrated, which form when the deposition rate is too high for the muddy deposits to consolidate, e.g., when water circulation slows down (e.g., Aller, 1998). Liquid sediments in lagoon-type systems, however, are mainly an outcome of peptisation of regularly textured muddy sediments, brought about by disintegration of micelles formed by organic particles and ions, which gelate the mud and fluidise it. This effect is associated with salinity changes (Poleszczuk, 1996; Wolnomiejski and Poleszczuk, 2003).

Over the period of study, sediment heavy metal pollution in the Great Lagoon was rather low. According to unpublished 2001 data of Poleszczuk, concentrations of three heavy metals (Zn, Cu, and Pb) in the topmost 5 cm mud layer were 55, 37, and 45 μg $g_{d.w.}^{-1}$. These values are several times lower than the critical concentrations reported by Skowronek (2011). However, the metal concentrations in samples collected from the same site, but on different days, were extremely variable: concentrations of Zn, Cu, and Pb were found to vary 28-, 12-, and 2.6 fold, respectively.

1.2.2.3 Water Mass Dynamics and Thermal Regime

Due to the large size and exposure of the Szczecin Lagoon, the wind action in the area is substantial, stronger than that in other inland water bodies. Furthermore the wind action is reinforced by the sea breezes. Air masses movements over the Lagoon remain unhampered in all directions (Robakiewicz, 1993). The Lagoon's water masses respond rapidly to changes in wind direction and speed. On the other hand, when not driven by the wind, the wave action ceases rapidly. As shown by Paszkiewicz and Jaworski (1980), the wind-driven wave action is accompanied by the so-called wave currents which enhance the drift current dynamics in the entire water layer. Poleszczuk (1997; 2007) discussed various types of post-wind water circulation patterns (including the Langmuir circulation) from the standpoint of their effect on the water column mixing. Robakiewicz (1993) and Tejakusuma (2004) described formation of large-scale eddies in the surface circulation in relation to wind direction and basin infilling. Podsetchine et al. (2004) published results of model studies on flow rates and currents in the Szczecin Lagoon as they relate to wind direction and speed. This is of a particular importance for, i.a., movement of the Odra water through the Lagoon (Radziejewska and Schernewski, 2008). Seawater incursions, changes of atmospheric pressure, and wind action induce local surges which in turn generate strong drift currents. They are an extremely important hydrodynamic factor playing a key role in mixing, disintegration of water column stratification, and intensification of sediment resuspension (Mutko, 1986; Buchholz, 1990).

The flood states of the Odra are of a minor importance for the Szczecin Lagoon. According to Buchholz (1990), the area between the downstream Odra branches, called the Międzyodrze, filled with polders and canals connecting river branches, intercept the peak flows and flatten the flood wave.

As shown in Table 1.2, the average water temperature near Świnoujście varies over time, the years 1999 and 2000 being the warmest. However, the ice cover duration showed little correlation with thermal conditions presented in the table. Ice conditions in the Lagoon vary extensively. Detailed data on ice conditions and local ice impacts were given by Girjatowicz (2005). Drifting ice fields are occasionally thrust onto the shore and change the shoreline of the low-lying areas. The ice-caused shore erosion has been observed to intensify in the recent years (Girjatowicz 2005). The shipping route along the channel is kept ice-free by icebreakers. There have also been years without ice. The general parameters of the Szczecin Lagoon ice cover are summarised in Table 1.2.

Table 1.2 Water temperature in the area of Świnoujście and general ice conditions in the Szczecin Lagoon in 1998-2001 (after Anon., 1998-2001).

Parameter / Year	1998	1999	2000	2001
Annual mean water temperature in the area of Świnoujście [°C]	9.5	10.4	10.5	9.7
Mean water temperature in the growing season (Apr.-Oct.) in the area of Świnoujście [°C]	14.4	15.7	15.1	14.8
Duration of ice season in the Szczecin Lagoon [days]	58	84	5	25
Number of days with ice cover in the Szczecin Lagoon	20	44	5	16

1.2.2.4 Suspended Particulate Input to the Lagoon

In the 1950s, the suspended particulate load carried by the Odra was 300 thou. t a year (Nowak, 1980). More recent estimates (e.g., Leipe et al., 1998 in Tejakusuma, 2004) set the annual load at 425 thou. t (assuming the Odra flow rate of 17 km^3 yr^{-1} and average suspended particulate concentration of 25 mg dm^{-3}). Tejakusuma (2004) published a table detailing the chemical composition of suspended particulates and proportions of various substances (including heavy metals) they contain. Periodic wind-induced eddies trap the suspension which is then gradually deposited on the bottom. Concentrations of suspended particulates in the Lagoon's water are strongly dependent on the wind speed.

Model simulations (Tejakusuma, 2004) showed deposition to dominate over transport at wind speed < 4.3 m s^{-1}, the two processes remaining in equilibrium at 4.3 m s^{-1}; at > 4.3 m s^{-1}, suspended particulates are not deposited, but are mainly transported away into the Pomeranian Bay instead. The overall budget shows the Szczecin Lagoon to be more of an "exporter" of the suspended particulates than a sink for them.

1.2.2.5 Physico-chemical Properties of Water

Generally, the water chemistry of the Lagoon and the dynamics of individual compounds over the year depend on the generally variable loads carried by the Odra, the basic factor functionally responsible for the trophic status and pollution of the area. The variability of nutrient concentrations and loads in the Odra in 1988-2006 was described by Pastuszak and Witek (2009a). Nutrient loads in 1998-2002 are shown in Table 1.3.

Table 1.3 Annual nutrient discharge by River Odra in 1998-2002 (after Anon., 1991-2003).

Nutrient \ Year	**1998**	**1999**	**2000**	**2001**	**2002**	**Average**	**Coeff. of variation**
Total phosphorus [t]	5 619	4 157	3 618	4 229	4 492	**4 423**	**0.167**
Total nitrogen [t]	87 335	77 758	56 142	67 233	79312	**73 556**	**0.164**

A certain role is played also by diffuse sources of chemical compounds in the Lagoon itself, seawater incursions, chemical processes affecting the water-sediment exchange (deposition and recycling), and excretion by aquatic organisms.

Owing to the highly dynamic mixing regime, the chemical stratification in the water column is very weak. Noticeable differences between the surface and the near-bottom layer are observable only during calm weather lasting longer than 3 days (Poleszczuk and Piesik, 2000) and during seawater incursions which inhibit the riverine through-flows; such conditions are, however, extremely rare. Consequently, differences in chemical parameters between the surface and the near-bottom layers, although analytically detectable (e.g., Poleszczuk et al., 2005), are in practice of a minor importance. The average values of chemical parameters and the Secchi depth during the period covered by this work are summarised in Table 1.4. Samples were collected from April to October (growing seasons) in monthly intervals using 2 dm^3 volume Ruttner sampler. Analytical procedures were those described by Poleszczuk and Sitek (1995).

Table 1.4 Average values of hydrochemical parameters and Secchi depth in the main basin of the Great Lagoon in growing seasons (April-October) of 1998-2002: surface layer, 0.5 m below water surface; bottom layer, about 0.5 m above bottom. s, standard deviation; v, coefficient of variation (s and v refer to growing season means). Range 1 refers to growing season mean values; Range 2 refers to values of individual samples.

Parameter	Layer	Mean	s	v	Range 1 min.	Range 1 max.	Range 2 min.	Range 2 max.
Salinity [PSU]	surface	0.8	0.18	0.22	0.6	0.9	0.2	1.6
	bottom	1.0	0.16	0.16	0.6	1.1	0.3	2.0
Oxygen [mg dm^{-3}]	surface	10.5	0.6	0.06	9.7	11.2	7.3	15.6
	bottom	9.6	0.6	0.06	9.1	10.6	7.1	12.8
Oxygen saturation (%)	surface	105	4.5	0.04	99	111	77	159
	bottom	95	5.6	0.06	89	104	67	122
Nitrates [μmol dm^{-3}]	surface	62	15.5	0.25	41	81	1.2	245
	bottom	60	17.3	0.29	37	80	1.0	242
Total nitrogen [μmol dm^{-3}]	surface	141	17.4	0.12	110	154	68	310
	bottom	139	19.1	0.14	106	154	63	324
Phosphates [μmol dm^{-3}]	surface	3.5	1.11	0.32	1.9	4.8	0.4	9.4
	bottom	3.6	1.22	0.34	1.9	4.9	0.4	9.5
Total phosphorus [μmol dm^{-3}]	surface	6.6	1.07	0.16	5.2	7.6	2.4	14.6
	bottom	7.1	1.05	0.15	5.6	8.0	2.0	12.8
BOD_5 [mg O_2 dm^{-3}]	surface	2.7	0.35	0.13	2.2	3.1	0.5	6.6
	bottom	2.2	0.34	0.15	2.0	2.8	0.9	4.6
pH	surface	8.5	0.11	0.01	8.3	8.6	7.8	9.0
	bottom	8.4	0.16	0.02	8.2	8.6	7.7	8.9
Secchi depth [m]		1.0	0.11	0.11	0.8	1.1	0.6	1.5

Table 1.5 provides data on chemical parameters, the Secchi depth, and the phytoplankton biomass (chlorophyll *a* concentration) for three areas of the Great Lagoon in 2001. The data illustrate direct relationships between some parameters (oxygen content, nitrate concentrations, BOD_5, Secchi depth) and the local phytoplankton concentrations.

Table 1.5 Water chemistry and phytoplankton biomass in three parts of the Great Lagoon (average values from the growing season (April-October) of 2001).

Parameter \ Part of the Great Lagoon	Main basin	Lake Nowowarpieńskie	Skoszewska Cove
Salinity [PSU]	0.7	1.0	0.5
Oxygen [mg dm^{-3}]	9.8	11.4	9.3
Oxygen saturation (%)	97	113	92
Nitrates [µmol dm^{-3}]	81	3	82
Total nitrogen [µmol dm^{-3}]	149	142	149
Phosphates [µmol dm^{-3}]	1.9	2.0	2.3
Total phosphorus [µmol dm^{-3}]	5.4	8.1	4.7
BOD_5 [mg O_2 dm^{-3}]	2.4	6.3	2.2
pH	8.2	8.9	8.1
Secchi depth [m]	1.0	0.6	2.0
Phytoplankton biomass [$g_{w.w.}$ dm^{-3}]	11.5	32.4	5.0

Salinity. During the period covered by this work (years 1998-2002), the salinity in the Lagoon was low, both in terms of average and maximum values. The salinity in the main basin averaged (over the period of study and across the water column) 0.9 PSU. Individual data points support the opinion of Mutko et al. (1994) about most of the salinity measurements showing the Lagoon to be limnic or oligohaline. It should be emphasised that the data referred to were collected during the growing seasons only. The salinity is usually at its lowest in spring (during the intensified riverine discharge), to increase in summer and, substantially so, in autumn. The highest salinity is observed in winter (Bangiel et al., 2004).

Oxygen concentration. Mean dissolved oxygen concentrations (10.5 and 9.7 mg dm^{-3} at the surface and near the bottom, respectively), oxygen saturation (105 and 95%, respectively), and ranges of the minimum and maximum values of those parameters across all the samples evidence good oxygen conditions in the Great Lagoon's water column. Particularly illustrative in this respect are the minimum and maximum values: the heaviest oxygen deficiency amounted to 67% saturation, the highest saturation being 159%. None of those values posed a threat to aquatic organisms: neither those requiring good oxygen conditions nor those sensitive to secondary effects of oversaturation were affected.

Nitrogen compounds. During the period of study, average concentrations of nitrates, the basic nitrogen source for the phytoplankton, were high, varied widely from season to season. According to Grelowski and Pastuszak (1996), the nitrogen discharge by the Odra is dominated by inorganic nitrogen compounds (mainly nitrates). For this reason, nitrates accounted for 65% of the total nitrogen discharge to the Lagoon during the intensified runoff in spring (April-May) (Fig. 1.3). In the remaining months of summer and autumn, nitrates made up as little as 23%, on the average, of the total nitrogen load. In addition to the key role of the Odra-borne loads in the nitrate discharge, the dynamic of nitrate concentrations in the Lagoon is affected by many other factors, including seawater incursions mentioned above, nutrient uptake by the phytoplankton, denitrification (Wielgat and Witek, 2004), and atmospheric transport. The latter process is increasingly frequently referred to as a significant pathway of input of chemicals. For example, Falkowska (1985) found 5.4 t nitrogen km^{-2} to be deposited by precipitation and atmospheric dust per year in the coastal zone (up to 20 nautical miles offshore) of the Baltic. Atmospheric deposition normalised to the Great Lagoon surface area amounts to about 3% of the riverine supply. The high contribution of atmospheric deposition to pollution of the southern Baltic was highlighted by data reported by Pęcherzewski (1991). Nitrate concentrations in successive months of the growing season were much more variable (Fig. 1.3) than the average concentrations in different years. The highest concentrations were recorded in spring, when nitrates were supplied by the enriched riverine runoff. In summer, particularly in August, the concentrations

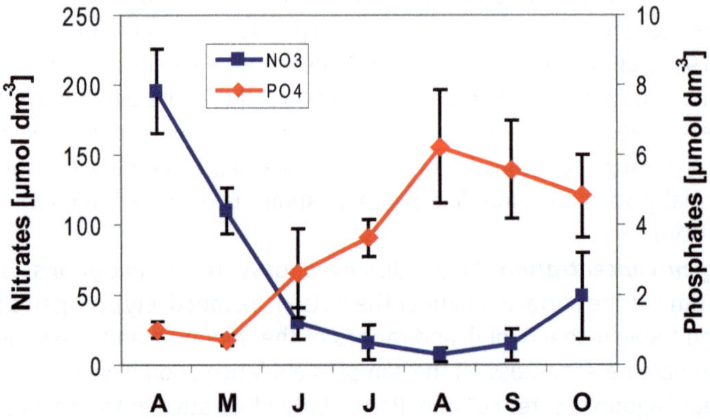

Figure 1.3 Seasonal changes of nitrate and phosphate concentrations [µmol dm^{-3}] in the Great Lagoon in 1998-2002 (average values in the surface and near-bottom layers and 95% confidence interval limits).

dropped to the minimum level as a result of nitrate concentration reduction in the Odra water and uptake by the phytoplankton. The intensive phytoplankton growth throughout the whole season (e.g., in Lake Nowowarpieńskie; Table 1.5) kept nitrate concentrations at a very low level. Judging by the critical levels identified by Vollenweider (1970), nitrogen concentration was doubtless a major limiting factor for diatoms and chlorophytes in the Lagoon in summer. The beneficiaries of this situation were only cyanobacteria, capable of tapping other nitrogen sources.

Phosphorus compounds. The average orthophosphate concentration over the period of study was 3.6 µmol dm^{-3}, the individual growing season averages ranging from 1.9 µmol dm^{-3} in 2001 to 4.8 µmol dm^{-3} in 2000. As seen in Fig. 1.3, the orthophosphate concentrations varied considerably during the growing season. A similar pattern of orthophosphate concentration variability in the Szczecin Lagoon was described by Bangiel et al. (2004).

Grelowski and Pastuszak (1996) found the phosphorus pool in the Odra water discharged into the Lagoon to be dominated by organic forms. Therefore, during the intensified riverine discharge in spring (April-May), orthophosphates in the central part of the Lagoon accounted for 26% of total phosphorus, but contributed more (56%) in summer and autumn. Seasonal variations in the Lagoon's orthophosphate concentrations are not significantly correlated with the orthophosphate load in the riverine runoff (Pastuszak, pers. inf.). The distinct increase in summer concentrations are strongly related to recycling of the sediment-bound phosphorus (Kaczewiak and Wolnomiejski, 1977; Wolnomiejski, 1994; Poleszczuk, 1997; 1998; 2007) and to orthophosphate pool remaining in the Lagoon unused by the phytoplankton due to the deficiency of mineral nitrogen (Wielgat and Witek, 2004; Pastuszak et al., 2005). Bloesch (1995) found almost 50% of the internal phosphorus load in a shallow lake to be generated by the release from bottom sediment. Elevated orthophosphate concentrations were observed during periods of strong seawater incursions, orthophosphate release from the sediment being enhanced by the higher ionic strength of the incoming seawater and periodic stagnation which intensifies hypoxia and anoxia in the sediment. The orthophosphate release from the sediment results in elevated phosphorus concentrations in the Lagoon's water in both the summer and early autumn. Thus, phosphorus does not limit the phytoplankton growth during this time. According to Vollenweider (1970), for the phytoplankton to form blooms, the readily available phosphorus species (i.e., orthophosphates) should occur at concentrations exceeding 3 µmol dm^{-3}. It was only in spring that orthophosphate concentrations in the Lagoon's main basin were lower than that critical level.

Silicate. Silicate concentrations were not determined in this study. However, they were measured in 2000-2005 by Pastuszak et al. (2008). The highest concentrations of soluble silicates (200-250 μmol dm^{-3}) occurred in winter, the lowest (< 1 μmol dm^{-3}) being typical of spring, primarily due to the mass occurrence of diatoms. The second silicate peak (up to 100 μmol dm^{-3}, a result of biogenic silicate recycling) was recorded in summer, followed by a decrease in autumn caused by another peak in diatom occurrence. Bangiel et al. (2004) showed silicate concentrations to be at their lowest in May, the concentrations being low enough to significantly limit the development of diatom populations.

BOD_5. The biochemical oxygen demand (BOD_5) was observed to vary substantially, from 0.47 to 6.67 mg O_2 dm^{-3}, i.e., within a range found also by other authors in the Great Lagoon (e.g., Poleszczuk, 1998). BOD_5 is primarily affected by the magnitude of phytoplankton biomass, as illustrated by data in Table 1.5. This relationship prompted Poleszczuk (1998) to contend that BOD_5 may be used as a proxy of phytoplankton biomass in the Lagoon.

Water pH. All the samples analysed during the study showed the Lagoon water to be slightly alkaline. In the main basin, water pH ranged within 7.7-9.0. Thus, at a moderate level of phytoplankton biomass, photosynthesis carried out by microalgae did not lead to excessive and prolonged overoxygenation of water, and did not disturb the carbonate equilibrium (such a disturbance leads to an excessive pH increase which is harmful for aquatic organisms).

Water transparency. The Secchi depth in the Lagoon's main basin across the five growing seasons averaged 1 m. In individual measurements, the Secchi depth ranged within 0.60-1.45 m. This range becomes considerably wider when the data collected in embayments are included: the lowest Secchi depth was 0.5 m (September 2001, Lake Nowowarpieńskie), the highest amounting to 4.0 m (October 2001, the Skoszewska Cove). In the growing season, light attenuation in water is closely related to the magnitude of phytoplankton biomass. Data from 28 sampling events in the 2001 growing season collected in the main basin, the Skoszewska Cove, and Lake Nowowarpieńskie evidence a statistically significant negative correlation between the Secchi depth and the chlorophyll *a* ($R = -0.787$). This correlation is illustrated, in a simplified, direct manner, by data in Table 1.5.

1.2.2.6 Assessment of the Lagoon's Trophic Status and Pollution Level

The Great Lagoon chemistry depends primarily on the chemical load discharged with the riverine runoff, atmospheric deposition, point and diffuse sources within the Lagoon, and authigenic pollution processes.

The Odra catchment is ihabited by about 14 million people and includes the most industrialised regions of Poland. Therefore, the load of pollutants it carries is potentially very high. The catchment supports most of Poland's heavy industry, black and brown coal mines, metal ore mines and processing plants, iron and steel industry, textile industry, and major food industry facilities. According to Knasiak et al. (1990), industrial areas of the Czech Republic and Germany contribute to the Odra pollution load as well. In the vicinity of Odra's discharge into the Lagoon, pollutants derive from agriculture, from the city of Szczecin (in the period of study, 1998-2002, Szczecin had no municipal water treatment plant), from a chemical plant manufacturing primarily phosphate fertilisers, and from wastewater and wastes originating in the harbour. Extremely important for the assessment of the Lagoon's pollution level is the compilation published by Grelowski and Pastuszak (1996), based primarily on data, collected by the Institute of Meteorology and Water Management, on pollutant supplies to the Odra within 1988-1994. The general conclusion is that the supply of heavy metals (particularly those most hazardous) and the load of organic pollutants were substantially or very substantially reduced in the 1990s. After the wastewater treatment plants in Międzyzdroje (1996) and Świnoujście (1997) became operative in 1996 and 1997, respectively, a marked improvement of the water quality in the northern part of the Lagoon was observed.

Changes in water chemistry, usually detrimental or unfavourable for aquatic organisms, may result from disturbances in ecosystem functioning. They may lead to authigenic pollution, with potentially toxic effects in the water body affected. These take place mainly when excessive amounts of hydrogen sulphide or ammonium are formed in the sediment and in the near-bottom water layer, when harmful substances are recycled, or when toxins are released during mass development and senescence of cyanobacteria. During the period of study (1998-2002), the main basin of the Lagoon did not show bioceonotic effects or symptoms of such processes.

According to the widely held opinion, the Lagoon is extremely heavily polluted. It has to be, however, remembered that the Lagoon is a flow-through water body with a large surface area and large volume, characterised by intensive water dynamics. Therefore, the chemical loads present in the runoff do not accumulate to an extent they would do in a water body that acts purely as a sink. By conditioning, metabolising, and temporarily retaining the pollutants brought in with the Odra runoff, the Szczecin Lagoon functions as a natural biological and dynamic wastewater treatment facility for the Odra water (Jakuczun and Nowacki, 1994; Wielgat and Witek, 2004; Minning, 2004; Pastuszak et al., 2005). Pollutant retention is particularly heavy in the shipping channel. According to Minning (2004), channel dredging removes 4-6 and 17-28% of the nitrogen and phosphorus loads brought in by the Odra, respectively. Even higher, on the order of 50%, is the removal of heavy metal (Cu, Pb, Zn) loads by dredging. Therefore, shipping channel dredging, with dredged material being deposited on land, plays a substantial role in the chemical budgets of the Lagoon. In addition to the effect of water dynamics, which enhance pollutant inactivation, the Lagoon supports abundant assemblages of organisms constituting the so-called biofilter. In this respect, particularly important are beds of the zebra mussel (*Dreissena polymorpha*). The entire zebra mussel population in the Lagoon is theoretically capable of filtering the entire water volume during less than 40 days (Wiktor, 1969).

Generally, no symptoms of saproby (sensu Olszewski, 1971), i.e., a state when biological production is hampered by an excessive input and decomposition of organic matter, were recorded in the Lagoon. Most often, the Lagoon is referred to as polytrophic (e.g., Radziejewska and Schernewski, 2008) or intermediate between eutrophic and polytrophic (Mutko, 1986; Knasiak et al., 1990; Poleszczuk and Sitek, 1993; Poleszczuk, 1998). In 2000, the Trophic State Index (TSI) of Carlson (1977) averaged 69 (Wolnomiejski and Woźniczka, 2008).

The distinct progress in reduction of the pollution load and improving the trophic status of the Lagoon, started in the 1990s, was to a large extent a result of environmental initiatives, particularly those implemented within the frameworks of international agreements and action plans (it has to be remembered that an almost 180 km-long stretch of the Odra constitutes a transboundary water area). One of the major aspects of transboundary cooperation is the protection and conservation of nature in the Odra valley. It involves, i.a., establishment of the International Lower Odra Valley Park. The transboundary stretch of the river is being monitored by the relevant Polish and German bodies acting within the framework of a Polish-German collaboration agreement. The Odra mouth, the Szczecin Lagoon, and adjacent waters have been granted protection by the EU NATURA 2000 Network as the special area of conservation of habitats and a special area of protection of birds. Water-covered areas on both sides of the border are the major ecosystem components in both types of NATURA 2000 protection. Moreover, in 1992, the Baltic System of Protected Areas (BSPA) was established, based on the premises of the (new) Helsinki Convention. The BSPAs areas along the Polish coast included, since 1996, the coastal part of the Pomeranian Bay within the Wolin National Park, including the area in the vicinity of the Lagoon's discharge to the Bay. Following Poland's accession to the European Union, the country was obliged to implement the joint, integrated policy of water management, based on the European Parliament's European Water Framework Directive (2000/60/WE) aimed at protecting and improving of the status of waters and aquatic ecosystems. The overarching aim of the Directive is the good ecological state of waters in all the EU countries, to be achieved by 2015. Efforts are also underway for the Szczecin Lagoon to be granted special protection as a biosphere reserve.

1.3 Phytoplankton

1.3.1 Composition and Biomass

The phytoplankton sampled in 1998-2002 was found to consist of 315 taxa (Luścińska, 2005). Diatoms (Bacillariophyceae), chlorophytes (Chlorophyta), cyanobacteria (Cyanoprocaryota), pyrrophytes (Pyrrophyta), cryptophytes (Cryptophyta), chrysophytes (Chrysophyceae), and euglenophytes (Euglenophyta) were represented by 168 (53.3% of the total number of taxa), 84 (26.7%), 48 (15.2%), 7 (2.2%), 6 (1.9%), and 2 species (0.7%), respectively. By far, most of the taxa were freshwater forms; brackish species usually appeared in autumn, when seawater incursions were intensified. Brackish forms were represented mainly by the diatom *Skeletonema subsalsum*. The brackish species' presence in the Lagoon was very short-lived, as the Lagoon did not provide appropriate conditions for their persistence.

The phytoplankton was sampled monthly during growing seasons, i.e. from April to October (concurrently with recording chlorophyll *a* concentrations), at two stations in the main basin as well as at two additional stations in the Lagoon's embayments. Water samples were taken with a Patalas sampler from three horizons: 0.5 m below the water surface, 0.5 m above the bottom and from the intermediate level, then mixed and preserved with Lugol solution. Samples were analysed by Dr. Marta Luścińska, the phytoplanktologist at the Nicolaus Copernicus University in Toruń. The phytoplankton was analysed using the inverted microscope technique and biomass was calculated by the volumetric method, assuming that 1 mm^3 is equivalent to 1 mg of fresh mass. The results, averaged across the two main-basin stations, are shown in Table 1.6.

Table 1.6 Mean phytoplankton biomass [$mg_{w.w.}$ dm^{-3}] and chlorophyll *a* concentration [μg dm^{-3}] in the main basin of the Great Lagoon in growing seasons.

Taxonomic group \ Year	1998	1999	2000	2001	2002	Average	Coeff. of variation
Cyanoprokaryota	12.6	14.8	8.9	2.4	3.4	8.4	0.65
Bacillariophyta	4.6	7.4	3.3	7.3	5.1	5.6	0.33
Chlorophyta	0.7	0.6	0.9	1.2	0.4	0.8	0.37
Remaining groups[a]	0.6	0.4	1.1	0.6	1.0	0.7	0.43
Phytoplankton together	18.5	23.2	14.2	11.5	9.9	15.5	0.35
Chlorophyll *a*	52.3	47.5	68.6	48.8	45.5	52.5	0.18

[a] mainly Dinophyta and Cryptophyta

The phytoplankton biomass over 1998-2002 averaged 15.5 mg dm^{-3}. The maximum biomass (60 mg dm^{-3}) was recorded in August 1999. The chlorophyll *a* concentration in the main basin in the growing seasons of 1998-2002 averaged 52.5 mg m^{-3}. The maximum chlorophyll *a* concentration (86.5 mg m^{-3}) was recorded in September 2000.

In August 2000, masses of algae with the biomass of up to 165 mg dm^{-3}, pushed by the wind were observed in the littoral zone of the eastern shore of the Lagoon. The masses consisted primarily of the cyanobacteria species *Microcystis aeruginosa* (87% of the biomass), *Anabaena flos-aquae* (5%), and *Planktothrix agardhii* (3%). However, those wind-driven algal accumulations should not be treated as phytoplankton blooms.

On account of a lack of mass cyanobacterial blooms during the period of study, the algal biomass was distributed uniformly in the water column. Such distribution was recorded also in the 1980s by Mutko et al. (1994) who compared data for the surface and near-bottom water layers. In addition, they frequently (e.g., throughout the entire 1984) observed higher chlorophyll *a* concentrations and algal biomass near the bottom than at the surface. The Provincial Inspectorate of Environmental Protection (PIEP) data summarised by Mutko et al. (1994) showed also instances of identical chlorophyll *a* concentrations at the surface and in the near-bottom water layer. For example, in 2001 and 2002, in the central part of the main basin, the chlorophyll *a* concentrations were higher near the bottom than at the surface in 40% of the sampling events (in 50% in 2001 alone). At one of the stations, even the average chlorophyll *a* concentration in the near-bottom water layer was higher than that near the surface throughout the growing season.

As shown by data in Table 1.6, the phytoplankton in the main basin of the Great Lagoon consisted mainly of cyanobacteria and diatoms throughout the period of study. On the average, cyanobacteria, diatoms, and the remaining major phytoplankton taxa contributed 54, 36, and 10% of the phytoplankton biomass, respectively. The cyanobacterial biomass was much more variable than that of other phytoplankton taxa. In 1998-2000, when the phytoplankton biomass was generally high in the Great Lagoon's main basin, cyanobacteria were the phytoplankton dominants. In 2001 and 2002, however, the diatom biomass was higher than that of cyanobacteria.

The cyanobacteria were dominated by *Microcystis aeruginosa*. Other prominently represented species included *Planktothrix agardhii, Pseudoanabaena limnetica, Aphanisomenon flos-aquae*, and *Limnothrix (redekei et planctonica)*.

Among diatoms, the most abundant species was *Actinocyclus normani*. Relatively abundant were also *Aulacoseira granulata, Cyclotella* sp. div., and *Diatoma tenuis*.

Chlorophytes were dominated by two taxa: *Scenedesmus* sp. div. and *Pediastrum boryanum, Cryptomonas* sp. div. dominating among the remaining phytoplankton groups. The most abundant phytoplankton taxa listed above accounted for about 90% of the total phytoplankton biomass.

A clear domination of *Microcystis aeruginosa* in the phytoplankton was recorded only in 1999. In 1998, *M. aeruginosa* co-dominated with *Planktothrix agardhii* (Niedźwiedzka, 2000). In 2000, two cyanobacterial species: *Planktothrix agardhii* and *Pseudoanabaena limnetica*, gained prominence, but their domination was not particularly pronounced. In the two subsequent years, the phytoplankton was dominated by the diatom *Actinocyclus normani.*

Seasonal changes in the phytoplankton biomass and domination structure were similar to those recorded in most temperate eutrophic water bodies, i.e., diatoms peaked in spring, while cyanobacteria were responsible for the summer or early-autumn peak in the phytoplankton biomass (Table 1.7).

Table 1.7 Average biomass [$mg_{w.w.}$ dm^{-3}] of dominant phytoplankton groups in different seasons of the year in 1999-2002.

Taxonomic group / Season of the year	Spring Apr-May	Summer June-Aug	Autumn Sep-Oct
Cyanoprokaryota	0.41	11.19	9.44
Bacillariophyta	6.75	4.64	6.67
Chlorophyta	0.41	1.43	0.34
Remaining groups[a]	0.56	1.02	0.65
Phytoplankton together	8.13	18.28	17.10

[a] mainly Dinophyta and Cryptophyta

The mass development of cyanobacteria was pronounced primarily in August and September, and was accompanied by a marked nitrate deficiency; the N_{tot}/P_{tot} ratio dropped below 10 (the ratio of mineral forms of nitrogen and phosphorus dropped even below 1; compare Fig. 1.3). On the other hand, differences between diatom densities during each growing season were relatively small.

The phytoplankton in the Lagoon's embayments showed a different pattern of changes in biomass and domination structure, compared to that in the main basin. This is exemplified by the 2001 data (Table 1.8).

The paucity of phytoplankton in the Skoszewska Cove resulted from a strong feeding pressure of *Dreissena polymorpha*, the bivalve being particularly abundant in the area Wolnomiejski and Woźniczka 2007). In contrast, the zebra mussel was virtually absent from the isolated and shallow Lake Nowowarpieńskie. The lake received an additional nutrient supply from the town of Nowe Warpno located in its vicinity.

Table 1.8 Average phytoplankton biomass [$mg_{w.w.}$ dm^{-3}] in three parts of the Great Lagoon in growing season of 2001.

Taxonomic group \ Part of the Great Lagoon	Main basin	Lake Nowowarpieńskie	Skoszewska Cove
Cyanoprokaryota	2.4	16.9	1.3
Bacillariophyta	7.3	11.8	3.0
Chlorophyta	1.2	3.6	0.5
Remaining groups[a]	0.6	0.2	0.2
Phytoplankton together	11.5	32.5	5.0

[a] mainly Dinophyta and Cryptophyta

It has to be borne in mind that the general picture of biodiversity, domination structure, abundance, and productive potential of the phytoplankton in the Lagoon depended primarily on the phytoplankton in the main basin which accounts for 92% of the total water volume in the Great Lagoon.

1.3.2 Phytoplankton Primary Production

1.3.2.1 Assumptions Adopted in Calculations

Phytoplankton primary production is most adequately measured with the ^{14}C technique. However, the technique was applied in the Great Lagoon five years before this study, in 1993-1997, by Westphal and Lenk (1998) according to whom the gross production of the phytoplankton averaged 648 gC m^{-2} yr^{-1}. Based on measurements of respiration and production of different size fractions of phytoplankton from the Puck Bay, conducted by Latała (1993), it may be estimated that on average the net primary production accounts for about 80% of the gross production. Based on data collected from numerous lakes, Lewis (2011) concurs with that estimate. Therefore, the net primary production estimated as 80% of the gross production would equal 518 gC m^{-2} yr^{-1}. However, a marked reduction in nutrient loads transported by the Odra was recorded in the Lagoon in the late 20th and the early 21st centuries (e.g., Pastuszak and Witek, 2009b). The phosphorus loads discharged in 1998-2002 (Table 1.3) were by 25% lower than those in 1993-1997, the nitrogen loads being lower by more than 13%. Starting with the primary production estimated for the Szczecin Lagoon in 1993-1997 and considering the changes in nutrient inputs mentioned, the net phytoplankton primary production in 1998-2002 can be estimated at about 400 gC m^{-2} yr^{1}. According to the dynamic biogeochemical model of Wielgat (2003), the net phytoplankton production in 1993-1997 amounted to 404-503 gC m^{-2} yr^{-1}, i.e., close to the empirical data of Westphal and Lenk (1998). Therefore, for the purpose of this work, the annual net primary production of the Great Lagoon phytoplankton in 1998-2002 is estimated at about 400 gC m^{-2}.

1.3.2.2 Primary Production Estimated With Other Methods

a. The annual primary production of the phytoplankton was also estimated based on the empirical biomass data, production rate (P/B) calculated for the growing season, and a correction factor for the annual production calculations.

- The published phytoplankton P/B data for the growing season span a relatively wide range. Most authors, however, indicated the range of 60-90, found in:

  - natural freshwater reservoirs and dam lakes in temperate latitudes (Winberg, 1972; 1979; Gak et al., 1972; Bohr et al., 1975 in Mikulski et al., 1975; Adamczak et al., 1979; Brylinsky, 1980; Winberg, 1985 in Bulon, 2004; Lacroix et al., 1999; Wetzel, 2001; Kazantseva, 2004);
  - coastal areas of the southern Baltic with low salinity gradients: the Puck Bay, the Curonian Lagoon, the coastal part of the Gulf of Riga, and the Pärnu Bay (Tomczak et al., 2009) ;
  - estuaries with steep salinity gradients (Mann, 2000; Rybarczyk et al., 2003; Rybarczyk and Elkaim, 2003; Libralato and Solidoro, 2009; Duan et al., 2009).

- The phytoplankton biomass in the Great Lagoon in 1998-2002 averaged 15.5 $mg_{w.w.}$ dm^{-3}, i.e., 58.9 $g_{w.w.}$ m^{-2} (the mean depth of the Great Lagoon is 3.8 m). Under continuous mixing of the entire water column of the Lagoon, the phytoplankton production is effected by algae distributed throughout the water column. Such conclusion was also drawn by Bucior (2009).
- At P/B = 60-90, the primary production amounts to 3534-5301 $g_{w.w.}$ m^{-2} in the growing season;
- The production effected during the growing season (April-October) was assumed to account for 83% of the annual production. This assumption was based on data presented by Wetzel and Likens (1995) who followed phytoplankton production during the entire year in the temperate Lake Lawrence (USA). In their study, 17% of the primary production occurred from November to March, outside of the growing season. A similar pattern emerges from a diagram plotted by Kajak (2001) for the annual production of phytoplankton in Lake Balaton. In Canadian lakes, Schindler (1972) calculated the off-season production to account for about 16% of the annual value.

Thus, the annual net production of the Great Lagoon phytoplankton ranged within 4258-6387 $g_{w.w.}$ m^{-2}.

- Calculations of the energy-based production indicators involved the use of conversion factors of Winberg (1971). The dry weight fraction of wet weight was assumed to be 0.20 and 0.15 for diatoms and for the remaining algae, respectively; the ash fraction in dry weight was assumed to be 0.5 and 0.07 for diatoms and for the remaining phytoplankton, respectively. Assuming that, on the average, diatoms and the remaining algae contributed 36 and 64%, respectively, to the averaged phytoplankton biomass, the dry weight fraction amounts to 0.168 (0.17 for all practical purposes) of the wet weight, 0.225 being the ash fraction in the dry weight. The dry weight contribution to the wet weight equal to 0.17 resulted also from data reported by Vinogradov and Shushkina (1987). Thus, the ash-free dry weight contributed 0.775 to the total dry weight. The ash-free dry weight (afdw) to carbon (C) conversion is: 1 g_{afdw} = 0.526 gC.
- As expressed in carbon units, the annual phytoplankton production was: 4258-6387 $g_{ww.}$ m^{-2} yr^{-1} × 0.17 $g_{d.w.}$ $g_{w.w.}^{-1}$ × 0.755 g_{afdw} $g_{d.w.}^{-1}$ × 0.526 gC g_{afdw}^{-1} = 295-443 gC m^{-2} yr^{-1}

b. When using formulae developed by Boulion (2004a) and, based on mean chlorophyll *a* concentration (52.5 mg m^{-3}) and a 210-day long growing season, the net phytoplankton production would amount to 277 gC m^{-2}, i.e., 334 gC m^{-2} yr^{-1}.

c. When using the empirical biomass data of this study and separate conversion factors for diatoms (after Strathmann, 1967) and the remaining phytoplankton (after Edler, 1979), the net phytoplankton production would amount to 426 gC m^{-2} yr^{-1}.

Results of the three compilation-based calculations presented above show that it is rational to assume the annual net phytoplankton production to amount to 400 gC m^{-2}, as shown by the functionally modified data of Westphal and Lenk (1998).

According to the empirical model of Smith (1979) which makes it possible to estimate primary production in the water body from the total phosphorus (P_{tot}) concentration and the Secchi depth, for a mean P_{tot} concentration in the Lagoon of 212 µg dm^{-3} (6.85 µmol dm^{-3}) and the Secchi depth of 1 m (Table 1.4), the net phytoplankton production over a 210-day long growing season would amount to 857 gC m^{-2}. However, Smith's model assumes primary production to be phosphorus-limited. This is actually not the case in the Szczecin Lagoon. As already mentioned in Section 1.2.2.5, only a part of the phosphorus pool is utilised in the growing season, as it is nitrogen that becomes a limiting factor in summer (Fig. 1.3).

1.4 Mesozooplankton

1.4.1 Taxonomic Composition

During the 30 years preceding the present study, there were no detailed studies on the Great Lagoon mesozooplankton, the quantitative aspects of its occurrence in the Lagoon being particularly neglected. The information presented here was obtained from monthly samples collected from two stations in the central basin of the Lagoon in the growing seasons of 1998, 2001, and 2002. Sampling was conducted in a standard manner: water samples were collected with a 5 dm^3 Patalas sampler from three levels: 0.5 m below the water surface, 0.5 m above the bottom, and in the intermediate part of the water column. The water was screened through a 55-60 μm mesh size net. The zooplankters retained on the net were identified by Dr. Paweł Napiórkowski of the Department of Hydrobiology, Nicolaus Copernicus University in Toruń; results of laboratory analyses are contained in an unpublished report. Biomass of the cladocerans and copepods was calculated using the formula of McCauley (1984), while the biomass of rotifers – using equations of Ejsmont-Karabin (1998), both based on mean body length of 50 individuals measured in each sample.

The samples were found to contain a total of 65 taxa: rotifers were represented by 40 taxa, cladocerans by 11, copepods by 13, and bivalves by 1 (larval stages of *Dreissena polymorpha*). An example of the list of taxa (year 2002 samples) is given below (the taxa most abundant or dominant in the group are underlined):

Rotatoria:
Ascomorpha ecaudis (Perty), *Asplanchna priodonta* Gosse, Bdelloidea indet., <u>*Brachionus angularis* Gosse</u>, <u>*B. budapestiensis* Daday</u>, *B. calyciflorus* Pallas, *B. diversicornis* (Dad.), *B. quadridentatus* Herm., *B. urceolaris* (O.F.M.), *Colurella adriatica* Ehrb., *C. obtusa* (Gosse), *Conochilus* sp. Ehrenberg, *Filinia longiseta* (Ehrb.), *F. terminalis* (Plate), *Kellicottia longispina* (Kell.), <u>*Keratella cochlearis* var. *typica* (Gosse)</u>: var. *cochlearis* (Gosse) et <u>var. *tecta* (Gosse)</u>, <u>*K. quadrata* (O.F.M.)</u>, *Lecane closterocerca* (Schmarda), *L. lunaris* (Ehrb.), *Lepadella patella* (O.F.M.), *Notholca acuminata* (Ehrb.), *N. squamula* (Ehrb.), *N. labis* Gosse, *N. limnetica*, *Polyarthra dolichoptera* Id., *P. maior* Burekh., *P. remata* Skor., *P. vulgaris* Carl., *Pompholyx sulcata* (Husd.), *Proales* sp. (Rouss.), *Ptygura* sp., *Synchaeta* sp. Ehrenberg, *Trichocerca capucina* Wierz. et Zach., *T. pusilla* (Lauterborn), *T. rousseleti* (Voight), *T. similis* (Wierz.), *T. rattus*, *T. inermis*.

Cladocera
Alona elongata (O.F.M.), *A. rectangula* Sars, <u>*Bosmina coregoni* Baird</u>, *B. longirostris* (O.F.M.), <u>*Chydorus sphaericus* (O.F.M.)</u>, <u>*Daphnia cucullata* Sars</u>, *D. hyalina* Leydig,

D. longispina (O.F.M.), *D. pulex* (De Geer), *Diaphanosoma brachyurum* (Liev), *Leptodora kindti* (Focke).

Copepoda: Calanoida
Eudiaptomus graciloides (Lilljeborg), *Eurytemora lacustris* (Poppe).

Copepoda: Cyclopoida
Acanthocyclops robustus (Sars), *A. vernalis* (Fisher), *Cyclops kolensis* Lillj., *C. scutifer* G.O.Sars, *C. strenuus* Fischer, *C. vicinus* Uljanin, *Diacyclops bicuspidatus* (Claus), *Eucyclops serrulatus* (Fischer), *Mesocyclops leucartii* Claus, *Thermocyclops oithonoides* (Sars).

Copepoda: Harpacticoida (not identified)

Bivalvia
Dreissena polymorpha (veligers).

The taxonomic composition of zooplankton may change somewhat over time as a result of episodic inflows of the Baltic water into the Lagoon which bring in certain species from the Baltic Sea. The changes, however, are restricted to the northern part of the Lagoon only (Wiktor, 1980; Chojnacki, 1989). In this study, no brackishwater forms were recorded.

1.4.2 Quantitative Variability of the Mesozooplankton

The abundance and biomass of the total mesozooplankton found in water samples collected with the Patalas sampler averaged, over the three years of study, 765 ind. dm^{-3} and 4.78 $mg_{w.w.}$ dm^{-3}, respectively (the biomass included 1.99 $mg_{w.w.}$ dm^{-3} contributed by cladocerans). The annual biomass averages were 5.54, 4.43, and 4.27 $mg_{w.w.}$ dm^{-3} in 1998, 2001, and 2002, respectively. It has to be borne in mind, however, that the standard sampling equipment (water samplers or vertical net hauls) provide no representative data on the presence of the largest (the size bordering on the macroplankton) zooplankter, the predatory cladoceran *Leptodora kindti*. Standard samples contain single individuals of the species only, whereas it is known to be one of the major and basic food items of obligatory and facultative plankton feeders among fish. Therefore, it was considered necessary to correct the zooplankton abundances and biomasses produced by standard analyses. Toward this end, results of 5 horizontal tows conducted with a Bongo net (61 cm diameter; ≈ 5.5 km h^{-1} towing speed) in spring and summer 2007 in the Great Lagoon were used to develop appropriate correction factors. A single relative parameter, the ratio between the wet weight of *L. kindti* and the total cladoceran wet weight was determined for this purpose.

The ratio, expressed as a percentage, was 19.8%. A very similar ratio (19.2%) was reported by Tyluś (2006) from samples collected in the Lagoon in 2003 (a 20 cm diameter Bongo net, ≈5.5 km h^{-1} towing speed). Over the entire growing season (including autumn), the ratio was 15.06%. Thus, the biomass resulting from the assumed 15% contribution of *L. kindti* (0.351 $mg_{w.w.}$ dm^{-3}) was added to the biomass calculated from the data collected in 1998, 2001, and 2002. The corrected biomass averaged, over the three years, 5.127 $mg_{w.w.}$ dm^{-3}. When converted to unit area (1 m^2), the mesozooplankton biomass averaged 19.5 $g_{w.w.}$ (assuming the mean depth of 3.8 m) (Table 1.9).

Table 1.9 Average mesozooplankton biomass in growing seasons of 1998, 2001 and 2002.

Taxa	$mg_{w.w.}$ dm^{-3}	$g_{w.w.}$ m^{-2}	%
Rotatoria	0.522	1.984	10.2
Leptodora kindti	0.351	1.334	6.8
Other Cladocera	1.989	7.558	38.8
Copepoda	2.255	8.569	44.0
Veligers of *Dreissena*	0.010	0.038	0.2
Total	5.127	19.483	100.0

The total average biomass can be broken down into contributions of individual major taxa: cladocerans, copepods, rotifers, and bivalve veligers which accounted for 45.6; 44.0; 10.2; and 0.2% of the total average biomass. Predators among the mesozooplankton showed an average biomass of 2.95 $g_{w.w.}$ m^{-2} (15%), the biomass of suspension feeders amounting to 16.53 $g_{w.w.}$ m^{-2} (85%).

1.4.3 Estimation of Zooplankton Production

The net mesozooplankton production in the Great Lagoon was estimated based on the empirical biomass data and the published P/B ratios.

Estimates of the production rate (P/B) for individual freshwater zooplankton species or the higher taxa, reported by various authors (Brylinsky, 1980; Hillbricht-Ilkowska et al., 1975; Kazantseva, 2004; Morgan et al., 1980; Wetzel, 2001; Winberg, 1979; Winberg et al., 1972; Vladimirova, 1974; Waters, 1977; Zawiślak, 1977 and references therein) differed widely. The P/B ratios of individual major mesozooplankton taxa covered wide ranges of 28.0-73.0 (rotifers), 13.8-36.9 (non-predatory cladocerans), 12.0-26.2 (*Leptodora kindti*), and 8.9-28.8 (copepods).

Production estimates in this study were based on the relevant zooplankton data reported by Vladimirova (1974) and collected during a 12-yr-long study in the Rybinsk Dam Reservoir in growing seasons extending from May through October. She arrived at the following P/B ratios: 60.0 for rotifers, 26.9 for cladocerans, and 10.0 for copepods. The P/B ratio of 15, calculated for Polish lakes from data reported by Hillbricht-Ilkowska et al. (1975), was used for estimating production of the predatory cladocerans *Leptodora* sp. The zebra mussel veliger P/B was assumed to be 2.0, i.e., the minimum P/B value of univoltine forms (Kajak, 1988).

The mesozooplankton production parameters and the production estimates themselves are shown in Table 1.10. The values apply to a 210-day-long growing seasons and were assumed to represent the annual production.

Table 1.10 Mesozooplankton production in the Great Lagoon in the growing season.

Taxon \ Parameter	Biomass [$g_{w.w.}$ m^{-2}]	P/B	Production [$g_{w.w.}$ m^{-2}]	Carbon conversion factor [gC $g_{w.w.}^{-1}$]	Biomass [gC m^{-2}]	Production [gC m^{-2}]
Rotatoria	1.984	60.0	119.04	0.050[a]	0.10	5.95
Leptodora sp.	1.334	15.0	20.01	0.064[b]	0.09	1.28
Other Cladocera	7.558	26.9	203.31	0.064[b]	0.48	13.01
Copepoda	8.569	10.0	85.69	0.064[b]	0.55	5.48
Veligers of *Dreissena*	0.038	2.0	0.08	0.064	< 0.01	< 0.01
Total	19.483		428.13		1.22	25.72

[a] Parsons et al. (1977);
[b] Vinogradov and Shushkina (1987)

The P/B ratio of the entire zooplankton averaged 22 (when wet weight based) and 21 (when organic C based). The production of suspension feeders was estimated at 366 $g_{w.w.}$ m^{-2} (22.2 g C m^{-2}), the production of predators amounting to 62 $g_{w.w.}$ m^{-2} (3.5 g C m^{-2}). Individual major taxa contributed, (wet weight based; in brackets - C_{org} based): rotifers 28 (23) %, cladocerans 56 (56) %, copepods 20 (21) %, and veligers < 1 (< 1) %

The $P_{mesozoopl.}/P_{phytopl.}$ ratio was 0.0643.

1.5 Macrozooplankton

Occasionally, particularly during and just after summer and autumn storms, the Szczecin Lagoon may feature various incoming Baltic macroplankters, including even the jellyfish *Aurelia aurita*. Their presence in the Lagoon is ephemeral only and they die off after 1 or 2 days. There is, however, a permanent macroplanktonic zooplankton component, confined to the northern part of the Odra River mouth system: the mysid *Neomysis integer*. Due to their mass occurrence and large size, mysids are a very important dietary item for fish, particularly the small species (e.g., smelt) and juveniles of predatory fish.

Regrettably, there are no reliable data on the mysid abundance and biomass in the Szczecin Lagoon, which is the case also in other shallow Baltic areas and coastal water bodies. The only source of information on the abundance of *N. integer* in the Polish part of the Lagoon dates back to the 1950s (Wiktor, 1961). Those data were, however, based on hardly representative samples obtained with vertical net hauls.

Based on the data reported by Wiktor (1961), the mysid abundance may be estimated to have averaged then – over the entire Lagoon – 35 ind. m^{-2}. Our standard weights allow to calculate the average individual weight as 15 $mg_{w.w.}$, which yields an average biomass of 0.525 $g_{w.w.}$ m^{-2}.

The annual P/B ratio was assumed to be 3.3. The P/B values reported for other mysids were, e.g., 3.55 (*Neomysis americana*; Stallone 1979) or 3.0 (*Mysis mixta* in the Gulf of Gdańsk, Witek, 1995).

The wet weight to organic C conversion factor was assumed to be identical to that used for the crustacean mesozooplankton, i.e., 0.064 gC $g_{w.w.}^{-1}$ (Vinogradov and Shushkina, 1987).

Those assumptions allowed to estimate the mysid production in the Great Lagoon as 0.111 g C m^{-2} yr^{-1} (Table 1.11). However the value is a gross underestimate because of both the absence of reliable data on the mysid abundance and the lack of knowledge on their local developmental specificity, particularly the number of generations per year.

Table 1.11 Biomass and production of *Neomysis integer*.

Density [a] [ind. m^{-2}]	Average individual weight [$mg_{w.w.}$]	Average biomass [$g_{w.w.}$ m^{-2}]	P/B [b] [yr^{-1}]	Production [$g_{w.w.}$ m^{-2} yr^{-1}]	Carbon conversion factor [c] [gC $g_{w.w.}^{-1}$]	Production [gC m^{-2} yr^{-1}]
35	15	0.525	3.3	1.73	0.064	0.111

[a] after Wiktor (1961)
[b] after Stallone (1979) and Witek (1995)
[c] after Vinogradov and Shushkina (1987)

1.6 Macrophytes

1.6.1 Composition and Biomass

During the past 40 years, no dedicated specialist study was focussed on higher plants and macrophytic algae in the Polish part of the Szczecin Lagoon. The most important source of information are still the publications by Garbacik-Wesołowska (1969; 1973) describing the macrovegetation status in the 1960s. She used aerial photographs to estimate the surface area covered by individual macrophytic communities. In 2000 and 2001, the basic data reported by Garbacik-Wesołowska (1969; 1973) were re-assessed based on observations carried out from a boat and focussing mainly on the macrophyte diversity and a general pattern of their distribution. Generally, no substantial changes in the composition and distribution of the key macrophyte species were found. Compared to the inventory provided by Garbacik-Wesołowska (1969), some species turned out to be missing, e.g., river rush (*Bulboschoenus maritimus*) and floating heart (*Limnanthemum nymphoides*). In addition, an area (65 ha) which had been, when surveyed by Garbacik-Wesołowska (1969), overgrown by the Characeae, turned out to be devoid of them; instead, the area supported a lush growth of pondweeds, particularly *Potamogeton pectinatus.*

Individual macrophytes, particularly helophytes, occur in the Great Lagoon in bands or in patches. The key, most abundant helophytes, i.e., common reed (*Phragmites australis*), bulrush (*Schoenoplectus lacustris)*, and narrow leaf cattail (*Typha angustifolia)*, usually occur in monospecific bands or patches. In contrast, the submerged vegetation often consists of different co-occurring species.

The Great Lagoon supports lush, well-developed stands of higher plants. Common reed grows occasionally to the height of 4.5 m above the bottom. It usually occurs in the littoral down to ±1m depth (2 m at the maximum), the reed belts extending out to 200 m away from the shore. Macrophytes grow best in shallow embayments of the Lagoon and on the shallows off the eastern shore. Particularly important for the abundance of higher plants, especially elodeids, is the underwater ridge extending along almost the entire eastern (usually wind-exposed) shore, residing 1-3 km off from it. Like a near-shore reef, the alongshore vegetation belt considerably dampens the force of the wave action upon the shore. The shallow bottom between the ridge and the eastern shore features wide helophyte belts, the so-called lacustrine littoral, formed by reeds and bulrush. Pondweed patches found there extend to 1.5 km away from the shore into the Lagoon.

The above-bottom macrophyte biomass in the Great Lagoon was assumed to equal that calculated by Garbacik-Wesołowska (1973) (Table 1.12). The 410 km^2 area of the Great Lagoon featured macrophytes growing on 15.5% of the bottom surface. The total biomass (dry weight-based) of the above-bottom parts of the plants amounted to 53,500 t.

Table 1.12 Summer biomass of above-ground tissues of aquatic macrophytes in the Great Lagoon (according to data of Garbacik-Wesołowska, 1973).

Macrophyte assemblage	Area covered by assemblage [ha]	Total biomass [$t_{d.w.}$]	Biomass in area covered by assemblage [$g_{d.w.}$ m^{-2}]	Biomass in phytolittoral [$g_{d.w.}$ m^{-2}]	Biomass in entire reservoir [$g_{d.w.}$ m^{-2}]
Emergent with *Phragmites communis (australis)*	927	31 649	3414	498	77
Emergent with *Schoenoplectus lacustris*	553	11 417	2065	180	28
Emergent with *Typha angustifolia*	73	1 132	1551	18	3
Emergent total	1 553	44 198	2846	696	108
Floating-leaved (*mostly Nuphar luteum*) total	311	1262	406	20	3
Submersed with *Potamogeton* sp. div.	3 863	6 021	156	95	15
Submersed with *Myriophyllum* and *Ceratophyllum*	627	2063	329	32	5
Submersed together	4 490	8 084	180	127	20
Floating-leaved and submersed total	4 801	9 346	195	147	23
Total macrophytes	6 354	53 544	843	843	131

Helophytes occupied 1 553 ha of the bottom, 60% of the area supporting common reed. The largest part of the phytolittoral (4 800 ha) was occupied by submersed and floating-leaved macrophytes – elodeids and nymphaeids, the former occupying more than 94% of the area overgrown by the two groups. Elodeids occurred as communities of different pondweed species, notably *Potamogeton perfoliatus, P. lucens,* and *P. pectinatus.* In addition, important were also *Ceratophyllum demersum* and *Myriophyllum spicatum.* The most common nymphaeid was *Nuphar luteum.* Among other nymphaeids, particularly important were *Polygonum amphibium* and *Nymphaea alba.*

The helophyte biomass in the Great Lagoon was particularly high. When estimating the macrophyte biomass, Garbacik-Wesołowska (1973) did not separate off the vegetation of the inner belt closest to the shore and featuring, e.g., *Stratiotes, Acorus, Iris* and *Sparganium* sp. Thus, the overall estimates contain the biomass of those species included into the biomass of the basic helophytes in the area of their occurrence. Algae (mainly *Cladophora* sp., *Enteromorpha* sp., and *Hydrodictyon* sp.) attached to stems of higher plants, particularly close to the bottom, contributed somewhat to the biomass of certain individual communities, especially those of helophytes.

1.6.2 Production

To estimate the entire production potential of macrophytes, it was necessary to consider the biomass of their rhizomes and roots into calculations, in addition to the above-bottom parts. However, published data on the sub-bottom parts of aquatic macrophytes are very scant and hardly available. In view of the absence of relevant empirical data, production of the sub-bottom parts was estimated from the published information on the biomass ratio between the sub- and above-bottom parts of the plants.

Production of above-bottom parts

Production of macrophytes (above-bottom parts) was assumed to be equal to their maximum (summer) biomass expressed as dry weight, following Garbacik-Wesołowska (1973), multiplied by P/B_{max} = 1.25 (Winberg, 1972). According to Boulion (2004b), Raspopov (1985) used a similar factor (1.2) for the reed alone.

The ash content of helophytes, nymphaeids, and elodeids was assumed – after Bernatowicz and Pieczyńska (1965), Grabowski (1973), and Rejewski (1979; 1988) – as 7.8, 10.7, and 30.0% of the dry weight, respectively.

Following Winberg (1971), the organic C to afdw conversion factor was assumed to be 0.455 gC g_{afdw}^{-1}.

Production of sub-bottom parts

The sub-bottom part biomass was estimated from a compilation of data extracted from Gayevskaya (1966), Bernatowicz and Wolny (1974), Rejewski (1975; 1979), Ozimek (1983), Wetzel (2001), and Żurek (2007). The sub-bottom parts were assumed to account for 50, 75, and 30% of the total plant weight in helophytes, *Nuphar* sp., and elodeids (primarily pondweeds), respectively.

According to Bernatowicz and Wolny (1974), the sub-bottom biomass result from a long term deposition of assimilates and is hardly a measure of production intensity in any single year. Gayevskaya (1966) found about half of the sub-bottom part of a plant to grow during a single growing season. Assuming a stable spatial succession of helophytes in the Szczecin Lagoon, whereby there is a stiff competition for space between roots and rhizomes within the sediment, the sub-bottom biomass can be assumed to increase by one-third each year. This assumption is supported by data of Wetzel (2001) on the production of above- and sub-bottom parts of marsh vegetation. Thus, the annual production of the sub-bottom parts of the plants was assumed to amount to one-third of their biomass.

In view of the lack of relevant empirical data, energy conversion factors were assumed to be identical to those applied to the above-bottom parts. Data provided by Rejewski (1979) for nymphaeids corroborate the plausibility of such an assumption.

The calculated production data are shown in Tables 1.13 and 1.14.

Table 1.13 Annual production of above-ground tissues of aquatic macrophytes in the Great Lagoon (average per standard m^2 of the entire reservoir).

Macrophytes	Biomass $[g_{d.w.}\ m^{-2}]$	Ash content [% d.w.]	Carbon conversion factor[a] $[gC\ g_{afdw}^{-1}]$	P/B	Production		
					$[g_{d.w.}\ m^{-2}\ yr^{-1}]$	$[g_{afdw}\ m^{-2}\ yr^{-1}]$	$[gC\ m^{-2}\ yr^{-1}]$
Emergent	108	7.8	0.455	1.25	135.0	124.47	56.6
Floating-leaved	3	10.7	0.455	1.25	3.75	3.35	1.5
Submersed	20	30	0.455	1.25	25.0	17.50	8.0

[a] according to Winberg (1971)

Table 1.14 Annual production of underground tissues of aquatic macrophytes in the Great Lagoon (average per standard m^2 of the entire reservoir).

Macrophytes	Biomass $[g_{d.w.}\ m^{-2}]$	Ash content [% d.w.]	Carbon conversion factor $[gC\ g_{afdw}^{-1}]$	P/B	Production		
					$[g_{d.w.}\ m^{-2}\ yr^{-1}]$	$[g_{afdw}\ m^{-2}\ yr^{-1}]$	$[gC\ m^{-2}\ yr^{-1}]$
Emergent	108	7.8	0.455	0.333	36	33.19	15.1
Floating-leaved	9	10.7	0.455	0.333	3	2.68	1.2
Submersed	8.6	30	0.455	0.333	2.9	2.03	0.9

In the entire Great Lagoon, total macrophyte production was 83.3 g C $m^{-2}\ yr^{-1}$, of which 66.1 g C $m^{-2}\ yr^{-1}$ was the production of the above-bottom parts, and 17.2 g C $m^{-2}\ yr^{-1}$ (20.6%) - production of sub-bottom parts.

The caloric content of macrophyte dry weight, calculated from a comparison between production as expressed in dry weight (g) and in organic carbon (Table 1.13), using Winberg's (1971) calories to organic carbon conversion factor (10 kcal gC^{-1}) was 4.19, 4.0, and 3.20 kcal $g_{d.w.}^{-1}$ for helophytes, nymphaeids, and elodeids, respectively.

The energy content of lacustrine helophytes and nymphaeids (Rejewski, 1979; 1988) was 4.36 and 3.90 kcal $g_{d.w.}^{-1}$, respectively.

1.6.3 The Surface Offered by Aquatic Macrophytes to Other Organisms

In addition to benthic habitats, aquatic vegetation offers additional living space for assemblages of sessile microalgae and invertebrates, collectively termed the periphyton.

Based on unpublished data of Wolnomiejski on the length, diameter, and density of reed stems per 1 m^2 bottom surface, the submerged parts were found to offer an average of 2 m^2 for settling by the periphyton. Surfaces of a similar magnitude were assumed for the remaining helophytes. Common reed was a dominant helophyte which occupied 60% of the bottom area supporting all the helophytic communities.

Wolnomiejski's unpublished data on plant surface to weight ratios were used to calculate the submerged surface area of submersed and floating-leaved plants, pondweeds and water lilies. The wet weight of the plants was determined using the displacement volume technique, assuming the density of 1 kg dm^{-3}. The surface area was determined from contouring the leaves on millimetre-ruled paper; the stems were assumed to be cylinders. In the case of pondweed-type plants, a 1 m^2 kg^{-1} conversion factor was used, 0.3 m^2 kg^{-1} being used for *Nuphar* and *Nymphaea*. For strongly branched plants (coontail, watermilfoil) the conversion factors developed by Czarnecka (2003) for artificial, plant-like branching structures, were used. In her study, a photographic image of Christmas-tree like "branches" was processed by a computer image analysis software to measure the surface area. The volume of the branches was determined in the measuring cylinder. As shown by the measurements, the surface area of 1 dm^3 of the entire "branches" averaged 7.7 m^2.

Considering the plant density converted to wet weight and the surface area of the bottom covered by individual communities as in Table 1.12, and using the plant weight to surface ratios given above, the surface area of the soft (i.e., the floating-leaved and submersed) vegetation covering a 48 km^2 bottom area averaged 148.5 km^2 (Table 1.15). Thus, a square metre of the bottom overgrown by the soft vegetation supported a 3.09 m^2 (rounded off to 3 m^2) wet surface area of plants.

Helophytes, having an underwater surface of 2 m^2 per m^2 of the bottom, covered 24.4% of the phytolittoral surface area (which totalled 175 km^2). Elodeids and nymphaeids, providing a submerged surface area averaging 3 m^2 per m^2 of the bottom, occupied 75.6% of the phytolittoral. Thus, generally, a square metre of the phytolittoral supported 2.76 m^2 of underwater macrophyte surface area.

Table 1.15 Underwater surface area of floating-leaved and submersed macrophytes in the Great Lagoon.

Assemblage	**Area covered by assemblage [km²]**	**Dry weight to wet weight conversion factor[a] [$g_{d.w.}$ $g_{w.w.}^{-1}$]**	**Fresh biomass [$g_{w.w.}$ m^{-2}]**	**Plant surface area per plant weight unit [m^2 $kg_{w.w.}^{-1}$]**	**Total plant surface area [km²]**
Floating-leaved (mostly *Nuphar luteum*) total	3.1	0.17	2.39	0.3	2.2
Submersed, with *Potamogeton* sp. div.	38.6	0.15	1.04	1.0	40.1
Submersed, with *Myriophyllum* and *Ceratophyllum*	6.3	0.15	2.19	7.7	106.2
Floating-leaved and submersed total	48.0				148.5

[a] values compiled from data in Gayevskaya (1966); Bernatowicz and Pieczyńska (1966); Penczak (1984); and Rejewski (1979; 1988)

1.7 Phytoperiphyton

1.7.1 Composition and Biomass

The diversity of periphytic algal assemblages was assessed from the analysis of four series of samples collected in July and August of the years 1998 and 1999 from submerged stem parts of reed growing at two sites in the north-eastern part of the Great Lagoon (Tymolewska and Wolnomiejski, unpubl. data). The reed stems collected were preserved in formalin; the periphyton layer was thoroughly removed from stem sections and all the macroscopic objects (mainly macroperiphytic invertebrates) were picked out. The phytoperiphyton wet weight-based biomass was determined by measuring the volume of algal mass sedimented twice for 24 h. To accelerate the sedimentation process, the algal suspension was treated with Lugol's solution. About one-third of the sedimented mass of algae was the external water remaining among the cells. The volumetric contribution of individual periphyton components to the sedimented mass was determined using the microscopic method of Szlauer (1996). The algal density was assumed to approximate 1 g cm^{-3}. The algae were identified to the genus level. Following the measurement of the surface area of the reed stem sections, the algal density was normalised to 1 m^2 stem surface area. The taxonomic composition and biomass of individual phytoperiphyton components are shown in Table 1.16.

Table 1.16 Taxonomic composition and biomass of peripyhtic algae growing on submerged parts of reed stems in the northern part of the Great Lagoon.

Taxa	**Biomass** [$g_{w.w.}$ m^{-2} substrate]	**Percentage**
Cyanoprokaryota:		
Lingbya sp.	5.3	
Planktothrix sp.	0.9	
other blue-greens	0.2	
Cyanoprokaryota total	6.4	9.7%
Bacillariophyceae:		
Navicula sp.	23.3	
Aulacoseira sp.	6.5	
Cocconeis sp.	6.1	
Gomphonema sp.	5.7	
Cymbella sp.	4.3	
Fragilaria sp..	4.3	
Rhoicosphaenia sp.	2.8	
Gyrosigma sp.	2.2	
other diatoms	2.9	
Bacillariophyceae total	58.1	88.6%
Chlorophyta:		
Cladophora sp.	0.5	
Oedogonium sp.	0.3	
other green algae	0.3	
Chlorophyta total	1.1	1.7%
Phytoperiphyton total	65.6	

Following Bohr (1962), the following layers (strata) and biological groups were identified in the periphyton growing on reeds in the Lagoon:

the primary (basic) layer, formed by filamentous, branched algae, most commonly attached to the substrate. They were represented by large filamentous *Oedogonium* held in place by their attachment cells; *Cladophora* attached to the substrate with rhizoids; certain *Aulacoseira* diatoms forming filamentous colonies; *Fragilaria* diatoms growing in ribbon-like colonies; "zigzagging" colonies of *Diatoma*; and *Cymbella* forming gelatinous rope-like structures. These constitute a secondary substrate amenable for colonisation by other forms of the periphyton;

epiphytic forms, growing on the basic layer algae, and – more seldom – including the algae settled directly on the original substrate; they are represented primarily by *Cocconeis, Gomphonema, Cymbella,* and some species of *Navicula;*

free-living forms, loosely associated with the periphyton or merely trapped in it, e.g., the filamentous diatom *Aulacoseira granulata* and some species of the genera *Navicula* and *Lingbya.* There are also typically planktonic algae, accidentally trapped in the periphyton, particularly the planktonic cyanobacteria (e.g., *Planktothrix)* or ubiquitous forms that may even occur on the bottom (e.g., *Fragilaria*). Some of them, according to Oleksowicz (1986) and the references therein, are ecotypes, i.e., varieties of a species that are physiologically adapted to dwelling in a habitat different than that typical of the species.

1.7.2 Assumptions Adopted for Generalisation of the Quantitative Characteristics of the Phytoperiphyton

It was assumed that the periphyton occurring on reed stems and on other macrophytes was similar in its composition and abundance. The findings of Bohr (1962) based on surveys of numerous lakes were extremely helpful in this regard. He found the composition of the phytoperiphyton on stems of reeds grown during a given year and a year before, on *Schoenoplectus lacustris,* and on stones in the same area to be similar. The reed periphyton in different areas situated far apart in the littoral of different lakes usually represents algal assemblages similar in composition and structure. In general, Bohr (1962) contended that the periphyton species composition is closely related to the ecological properties of a site, substrate effects playing a secondary role. Using those findings as the starting point, it was assumed that the pattern of periphyton variability emerging from examination of reed stems can be extended onto the substrates offered by other macrophytes.

The qualitative and quantitative characteristics of the phytoperiphyton described here are based on the analysis of samples collected in July and August, considered representative of the entire growing season. The feasibility of such an assumption stems from long-term observations reported on by Bohr (1962) and showing that the phytoperiphyton components which are most abundant and most persistent occur throughout the year, and it is only their abundance that changes. A community of periphytic algae attains its peak diversity in August. On the other hand, Kuczyńska-Kippen et al. (2004) recorded the highest diversity in spring, but the biomass of periphytic algae was at its lowest then.

Data provided by other authors on the energy content (as a biomass equivalent) of periphytic algae also show the summer samples to be adequate for estimation of quantitative aspects of the periphyton throughout the growing season. This is evidenced by data collected by Adamczak et al. (1979) from a eutrophic lake where the phytoperiphyton biomass (energy content) in spring, August, and October amounted to 43, 57, and 111 kcal m^{-2} reed stem surface, respectively (the average was 70 kcal m^{-2}, i.e., somewhat more than the energy content in August). In the same lake, but in a different year, Bohr et al. (1975) revealed a similar pattern: the periphyton biomass (energy content) in May, July, and October amounted to 11, 60, and 84 kcal m^{-2} reed stem (averaging 52 kcal m^{-2}). According to Szczepańska (1970), the chlorophyll *a* content in the periphyton of four Masurian lakes changed from low in spring to intermediate in summer to high in autumn. Changes in the energy content followed the same pattern.

1.7.3 Phytoperiphyton Production

Phytoperiphyton production was based on the empirical biomass data and P/B values extracted from the literature.

- The literature contains only scant data on the production rates of phytoperiphyton growing on natural substrates. According to Adamczak et al. (1979), the net P/B of the periphyton growing on reed in a eutrophic lake was 11.8 for the growing season. Bohr et al. (1975) reported the P/B of 15 (for 240-day-long growing season, i.e., 13.1 for 210 days, the turnover time amounting to 16 days). Wetzel (2001) quoted the data of other authors on the turnover time of the phytoperiphyton growing on macrophytes, the turnover amounting to 21.7 and 14.4 days on *Equisetum* sp. and *Sagittaria* sp., respectively. When normalised to 210 days, the respective P/B values amount to 9.7 and 14.6. For epiphytic algae of Lake Esrom, Wetzel (2001) reported P/B of 14; this value was adopted for this study.
- The net phytoperiphyton production per phytolittoral unit area was then: 65.6 $g_{w.w.}$ phytoperiphyton m^{-2} substrate × 2.76 m^2 m^{-2} (average surface area of macrophytes per m^2 phytolittoral surface area, cf. Section 1.6.3) × 14 (P/B) = 2534 $g_{w.w.}$ m^{-2} yr^{-1}.
- Phytolittoral accounted for 15.5% of the Great Lagoon area; therefore, the phytoperiphyton production over the entire Great Lagoon was:

$$2534\ g_{w.w.}\ m^{-2}\ yr^{-1} \times 0.155 = 393\ g_{w.w.}\ m^{-2}\ yr^{-1}.$$

- Conversion factors of Winberg (1971) were used for calculating the energy-based production: the dry weight to wet weight ratios of diatoms and the remaining algae were 0.20 and 0.15, respectively; the ash to dry weight ratios of diatoms and the remaining phytoplanktonic algae were 0.5 and 0.07, respectively. Assuming that diatoms and the remaining phytoplankters accounted for 88.6 and 11.4% of the phytoplankton biomass, respectively, the average dry weight to wet weight ratio and the ash to dry weight ratio amount to 0.194 and 0.451, respectively. Thus, the ash-free dry weight accounted for 0.549 of the dry weight. The organic carbon content in 1 g ash-free dry weight was, using Winberg's (1971) conversion factors, 0.526 g.
- Consequently, the phytoperiphyton carbon-based production in the entire Great Lagoon was calculated as:

$393\ g_{w.w.}\ m^{-2}\ yr^{-1} \times 0.194\ g_{d.w.}\ g_{w.w.}^{-1} \times 0.549\ g_{afdw}\ g_{d.w.}^{-1} \times 0.526\ gC\ g_{afdw}^{-1} = 22.02\ gC\ m^{-2}\ yr^{-1}$, or, when rounded up, $22\ gC\ m^{-2}\ yr^{-1}$.

Boulion (2004a; b), using a formula based on research conducted in some tens of lakes, found the periphyton biomass to account for 1.5% of the macrophyte biomass, and used the P/B ratio of 30. Should that formula and the P/B value be applied to our data, the Great Lagoon phytoperiphyton production would amount to 23.8 gC $m^{-2}\ yr^{-1}$, i.e., close to the value produced by our own calculations.

1.8 Mesozooperiphyton

Information on the mesozooperiphyton was extracted from the 1998 and 1999 samples used for the phytoperiphyton analyses. The methods used for obtaining quantitative information were identical as well.

In view of the lack of any comprehensive references, the mesozooperiphyton composition and biomass data collected from reed were regarded as representative for the formation occurring on all the macrophytes. The average surface offered by macrophytes in the phytolittoral was 2.76 $m^2\ m^{-2}$ bottom surface area (cf. Section 1.6.3). As already mentioned, the phytolittoral accounted for 15.5% of the entire Great Lagoon surface area.

More than 90% of the mesozooperiphyton were contributed by rotifers and nematodes. The mean biomasses of rotifers and nematodes were 1.13 and 2.20 $g_{w.w.}\ m^{-2}$ reed stem surface area, respectively.

The production rate (P/B) of periphytic rotifers was assumed to be identical with that of the planktonic forms (60). Piesik and Wawrzyniak-Wydrowska (2003) reported rotifers found in the periphyton to be meroplanktonic. The nematode P/B was assumed to be 20. Galtsova (1991) found the published nematode P/B values to vary within a wide range, from 3 to more than 50.

The way the mesozooperiphyton production was calculated is shown in Table 1.17.

Table 1.17 Biomass and production of mesozooperiphyton in the Great Lagoon. Organic carbon conversion factor (Cc) for rotifers after Parsons et al. (1977); for nematodes after Galtsova (1991); 1 kcal assumed to be equivalent to 0.08 gC (Vinogradov and Shushkina, 1987).

Component	Biomass		P/B	Cc	Production		
	[$g_{w.w.}$ m^{-2} substrate]	[$g_{w.w.}$ m^{-2} phytolittoral]		[gC $g_{w.w.}^{-1}$]	[$g_{w.w.}$ m^{-2} yr^{-1} phytolittoral]	[$g_{w.w.}$ m^{-2} yr^{-1} in entire Lagoon]	[gC m^{-2} yr^{-1} in entire Lagoon]
Rotatoria	1.13	3.12	60	0.05	187.20	29.01	1.451
Nematoda	2.20	6.07	20	0.069	121.40	18.82	1.298
Total	3.33	9.19			241.83	47.83	2.749

1.9 Macrozooperiphyton

The data on biomass of the reed epifauna were extracted from Czarnecka and Tymolewski (2003) who collected their samples in July and August of the years 1998-2000 in the north-eastern part of the Great Lagoon. In addition, the unpublished 2003 data of Wolnomiejski (from a total of 13 samples collected in June-August) concerning different parts of the Lagoon were used as well. The pondweed epifauna was sampled (a total of 14 samples) in the central-eastern part of the Lagoon in July and August 2002.

Sections of submerged reed stems and pondweeds were rinsed over a 0.05 mm mesh size sieve. After rinsing, plants were additionally examined to collect any remaining organisms. The fauna was preserved in formalin or in the Dorogostayskiy fluid (Żmudziński, 1990). Preservation effects were corrected for using experimentally developed factors.

1.9.1 Taxonomic Composition

The macrozooperiphyton was found to be represented by the Coelenterata (*Cordylophora caspia* and *Hydra* sp.); Bryozoa (*Plumatella repens* and *P. fungosa*); Oligochaeta (almost exclusively naidids, mainly *Nais variabilis* and *Stylaria lacustris*); Chironomidae (more than 90% chironomid larvae were the ubiquitous periphytic forms: *Glyptotendipes* sp., *Dicrotendipes tritomus, Cryptochironomus* e.g. *pararostratus, Cricotopus (Isocladius)* sp., *Tanytarsus lauterborni* and non-identified Pelopiinae); Amphipoda (mainly *Gammarus tigrinus* and *Corophium curvispinum*; in addition, low abundances of *Obesogammarus crassus* were recorded as well); Cirripedia (*Balanus improvisus*); Hirudinea (*Glossiphonia heteroclita, G. complanata, Helobdella stagnalis*, and *Herpobdella octooculata*); Gastropoda (mainly *Bithynia tentaculata* and, in lower abundances, *Lymnaea peregra* and *L. auricularia, Theodoxus fluviatilis* and *Acroloxus lacustris*); Bivalvia (only *Dreissena polymorpha,* mainly as post-veligers and small young-of-the-year individuals); and remaining taxa (some copepods, cladocerans, beetle larvae –mainly *Donacia* sp., and larvae of the Ceratopogonidae and Trichoptera, represented mainly by the Hydroptilidae).

1.9.2 Assumptions Adopted in Calculations

- In view of the lack of relevant studies concerning other rush plants, data obtained for reed were regarded as representative of all the helophytes (almost 72 and 60% of the biomass and coverage, respectively, of which being accounted for by reed).

- Similarly, data on the motile epifauna of pondweed were considered representative of the entire submerged and floating-leaved vegetation. A cursory examination of quantitative samples, however, showed colonial and sessile macrozooperiphyton to be virtually absent from the *Ceratophyllum, Myriophyllum* and *Elodea*-dominated plant communities. Therefore, when assessing quantitative aspects of macroepifaunal colonisation of the floating-leaved vegetation, an appropriate correction factor was used.

- The macrozooperiphyton biomass in the entire Great Lagoon was calculated based on data on the surface area offered as substrate by different macrophyte communities (cf. Section 1.6.3). It was assumed that helophytes occupying a square metre of the Great Lagoon bottom offer 2 m^2 of substrate, an average of 3 m^2 being offered by submersed and floating-leaved plants of all the relevant species.

1.9.3 Macrozooperiphyton Abundance and Biomass

The most abundant components of the macrozooperiphyton were oligochaetes and crustaceans (primarily amphipods). For example, a square metre of reed stem surface was found to support, on the average, 5440, 5160, 3860, and 78 individuals of oligochaetes, amphipods, larval chironomids, and gastropods, respectively. Taken together, oligochaetes, chironomids, crustaceans, and gastropods made up more than 85% of the motile macroperiphyton biomass (Table 1.18).

As already mentioned, the phytolittoral accounted for 0.155 of the surface area of the entire Great Lagoon. To express the epifaunal biomass in organic carbon units, conversion factors identical to those adopted for the benthos

Table 1.18 Macrozooperiphyton biomass in the phytolittoral of the Great Lagoon.

Taxa	Biomass per plant surface area [$g_{w.w.}$ m^{-2} substrate]		Biomass in macrophyte assemblage [$g_{w.w.}$ m^{-2}]		Average biomass in phytolittoral [$g_{w.w.}$ m^{-2}]
	Phragmittes	***Potamogeton* sp. div.**	**Emergent**	**Floating-leaved and submersed**	
Cordylophora caspia	64.3	42.6	128.6	37.0	59.4
Bryozoa	0.9	2.4	1.8	2.1	2.0
Balanus improvisus	0.1	0.6	0.2	0.5	0.4
Dreissena polymorpha	3.1	4.6	6.2	4.0	4.5
colonial and sessile forms, total	**68.4**	**50.2**	**136.8**	**43.6**	**66.3**
Oligochaeta	2.7	0.8	5.4	2.4	3.1
Chironomidae	1.7	2.2	3.4	6.6	5.8
Amphipoda	4.6	3.4	9.2	10.2	10.0
Hirudinea	0.3	1.2	0.6	3.6	2.9
Gastropoda	2.4	5.0	4.8	15.0	12.5
Remaining taxa	1.4	0.8	2.8	2.4	2.5
Motile forms total	**13.1**	**13.4**	**26.2**	**40.2**	36.8
Macrozooperiphyton total	**81.5**	**63.6**	**163.0**	**83.8**	**103.1**

(see the following section), i.e., 0.088, 0.077, 0.070, 0.026, and 0.064 gC $g_{w.w.}^{-1}$ for chironomid larvae, oligochaetes, hirudineans, molluscs (with shells), and the remaining motile forms, respectively. The conversion factor for *Balanus*, represented almost exclusively by the young-of-the-year individuals, was 0.026 gC $g_{w.w.}^{-1}$, i.e., identical to that of molluscs. A conversion factor of 0.04 gC $g_{w.w.}^{-1}$ was used for the two colonial forms, *Cordylophora caspia* and bryozoans having a chitinous external skeleton (zoecium, perisarc), the factor being identical to that of ostracods (Galtsova, 1991) whose thick, chitinous-calcareous carapace accounts for much of an individual weight.

The macrozooperiphyton biomass data (along with their organic carbon equivalents) averaged over the entire Great Lagoon are shown in Table 1.19.

Table 1.19 Macrozooperiphyton biomass in the entire Great Lagoon. Cc, organic carbon conversion factor (see text).

Taxon	Average biomass in phytolittoral [$g_{w.w.}$ m^{-2}]	Average biomass in the entire Lagoon [$g_{w.w.}$ m^{-2}]	Cc [gC $g_{w.w.}^{-1}$]	Average biomass in the entire Lagoon [gC m^{-2}]
Cordylophora caspia	59.4	9.21	0.04	0.368
Bryozoa	2.0	0.31	0.04	0.012
Balanus improvisus	0.4	0.06	0.026	0.002
Dreissena polymorpha	4.5	0.70	0.026	0.018
colonial and sessile forms, total	**66.3**	**10.28**		**0.400**
Oligochaeta	3.1	0.48	0.077	0.040
Chironomidae	5.8	0.90	0.088	0.079
Amphipoda	10.0	1.55	0.064	0.099
Hirudinea	2.9	0.45	0.070	0.031
Gastropoda	12.5	1.94	0.026	0.050
remaining taxa	2.5	0.39	0.064	0.025
motile forms total	**36.8**	**5.71**		**0.324**
Macrozooperiphyton total	**103.1**	**15.99**		**0.724**

1.9.4 Macrozooperiphyton Production

To calculate production (Table 1.20) of some macrozooperiphyton components, the P/B values identical to those of the macrobenthos (see the following section, Tab. 1.26) were used, as the two ecological categories are dominated by identical genera or even species. This was particularly the case with oligochaetes, gastropods and hirudineans. For all Chironomidae, the P/B of 7 was applied, like for *Glyptotendipes sp.*, the dominant genus among periphytic chironomids. The P/B of 3 was applied for all Amphipoda, basing on the value adopted for Gammaridae (Witek, 1995). Periphytic *Dreissena polymorpha* had higher P/B value than benthic representatives, because the periphytic forms were small and fast-growing. The young zebra mussels grow at the fastest rate and show a low mortality (Wiktor, 1969). For comparison, P/B of 2.54 and 0.48 was reported by Stallone (1979) for *Mya arenaria* aged 2 and 8 years, respectively. Estimation of the production rate of colonial organisms such as bryozoans or *Cordylophora* hydroids was difficult. P/B of 2 was applied to bryozoans, *Cordylophora*, and

Table 1.20 Biomass and production of macrozooperiphyton in the Great Lagoon.

Taxon	Biomass [gC m^{-2}]	P/B	Production [gC m^{-2} yr^{-1}]
Cordylophora caspia	0.368	2	0.736
Bryozoa	0.012	2	0.024
Balanus improvisus	0.002	2	0.004
Dreissena polymorpha	0.018	2	0.036
colonial and sessile forms total	**0.400**		**0.800**
Oligochaeta	0.040	5	0.200
Chironomidae	0.079	7	0.553
Amphipoda	0.099	3	0.297
Hirudinea	0.031	2	0.062
Gastropoda	0.050	2,2	0.110
remaining taxa	0.025	3	0.075
motile forms total	**0.324**		**1.297**
Zooperiphyton total	**0.724**		**2.077**

Balanus in view of Kajak's (1988) opinion that P/B of the species having a life span of 1 year usually ranges within 2-4.5. Most likely, however, such P/B values (usually 2) applied to species with a one-year-long life span are underestimates. As found out by Kamiński (1992), the common bryozoan *Plumatella fungosa* may produce 3 generations during a single growing season.

1.10 Macrozoobenthos

The part of the Great Lagoon macrobenthos that was studied most thoroughly in 1998-2002 was the pelophilous (muddy bottom) fauna. In addition, the distribution and abundance of *Dreissena polymorpha* throughout the Odra River mouth system was studied in detail in 2000-2004. In contrast, the littoral macrobenthos was investigated less intensely. Most of the data were obtained from samples collected in 2001 and 2002 on the eastern and western shores of the central basin and on Lake Wicko and the Skoszewska Cove. Earlier data (1983-1985) of Wolnomiejski and Grygiel (1994a) were used in this study as well.

The sampling and sorting methods used followed those described by Wolnomiejski and Grygiel (1989) and Wolnomiejski (1994). Samples were collected with a 7.5 kg Ekman-Birge grab and a standard 16 kg Van Veen grab;

the compact littoral sediments were sampled with a tube corer mounted on a pole. The sediment collected was washed on an 0.5 mm mesh size sieve. The animals were sorted and weighed live. In addition to the typical macrobenthos, the mud surface featured semi-benthic ostracods of a size transitional between the macro- and the meiobenthos. The macrobenthic fraction of ostracod biomass was estimated from separate samples, by extracting and sorting live ostracods from subsamples of the mud washed on the 0.5 mm mesh size sieve. The meiobenthic ostracods were studied using methods applicable to the meiobenthos (Radziejewska, pers. comm.), and are discussed in the section dealing with the Lagoon's benthic meiofauna.

1.10.1 Macrobenthic Habitats, Taxonomic Composition, and Biomass

The taxonomic composition, dominance structure, and quantitative parameters of the macrobenthic community structure are largely controlled by the sediment type. The basic sediment types in the Lagoon are mud (silts), sandy mud, muddy sands, and sands (cf. Section 1.2.2.2). The spatial distributions of those sediment types in the Lagoon are determined by the topography of the basin, depth, and shore exposure to wind-driven wave action.

The mud-type bottom (mud and sandy mud) occupies 50-55% of the Great Lagoon central basin's floor. The pelophilous macrofauna is species-poor. The major and constant components include larvae of *Chironomus* sp. and *Procladius* spp. as well as oligochaetes (Tubificidae). Larvae of *Chironomus* sp. typically dominate the biomass, the total fauna abundance being dominated by oligochaetes. The abundance and biomass of the pelophilous macrobenthos in successive growing seasons of 1998-2002 are shown in Table 1.21.

Noteworthy is a high variability of both the abundance and biomass of the macrozoobenthos observed during the period of study (the coefficient of variation ranging from 0.56 to 0.82) and the synchrony of quantitative changes in the major components, i.e., larval *Chironomus* sp. and oligochaetes. In 1998 and 2001, the fauna showed a particularly high biomass, its average exceeding 50 $g_{w.w.}$ m^{-2}, while a distinctly reduced biomass was observed in 2002.

Taken together, the two major macrobenthos components, i.e., oligochaetes and larval *Chironomus*, accounted for 97% of both the abundance and biomass of the pelophilous macrofauna. The dominance structure was similar in every year: oligochaetes dominated the abundance (55-82% depending on the year), larval *Chironomus* being a biomass dominant (62-81%).

The muddy sand on the offshore slope and at the foot of shallows, within the 2.5-4.5 m depth zone, support both pelophilous and littoral forms (Table 1.22). The occurrence of littoral forms was facilitated not only by the presence

Table 1.21 Average abundance and biomass of the pelophilous macrofauna in the muddy sediments of the central part of the Great Lagoon in the growing seasons of 1998-2002.

Taxon and taxonomic group	1998	1999	2000	2001	2002	Average	Coeff. of variation
				Abundance, ind. m^{-2}			
Chironomus sp.	1636	750	870	2730	410	1279	0.724
Procladius spp.	199	262	201	63	66	158	0.564
Other Chironomidae	13	19	68	32	17	30	0.755
Oligochaeta	8728	3385	1273	3778	1390	3711	0.815
Remaining taxa	33	16	24	10	5	18	0.633
MACROFAUNA total	10610	4432	2436	6613	1888	5196	0.684
				Biomass, $g_{w.w.}$ m^{-2}			
Chironomus sp.	37.7	12.2	12.7	44.1	7.7	22.9	0.730
Procladius spp.	0.5	0.7	0.4	0.2	0.1	0.4	0.628
Other Chironomidae	+	+	+	+	+	+	
Oligochaeta	18.9	6.8	3.2	10.0	3.0	8.4	0.782
Remaining taxa	0.1	+	0.4	+	+	0.1	
MACROFAUNA total	57.2	19.7	16.7	54.3	10.8	31.8	0.699

+ less than 0.1 g m^{-2}

Table 1.22 Abundance and total biomass of the benthic macrofauna in different littoral and sublittoral habitats, as exemplified by selected samples.

Main taxon and taxonomic group	Muddy sand in sublittoral, 3.5-4 m depth Jul. 2001	Sandy bank, 2 m depth Sep. 2000	Bottom with *Dreissena* bed, 3 m depth Jul. 1983	Bottom covered by plant remains within helophyte belt, 1 m depth. May 2000
Abundance [ind. m^{-2}]:				
Chironomus sp.	962		155	42
Procladius spp.	221			166
Glyptotendipes sp.	305	707	27 946	5395
Dicrotendipes gr. *nervosus*		49	14 606	83
Cladotanytarsus sp.	59	99	241	996
Other Chironomidae	778	222	2833	1826
Oligochaeta	1918	5491	1502	4233
Sphaeridae n.d.	68		10	322
Gastropoda sp. div.	397	191	386	581
Hirudinea sp. div.	7		1729	152
Crustacea sp. div.	113	49	126	42
Remaining taxa	100	343	896	1342
Total abundance	**4928**	**7151**	**50 430**	**15 180**
Total biomass [$g_{w.w.}$ m^{-2}]	**33.8**	**30.8**	**89.4**	**49.5**

of detritus and shell debris increasing the sediment compactness, but also by the proximity to the near-shore zone from which those organisms can be transported by water circulation. Important are also local beds of *Dreissena polymorpha* offering a particularly suitable habitat for numerous taxa typical of the littoral.

Beds of *Dreissena polymorpha*, occurring both in the littoral and in the sublittoral, create a suitable spatial and/or trophic niche for numerous benthic organisms, particularly for hirudineans (mainly *Herpobdella* and *Helobdella*), gastropods (particularly *Valvata* and *Bithynia*), crustaceans (especially *Asellus aquaticus* and *Corophium* sp.), and chironomid larvae (*Glyptotendipes* sp., *Polypedilum nubeculosum*, and *Dicrotendipes* gr. *nervosus*). The littoral and sublittoral zebra mussel beds supported the most abundant macrobenthos.

The eu- and upper littoral were colonised by the taxonomically most diverse fauna (Table 1.22). This diversity was particularly evident where the bottom was covered by large macrophytic remains and mollusc shells. Such habitats supported a particularly high diversity of chironomid larvae. These were, however, mostly small-sized forms (e.g., the Tanytarsini, *Cryptochironomus* sp., and *Dicrotendipes* gr. *tritomus*), for which reason their biomass was relatively low. The bottom overgrown by macrophytes was inhabited by abundant gastropods (mainly *Valvata piscinalis*, *Bithynia tentaculata*, and *Potamopyrgus antipodarum*) and small sphaerid bivalves (particularly *Pisidium* sp. div.). Among the remaining taxa, those occurring most frequently included crustaceans as well as larval trichopterans and ephemeropterans. All the samples from the littoral yielded as few as three individuals of polychaete *Marenzelleria* sp. Like in other eutrophic water bodies, the littoral benthos was very patchy in its distribution, the patchiness being evident even in habitats situated close to one another.

The macrobenthic biomass supported by major habitats of the Great Lagoon, estimated from information provided by samples collected mainly in 1998-2002, is shown in Table 1.23.

A separate, very important aspect of the Szczecin Lagoon biota is the exceedingly abundant presence of the zebra mussel (*Dreissena polymorpha*). A detailed study on the distribution and abundance of the bivalve in the Polish part of the Szczecin Lagoon and the adjacent water bodies was carried out in 2000-2004 (Woźniczka and Wolnomiejski, 2005). The crude weight (with shells) of *D. polymorpha* was determined for each sample using the length (L, mm) - weight (W, mg) relationship calculated from the locally obtained material. The relationship took the form of

$$W = 0.1487 \times L^{2.946}$$

This relationship is very similar to the data obtained by Wiktor (1969) for the zebra mussel from the Szczecin Lagoon. The crude weight includes the water

Table 1.23 Average biomass [$g_{w.w.}$ m^{-2}] of macrobenthos in main bottom habitats of the Great Lagoon (zebra mussels, unionids and macrobenthic ostracods not included).

Bottom habitat / Taxon and taxonomic group	Muddy deposits of central basin; 4.5-6 m depth	Muddy sand bottom of shoal slope (sublittoral); 2-4 m depth	Sandy bottom; down to 2 m depth	Bottom with *Dreissena* bed; down to 4.5 m depth	Bottom covered by plant remains and overgrown with macrophytes; down to 2 m depth
Chironomus sp.	22.9	16.2	0.7	0.9	1.0
Procladius spp.	0.4	0.8	0.2	+	0.2
Glyptotendipes sp.	+	1.7	1.3	18.3	5.4
Dicrotendipes gr. *nervosus*	+	0.3	0.1	4.0	0.1
Other Chironomidae	0.1	1.3	0.8	2.7	1.6
Oligochaeta	8.4	5.5	4.0	2.4	3.8
Mollusca sp. div. [a]	+	9.3	13.0	24.5	45.2
Hirudinea sp. div.		0.1	0.4	12.6	3.7
Remaining taxa [b]	0.1	2.1	2.2	3.3	3.5
MACROFAUNA total	31.9	37.3	22.7	78.3	64.5
Number of samples	91	22	11	5	5
Habitat contribution to total bottom area	50%	20%	15%	10%	5%

[a] without Unionidae and *Dreissena* (estimated separately)
[b] without macrobenthic Ostracoda (estimated separately)
+ less than 0.1 g m^{-2}

contained in the mantle cavity which is not a part of any tissue. Analyses showed this water to account for about 20% of the weight. Thus, subtraction of this water produced the true wet weight of the bivalves (with shells).

The zebra mussel was very unevenly distributed. This inhomogeneity was primarily due to variation in depth, substrate type, and exposure of the bottom to wave action. The highest abundances of the bivalve were recorded on the slopes of shallows (in the sublittoral) within the 3-4.5 m depth zone. In contrast, the muddy bottom at depths exceeding 5 m and sandy shoals within the 1-2.5 m depth zone were apparently the most inhospitable habitats for the zebra mussel (Table 1.24).

Table 1.24 Depth-related variability of *Dreissena polymorpha* abundance in the main basin of the Great Lagoon (data 2000-2004).

Depth range [m]	Number of samples	Average abundance [ind. m^{-2}]	Average biomass (with shells) [$g_{w.w.}$ m^{-2}]
0 - 0.5	23	1653	122
0.5 - 1	25	1993	193
1 - 1.5	23	196	19
1.5 - 2	25	859	33
2 - 2.5	22	265	27
2.5 - 3	21	957	94
3 - 3.5	25	2096	278
3.5 - 4	32	1517	297
4 - 4.5	20	1807	384
4.5 - 5	23	476	105
5 - 5.5	28	95	34
5.5 - 6	24	1	1
> 6	31	≈ 0	≈ 0

The bottom surface area-scaled (between 1 m isobaths) estimates of the zebra mussel biomass showed the main basin of the Lagoon to support an average biomass of 114 $g_{w.w.}$ m^{-2}, which means that the total wet weight (with shells) of the bivalve in the main basin of the Lagoon amounted to 39 thou. t. Overall biomass of the zebra mussel in the whole Great Lagoon (including embayments) was 58 330 t (Table 1.25).

Data with which to estimate the biomass of unionid bivalves (mainly *Unio tumidus* and *Anodonta anatina)* were obtained from the 564 stations sampled for the zebra mussel. The estimated unionid biomass in the Great Lagoon averaged

Table 1.25 Biomass (with shells) of *Dreissena polymorpha* in different parts of the Great Lagoon in 2000-2004.

Part of the Great Lagoon	Number of stations	Surface area [km^2]	Average biomass [$g_{w.w.}$ m^{-2}]	Total biomass in the area [$tonnes_{w.w.}$]
Main basin	322	343	114	39 102
Roztoka Odrzańska	40	26	330	8580
Skoszewska Cove	143	20	493	9860
Lake Nowowarpieńskie	26	9	13	116
Lake Wicko	33	12	56	672
Total	564	410	142[a]	58 330

[a] weighted average

12.19 $g_{w.w.}$ m^{-2}, i.e., the entire Great Lagoon supported 5 thou. t (wet weight with shells) of those bivalves. The unionids were definitely most abundant in the Lagoon's embayments, particularly in the Roztoka Odrzańska (the River Odra mouth) where the Odra discharge ensured an intensive water flow.

Biomass of the macrobenthic ostracods was estimated following a protocol used in an earlier multi-year study (Wolnomiejski and Grygiel, 1989). For the three major zones of the bottom (their respective contributions to the entire bottom surface area are given in the parentheses below), the biomass – as estimated from the 11 samples collected in 1998-2000 was as follows:

littoral (near-shore shoals) (25%) – 0.7 $g_{w.w.}$ m^{-2},
sublittoral (slopes of shallows) (20%) – 4.5 $g_{w.w.}$ m^{-2},
muddy bottom in the central basin (55%) – 8.2 $g_{w.w.}$ m^{-2}.

The Lagoon-wide (weighted) macrobenthic ostracod biomass averaged 5.58 $g_{w.w.}$ m^{-2}.

1.10.2 Macrobenthos Production

The macrobenthos production was calculated based on the empirically derived biomass values and production rates (P/B) extracted from the literature. The P/B values applied were arithmetic means or modes of values reported by numerous authors (Table 1.26). The P/B values reported from areas outside of the temperate zone were disregarded; such values were usually lower at high latitudes, particularly with respect to chironomid larvae. For the zebra mussel,

Table 1.26 Selection of P/B values for representatives of macrobenthos.

Taxon	P/B range	References	Average P/B	Applied P/B
Chironomus sp.	4.7-14.15	Sokolova (1971); Winberg et al.(1972); Giziński and Wiśniewski (1971); Żytkowicz (1976); Wiśniewski (1976); Wolnomiejski and Grygiel (1989); Krylova and Ten (1992); Stead et al. (2005)	8.05	8
Procladius sp.	4.24-5.7	Andronikova et al. (1972); Winberg et al. (1972); Żytkowicz (1976)	4.91	4.5 [m]
Glyptotendipes sp.	5.5-8.88	Malej (1974); Morgan et al. (1980); Humphreys (1981)	7.56	7 [m]
Chironomidae without Tanytarsini and Orthocladiinae	5.3-12.88	Kajak and Rybak (1966); Sokolova (1971); Maitland et al. (1972); Winberg et al. (1972); Wolnomiejski et al. (1976); Giziński et al. (1976); Wiśniewski (1980)	9.06	9
Chironomidae with Tanytarsini and Orthocladiinae	5.3-35.0	Above-mentioned authors and Żytkowicz (1976); Kazantseva (2004); Stead et al. (2005)	10.27	10
Tubifex, *Limnodrilus* and *Potamothrix* sp.	3.0-5.4	Winberg (1979); Wetzel (2001)	4.83	
Oligochaeta overall (mostly Tubificidae)	3.4-8.4	Gak et al. (1972); Winberg et al. (1972); Wolnomiejski and Giziński (1975); Żytkowicz (1976); Morgan et al. (1980); Kazantseva (2004); Stead et al. (2005)	5.12	5
Hirudinea	2.0-4.1	Gak et al. (1972); Winberg et al. (1972); Wetzel (2001)	2.61	2 [m]
Ostracoda		Herman and Heip (1985)	2.7	2.7
Dreissena polymorpha		Walz (1978 in Dermott et al., 1993) for juvenile forms in Lake Constance	6.8	
Dreissena polymorpha	0.40-2.31	Mikulski et al. (1975); Stańczykowska (1976; 1977); Winberg (1979); Kajak (1988); Mackie et al. 1989)	0.88	
Dreissena polymorpha		Wiktor (1969), in Szczecin Lagoon	≈ 1.0	
Dreissena polymorpha	1.02-1.13	Based on *Dreissena* population model (Wolnomiejski, unpubl. data)		1.0
Unionidae	0.13-0.35	Mann (1971); Winberg (1979); Humphreys (1981); Wetzel (2001)	0.3	
Unionidae		Lewandowski and Stańczykowska (1975)	0.33	0.35
Sphaeridae and Gastropoda	0.52-4.44	Mann (1971); Mann et al. (1972); Mattice (1972); Winberg et al. (1972); Winberg (1979); Alimov (1981); Gaten (1986); Wetzel (2001)	2.2	2.2
Remaining taxa (Crustacea, Trichoptera, Ceratopogonidae, Ephemeroptera and other)	1.5-6.2	Beattie et al. (1972), Gak et al. (1972); Winberg et al. (1972); Morgan et al. (1980); Winberg (1979); Witek.(1995); Wawrzyniak-Wydrowska (1997); Wetzel (2001)	3.8	4.0

[m] modal value

the P/B value determined by Wolnomiejski (unpubl.) in the Great Lagoon was applied. Majority of the P/B values shown in Table 1.26 were determined for the growing season. The growing season (210 days) production was assumed to be an equivalent of the annual production.

Estimates of production rate of *Dreissena polymorpha* published by various authors differ widely. For an overall assessment of the zebra mussel production potential, estimates based on soft parts only (Dermott et al., 1993) or those concerning juvenile forms, are of low utility. Most authors arrived at P/B lower than 1. In contrast, Wiktor (1969) found that in the flow-through Szczecin Lagoon, the zebra mussel production rate is clearly higher than that reported for Polish lakes, and estimated it at ≈ 1. Estimates of Wolnomiejski (unpubl.) based on the population model relying on shell length distribution (samples from more than 300 stations sampled in 2001-2003) corroborated the conclusion drawn by Wiktor (1969).

The P/B values of the remaining macrobenthos were based on the following premises:

- the average P/B of herbivorous benthos is 3.7 (ranging within 0.6-12.6) (Brylinsky, 1980),
- P/B of an annual (1-yr-long life cycle) species is usually 2-4.5 (Kajak, 1988),
- P/B of univoltine (single-generation) insect larvae is 5 (Waters, 1979 quoted by Grzybkowska, 2002),
- the average P/B values of multi-generation, two-generation, single-generation, 2-3 yr-old, and long-lived forms are 8.3; 5.4; 4.5; 2.3; 1.7, respectively, the overall average being 4.44 (Wetzel, 2001).

The production of the Great Lagoon macrofauna was estimated (Table 1.27) based on the weighted average biomass of individual macrobenthos components (Table 1.23) and the above-mentioned (Section 1.10.1) data for *Dreissena*, unionids, and macrobenthic ostracods.

On the scale of the entire basin, *D. polymorpha* contributed more than 72% of wet weight biomass, 7.5% being contributed by *Chironomus* sp. The proportions levelled off in production: the zebra mussel and *Chironomus* sp. accounted for 38 and 32% of the total Great Lagoon macrobenthos production, respectively. Important contributors to the total macrobenthic productions were also oligochaetes (> 8%, mainly the pelophilous forms) as well as small littoral molluscs and larvae of *Glyptotendipes* sp. chironomids, whose contributions amounted to 5% each.

In the face of scarcity of data on carbon content of benthic invertebrates, the energy contents of the major macrobenthic components (Table 1.28) were extracted from the literature. The energy contents of molluscs were estimated based on the assumptions concerning *D. polymorpha*. The soft tissue weight

Table 1.27 Average biomass (wet weight) and production of the macrobenthos per standard m² of the entire Lagoon.

Taxon	**Biomass [$g_{w.w.}$ m^{-2}]**	**P/B**	**Production [$g_{w.w.}$ m^{-2} yr^{-1}]**
Chironomus sp.	14.94	8.0	119.5
Procladius spp.	0.40	4.5	1.8
Glyptotendipes sp.	2.64	7.0	18.5
Dicrotendipes spp.	0.48	9.0	4.3
Other Chironomidae	0.78	10.0	7.8
Oligochaeta	6.33	5.0	31.6
Hirudinea	1.47	2.0	2.9
Ostracoda (macrobenthic)	5.58	2.7	15.1
Dreissena polymorpha (with shells)	142.27	1.0	142.3
Unionidae (with shells)	12.19	0.35	4.3
Other Mollusca (with shells)	8.52	2.2	18.7
Remaining forms (Crustacea, Trichoptera, Ceratopogonidae, Ephemeroptera and other)	1.31	4.0	5.2
Macrobenthos total	**196.91**	**1.9**	**372.0**
Macrobenthos without *Dreissena* and Unionidae	42.45	5.3	225.4

Table. 1.28 Caloric value and average organic carbon content of macrobenthos components, assuming 1 kcal to be equivalent to 0.08 gC (Vinogradov and Shushkina, 1987).

Taxon	**Caloric value [cal $g_{w.w.}^{-1}$]**	**Data source**	**Average carbon content [gC $g_{w.w.}^{-1}$]**
Chironomus sp.	1100-1375	Based on data cited by Sokolova (1983)	0.098
Other Chironomidae	1010-1210	Rodgers and Qadri (1977); Penczak et al. (1982; 1984); Rumohr et al. (1987)	0.088
Oligochaeta	890-980	Penczak et al. (1982); Rumohr et al. (1987); Witek (1995)	0.077
Ostracoda	500	Rumohr et al. (1987); Galtsova (1991)	0.040
Mollusca with shells[a]	324	Wiktor (1969); Stańczykowska and Ławacz (1976); Alimov (1981); Penczak et al. (1984); Brey et al. (1988)	0.026
Remaining taxa	645-910	Rodgers and Qadri (1977); Penczak et al. (1982; 1984); Rumhor (1987)	0.064
Hirudinea	No data	Compilation of 0.077 (as for Oligochaeta) and 0.064 (as for remaining taxa)	0.070

[a] calculated based on published data (see text for explanations)

accounts for 60% of the wet weight (with shell) (Wiktor, 1969). The mollusc dry weight accounts for 12% of wet weight without shell (Penczak et al., 1984). The mollusc caloric content (4500 cal $g_{d.w.}^{-1}$) was adopted following Brey et al. (1988). Very similar data were reported by Alimov (1981): 4530 cal $g_{d.w.}^{-1}$ for dreissenids, and 4233 cal $g_{d.w.}^{-1}$ as a freshwater bivalve average.

Identical conversion factors were applied to other molluscs in which the contribution of soft tissues to the wet weight with shell is usually lower, but dry weight contributes more to the wet weight without shell (Stańczykowska and Ławacz, 1975; Rumohr et al., 1987). It is difficult to obtain unequivocal data for whole molluscs because of difficulties with accurate determination of the amount of external water and the weight of other external materials, e.g., sand or other particles adhering to the shell. In addition, the bivalve groups identified (unionids and other molluscs) consist of taxa differing widely in their structure and shell development.

When converted to organic carbon units (Table 1.29), *D. polymorpha* contributed 53% to the macrobenthos biomass in the whole Great Lagoon, *Chironomus* sp. contributing 21%. In contrast, *Chironomus* sp. contributed more than 52% to the total macrobenthos production, the zebra mussel accounting for 16.5%.

Table 1.29 Average biomass and production (in carbon units) of macrobenthos on standard m² of the Lagoon.

Taxon	**Biomass [gC m^{-2}]**	**Production [gC m^{-2} yr^{-1}]**
Chironomus sp.	1.46	11.71
Procladius spp.	0.04	0.16
Glyptotendipes sp.	0.23	1.63
Dicrotendipes spp.	0.04	0.38
Other Chironomidae	0.07	0.69
Oligochaeta	0.49	2.43
Hirudinea	0.10	0.20
Ostracoda (macrobenthic)	0.22	0.60
Dreissena polymorpha	3.70	3.70
Unionidae	0.32	0.11
Other Mollusca	0.22	0.49
Remaining forms (Crustacea, Trichoptera, Ceratopogonidae, Ephemeroptera and other)	0.08	0.33
Macrobenthos total	**6.97**	**22.43**
Macrobenthos without *Dreissena* and Unionidae	2.95	18.62

The organic carbon P/B of the entire macrobenthos was 3.2, the value increasing to 6.3 if large bivalves (the zebra mussel and unionids) are excluded.

Ecological efficiency of the macrobenthos relative to the phytoplankton primary production was high (5.6 and 4.7% for the entire macrobenthos and the macrobenthos without large bivalves, respectively). According to Kajak (2001), in freshwater reservoirs the corresponding efficiency ranges within 0.2-6.0%.

1.10.3 Emergence of *Chironomus* sp.

Sokolova (1983) reviewed dates of mass emergence of *Chironomus* sp. reported in the literature for different water bodies and obtained in her own study. Emergence in May and August is the rule in the temperate zone. In the Vistula Lagoon, the insects emerge in May and July, a third mass emergence occurring occasionally during hot summers (Krylova and Ten, 1992). In the Szczecin Lagoon, the mass emergence occurred usually in the second and third decades of May and in the first and second decades of August.

Energy losses and the removal of a part of the *Chironomus* production out of the Lagoon were estimated using unpublished data of Wolnomiejski on individual weights of pupae and imagines as well as *in situ* observations. Assuming the annual *Chironomus* production to be 100%, the estimates are as follows:

- Based on a year-long sampling in the Great Lagoon (Wolnomiejski and Grygiel, 1989), the mass of the two largest size classes (20-22 and >22 mm) of *Chironomus* sp. larvae beginning to pupate, in the two periods of pupation and mass emergence, was estimated as equal to about 25% of the annual production.
- It was assumed that about 25% of the larvae are lost during pupation due to a distinct increase in natural mortality of large larvae (insufficient weight of diseased and genetically impaired larvae) and fish predation on the largest larvae. The remaining 18.75% of the production enter pupation.
- About one third of the total weight of larvae are lost during pupation due to energy expenditure on destruction and remodelling of internal organs and external extremities (the individual weight of the largest larvae and pupae averages 34 and 23 mg, respectively; Wolnomiejski, 1994). Grzybkowska (2008) described a number of processes accompanying pupation: organ transformation, muscle degeneration, enzymatic decomposition, and appearance of additional morphological structures, including those used for respiration and swimming. The completely formed pupae account then for about 12.5% of the larval production.

- During migration of the pupae to the water surface and emergence of imagines, one third of the pupae was assumed to be lost in insuccessful emergence (fish and bird predation and sinking of the imagines due to wave action). The remaining 8.4% of the larval production are involved in transformation to the adult form.
- Assuming ca. 20% weight loss during metamorphosis into the adult form, 6.74% of the production remains.
- As exuvia account for ca. 60% of the weight of the metamorphosing pupae, the emerged adults leaving the basin account for about 2.7% of the larval production. The exuvia float on the water surface for about 2-3 days and sink thereafter; a substantial portion of them is removed from the Lagoon with the runoff.

Darwall et al. (2010) who described extremely large mass emergence of *Chaoborus edulis* in Lake Malawi, concluded – based on their own observations and published data – that up to 40% of the annual production can contribute to emergence, but only a small proportion is removed from the lake as adults.

1.11 Meiobenthos

Meiobenthos of the Great Lagoon are poorly identified. According to Drzycimski (1989), it consists of 14 major taxa, the most common of which are ostracods, nematodes, turbellarians, rotifers, harpacticoid copepods, gastrotrichs, tardigrades, and halacarid mites. Ostracods and nematodes are distinctly most abundant, as reported by Radziejewska (2009, and unpubl. data) as well as by Radziejewska and Drzycimski (1988; 1990). This study uses information supplied, prior to final processing of data, by Dr T. Radziejewska who sampled the meiobenthos monthly from April/May to October 2002, using a gravity corer at 4 stations in the Great Lagoon: two stations were located on the muddy bottom of the main basin, the remaining two on the sandy bottom of the Skoszewska Cove in its northern and southern part. The meiobenthic biomass for this study was calculated using standard individual weights of Arlt (1973).

As already mentioned, the meiobenthos was clearly dominated by ostracods and nematodes. The average wet weight-based biomasses of the meiobenthos occurring in the uppermost centimetre of the sediment were:

Station 1 (NE part of the main basin, 6 m depth)	-	34.0 $g_{w.w.}$ m^{-2}
Station 2 (central part of the main basin, 5.5 m depth)	-	12.0 $g_{w.w.}$ m^{-2}
Station 3 (Skoszewska Cove, 4 m depth)	-	39.2 $g_{w.w.}$ m^{-2}
Station 4 (Skoszewska Cove, 3 m depth)	-	32.3 $g_{w.w.}$ m^{-2}

The biomass of meiobenthic organisms occurring in the uppermost centimetre of the sediment averaged 29.4 $g_{w.w.}$ m^{-2}.

According to Radziejewska (pers. comm.), the uppermost centimetre of the sediment supported about 70% of the total meiobenthic biomass. Therefore, the total biomass averaged 42 $g_{w.w.}$ m^{-2}. Ostracods and nematodes accounted for about 70 and about 30%, respectively, the contribution of the remaining taxa not exceeding 1%. In terms of the contributions of individual trophic groups, detritivores, suspension feeders, predators, and herbivores accounted for 74, 25, 0.6, and 0.3%, respectively (Radziejewska, pers. comm.). Detritus feeders included almost all nematodes and a substantial proportion of ostracods. Some of the latter (particularly *Cypria ophtalmica*) were suspension feeders; halacarids and tardigrades were herbivores, while the few predators present were encountered mainly among ostracods.

The meiobenthos production rate (P/B) is most frequently assumed as 8-10 (McIntyre, 1969; Hillbricht-Ilkowska et al., 1975; Drgas 2000). According to the data reported by Kazantseva (2004) for a lake near Sankt-Petersburg (Russia), the meiobenthic P/B in two consecutive years amounted to 6.5 and 10.7. Usually, the highest production rate has been ascribed to nematodes: Gerlach (1971), Heip and Herman (1988), and Galtsova (1991) used the values of 9-10, 10.1-35.3, and 3-53, respectively. In contrast, low P/B values are typical of ostracods. Wetzel (2001) stated that ostracods produce usually one generation per year; their metabolism is slow and the period of maturity, when they do not grow, is long. Therefore the ostracod growth rate is low. According to Galtsova (1991), the annual P/B of marine ostracods is 5, whereas Drgas (2000) reported 2.9 for freshwater species. Similarly, Herman and Heip (1985) quoted 2.7 for a brackish-water ostracod species. Generally, it seems most appropriate to use 3 as the P/B value of ostracods (including their macrobenthic representatives), similarly to that of many benthic forms with a single generation during the year (Kajak, 1988; Wetzel, 2001). Herman and Heip (1985), too, mention that it is rational to apply the generalised P/B = 3 for ostracods.

Using their P/B estimates based on individual weights, Banse and Mosher (1980) revealed a general tendency towards reduced production rate with increasing body weight. However, they emphasised that the production rate of meiofaunal organisms, despite their small size, was 3-5 times lower than that expected from corresponding relationships found in for the macrobenthos. Herman and Heip (1985), too, mentioned a relatively low production rate of some members of the meiobenthos. Galtsova (1991) pointed to a generally low P/B of marine meiobenthos as well. However, production rate of some species, particularly among stream nematodes, may be much higher (Stead et al., 2005).

Considering the above-mentioned findings, the meiobenthos production was estimated using an overall generalised P/B of 8.0 (Hillbricht-Ilkowska et al., 1975). Such a production rate would be obtained assuming P/B values for ostracods and nematodes equal to 3 and 20, respectively, and the ostracod to nematode biomass ratio of 7:3. Thus, the wet weight-based production of the Great Lagoon meiobenthos was estimated at 42 $g_{w.w.}$ m^{-2} × 8 yr^{-1} = 336 $g_{w.w.}$ m^{-2} yr^{-1}.

The caloric content of the two meiobenthic dominants, nematodes and ostracods, differs markedly. According to Galtsova (1991), 1 g wet weight of nematodes contains 860 cal energy, 500 cal being contained in 1 g ostracods. Thus, assuming 7:3 ostracod to nematode biomass ratio in the Great Lagoon, the meiobenthos caloric content averages 608 cal $g_{w.w.}^{-1}$. The organic carbon content, assuming 1 kcal = 0.08 g C (after Vinogradov and Shushkina, 1987), amounts to 0.0486 g C $g_{w.w.}^{-1}$.

Thus, the Great Lagoon meiobenthos production, expressed in organic carbon units, was 336 $g_{w.w.}$ m^{-2} yr^{-1} × 0.0486 gC $g_{w.w.}^{-1}$ = 16.3 g C m^{-2} yr^{-1}.

1.12 Fish Fauna

1.12.1 Commercial Catches: Magnitude and Composition

The fish fauna in the Szczecin Lagoon is very diverse and highly abundant. According to unpublished data of Wolnomiejski, supplemented by information provided by Skóra (1996), the Great Lagoon supports 49 fish species, half of them being freshwater. There are also species that alternate between freshwater and marine habitats (mainly European whitefish, salmon, and eel), typically marine species (mainly herring and flatfishes), and a group of accidental species penetrating into the Lagoon from the entire catchment. Fifteen fish species have commercial importance and are reported in catch statistics (Table 1.30).

In 1998-2002, the major part (85%) of the reported commercial catches was made up by 3 species: roach, bream, and perch (Table 1.30). The mean yield at that time was 56 kg ha^{-1} yr^{-1}.

Roach catches were exceptionally stable from year to year (coefficient of variation v = 0.08). The total catch showed an identical coefficient of variation. The remaining major total catch components varied relatively little as well (v ranging from 0.22 to 0.36), although the variability was higher than that of roach.

In the 1960s and early 1970s, the Great Lagoon fish were considered unsuitable for human consumption due to the phenolic odour of fish meat. However, after that period fish muscle concentrations of pollutants did not exceed the critical level.

Table 1.30 Annual commercial fish catches/landings [tonnes] referred to 410 km² area of the Great Lagoon. Avg., average; v, coefficient of variation (data after catch statistics of the Maritime Office in Szczecin).

Fish species / Year	1998	1999	2000	2001	2002	Avg.	v
Roach (*Rutilus rutilus*)	738	860	778	818	905	820	0.08
Common bream (*Abramis brama*)	731	689	687	420	517	609	0.22
Perch (*Perca fluviatilis*)	696	601	356	496	463	522	0.25
Pike-perch (*Stizostedion lucioperca*)	76	138	107	96	89	101	0.23
Eel (*Anguila anguila*)	106	74	53	53	47	67	0.36
Blue bream (*Abramis ballerus*)	46	72	81	62	83	69	0.22
White bream (*Abramis bjoerkna*)	22	38	42	51	56	42	0.31
Burbot (*Lota lota*)	7	10	10	17	28	14	0.59
Pike (*Esox lucius*)	9	9	9	6	6	8	0.21
Whitefish (*Coregonus lavaretus lavaretus*)	1	4	5	9	4	5	0.62
Asp (*Aspius aspius*)	2	3	6	4	7	4	0.47
Trout (*Salmo trutta* m. *trutta*)	4	5	3	3	1	3	0.46
Wels (*Silurus glanis*)	2	2	0.1	1	0.1	1	0.91
Tench (*Tinca tinca*)	1	1	1	0	0.2	1	0.78
Herring (*Clupea harengus*)	5	0.1	36	15	1	11	1.31
Remaining species total	21	22	38	13	11	21	0.51
Total catch	**2467**	**2528**	**2212**	**2064**	**2218**	**2298**	**0.08**

1.12.2 Virtual Correction of Catch Magnitude and Composition

The catch data reported in Table 1.30 are not a fully plausible measure of exploitation potential of the Great Lagoon fish fauna. The data can be more aptly termed the "purchases of fish caught by professional fishermen in the Great Lagoon". Official catch/purchase data do not usually include the very important by-catch or consider a possibility of catching other species, particularly the very abundant ruffe and smelt. These are not considered to be species marketable for consumption, neither are they delivered to the purchase centres as the so-called fodder fish. They are usually thrown overboard together with under-sized fish of other species, or are taken on land and thrown away there, or are even used as fertiliser. Thus, they are virtually ignored by catch statistics which otherwise include "fish pest" or "other species" categories. As the ruffe and smelt are the most important diet items of the local predatory fish, particularly the pike-perch and perch (e.g., Wengrzyn, 1978; Wolnomiejski and Grygiel, 1994c) and black cormorant (Bzoma, 2008a; b), it was necessary to correct the Great Lagoon fish fauna abundance estimates and the fish role in the community functioning by complementing the annual catch data with contribution of these two species. The premises for the correction stemmed from:

- the magnitude of pre-war catches (Pęczalska, 1974; Kompowski and Neja, 1990); according to the German catch data, the annual ruffe catches in 1900-1924 averaged about 300 t. Before World War II, ruffe, stickleback, bleak and smelt taken together made up almost one-third of the total catch from the entire Lagoon and the adjacent waters which averaged 3555 t yr^{-1}. In both parts of the Szczecin Lagoon, the smelt catches averaged several hundred to 1000 t yr^{-1};
- the magnitude of Polish post-war catches (Pęczalska, 1974; Kompowski and Neja, 1990); the combined contribution of ruffe, stickleback, bleak, and smelt, to the post-war catches was reported as 8% by 1980 and a little above 4% later on. The reduction resulted from the cessation of purchasing those species for processing for fish meal. In the 1970s, the so-called "weeding" of the Great Lagoon yielded about 500 t of ruffe a year. Such activities were abandoned in later years;
- the magnitude of post-war German catches in the Kleines Haff (Pęczalska, 1974; Garbacik-Wesołowska et al., 1995); in 1967-1971, the annual ruffe catches of 170-375 t, i.e., 9 t per hectare, were reported. In 1972, the catches amounted to as much as 700 t. Later on, in 1982 and 1983, the catches were 425 and 580 t, respectively, a record-breaking catch of 1450 t being reported in 1990;

- results of control fishing carried out by the Sea Fisheries Institute, mainly by means of a small-mesh trawl (Koronkiewicz et al., 1997; Garbacik-Wesołowska et al., 1998; Psuty-Lipska et al., 2004; and other unpublished data sets). In 1997, the catches consisted primarily of perch (31% of the total catch), ruffe (21 %), roach (18%), bream (9%), and smelt (5%). In 1998, ruffe contributed 16% to the average catch per unit effort (the second most important species); smelt accounting for 3.6% (the fourth). In 2004, among almost 60 thou. fish individuals representing 22 species caught in 75 tows (Psuty-Lipska et al., 2004), ruffe made up almost 13% (the third most abundant species), smelt accounting for more than 10% (the fourth) of all the individuals in the catches. It has to be added that catches obtained by demersal tows hardly reflect the true abundance of the pelagic smelt. The abundance of smelt in the Great Lagoon varied from year to year, the highest abundance being recorded in, i.a., 1999. On the other hand, the high abundance of ruffe remained stable until 2004 and dropped markedly thereafter;
- a targeted research on ruffe and smelt by-catch in commercial fisheries in the Szczecin Lagoon (Wysokiński and Garbacik-Wesołowska, 1995a b; Garbacik-Wesołowska et al., 1998; Wesołowska and Boberski, unpubl. data). In the perch-roach gill net catches effected in 1995, ruffe were the second (following roach) most abundant species caught;
- a comparison with catches in other southern Baltic lagoons (Borowski and Dąbrowski, 1996; Borowski et al., 1998; Borowski, 2000; Pustelnikovas, 1998; Psuty-Lipska et al., 2004). For example, in the 1980s and 1990s, substantial amounts of ruffe and smelt were caught in the Curonian Lagoon, the two species accounting for up to 25 and 15% of the total catch, respectively.

Compilation of data from all the sources indicated above allowed to conclude that the official catch data should be corrected by 300 t of ruffe and 200 t of smelt, the amounts not reported in catch statistics. After correction, catches of adult fish (i.e., more or less the commercial stocks) from the 410 km^2 area of the Great Lagoon averaged 2798 t yr^{-1}, which – when broken down into the three major trophic groups – shows the catches of benthivorous (roach, common bream, ruffe, white bream, whitefish and tench), piscivorous/predatory (perch, pike-perch, eel, burbot, pike, asp, trout, and wels), and planktivorous (smelt, blue bream, herring and others, e.g. bleak, three-spined stickleback) species to amount to 1776, 721, and 301 t, respectively.

1.12.3 Additional Fishing Mortality of Undersized Fishes (by-catch)

1.12.3.1 Fishing Gear

Trawl catches are banned throughout the entire Polish part of the Szczecin Lagoon. In the Great Lagoon, the most common fishing gear are fyke nets (especially deep-water eel- directed fyke nets) and perch-roach gill nets. In 1998-2002, in the entire Polish part of the Szczecin Lagoon and in the adjacent waters, i.e., over the area of 514 km^{-2}, there were 152-158 (121-126 in the Great Lagoon) fishing boats; there were 2320-2410 (1850-1920) fyke nets and 5260-5630 (4195-4490) gill nets. The construction and application of those fishing gear types were described by Wysokiński and Garbacik-Wesołowska (1995 a, b.), Borowski and Dąbrowski (1996), Borowski et al. (1998), Garbacik-Wesołowska and Boberski (2000), and Psuty-Lipska (2006). In addition, some catches were also effected with boat seine nets the use of which requires special permission for defined localities; sporadically – hook and line fishing was carried out as well.

The perch-roach gill nets, and particularly the deep-water fyke nets, are non-selective with respect to numerous fish species, and catch and destroy substantial amounts of young, undersized individuals. This is particularly the case with respect to the fry and juveniles of pike-perch caught by fyke nets, although the cod-ends are equipped with protective sieves making it possible for the young fish to escape. The undersized fish (particularly roach and bream) as well as the so-called "undesired" species (particularly white bream) caught in fyke nets are usually thrown overboard, only few of them surviving. Gill nets retrieved from the water, together with fishes trapped in their meshes, are brought on land, thus all the fish are dead. The undersized pike-perch, perch, and eel caught are retained for consumption by fishermen or sold.

1.12.3.2 Undersized Fish in Fyke Net and Gill Net Catches

Analyses of harmful effects produced by fishing gear, described by authors referred to above, concerned the numerical magnitude of undersized fish by-catches. Here the extent of fry and juvenile fish biomass elimination from the Great Lagoon, as an additional fishing mortality associated with the fishing gear used, was estimated based on mass measurements performed mainly in 1994-2004 directly on board fishing boats (unpublished data made available by Dr. A. Garbacik-Wesołowska and Mr E. Boberski, M.Sc. of the National Marine Fisheries Research Institute, Branch in Świnoujście) and on unpublished data of the authors and opinions of persons associated with fisheries. The catch biomass was calculated with weight-length regression equations based on measurements of local fish (Wolnomiejski unpubl.). It was assumed that fyke net catches contributed 60% to the total annual catch, the perch-roach gill nets providing the remaining 40%. More than 30 thou. fish individuals (about 4 t) were measured altogether; undersized fish (fry and juveniles) accounted for 45 and 33% of the number of individuals and their total weight, respectively. The calculations performed for fyke net fisheries are shown in Table 1.31 as an example.

Table 1.31 By-catch of fry and under-sized fish of four major commercial species caught by fyke nets in the Great Lagoon, assuming 60% fyke net catch contribution to the total commercial catch (based on Garbacik-Wesołowska and Boberski, unpubl.). *l.t. – longitudo totalis.*

Fish species /minimum legal length ***l.t.*** **/ and size (age) categories**	**Data from analyses**				**Calculated data**
	Number (N) of fish caught and measured [indiv.]	**Weight (B) of fish caught [kg]**	**% N**	**% B**	**Average annual catch in the Great Lagoon [t]**
Rutilus rutilus /17 cm/:					
-fry	≈ 0				≈ 0
-below size limit (immature)	3107	122	34.3	17.9	107[a]
-above size limit (mature)	5960	561	65.7	82.1	491
Abramis brama /40 cm/:					
-fry	242	3.5	13.2	0.5	3
-below size limit (immature)	1292	307	70.6	42.5	272[a]
-above size limit (mature)	297	412	16.2	57.0	365
Perca fluviatilis /17 cm/:					
-fry	≈ 0				≈ 0
-below size limit (immature)	3753	160	50.5	28.6	125
-above size limit (mature)	3675	400	49.6	71.4	313
Stizostedion lucioperca /45 cm/:					
-fry	1557	65	39.5	6.5	10
-below size limit (immature)	2100	544	53.3	53.9	83
-above size limit (mature)	282	400	7.2	39.6	61

[a] ⅓ of released fish assumed to survive

The calculations involved adopting the following assumptions, as exemplified for pike-perch (*Stizostedion lucioperca*):

- the total annual catch (fish of the regular size as determined by the appropriate legal regulations) in 1998-2002 averaged 101 t; according to the reasoning presented above, the fyke net fishery contributed 60%, i.e., 61 t;

- individual fish length measurements and the length-weight relationships resulted in the estimated 39.6% contribution of regularly sized fishes to the total fyke net catches (Tab. 1.31);
- thus, the undersized fish by-catch consisted of 83 t of juveniles and 10 t of fry.

In the gill net catches, the largest by-catch comprised juvenile (undersized) pike-perch (24 t yr^{-1}) and bream (18 t yr^{-1}). Gill nets were also responsible for substantial catches of ruffe, including juveniles (ruffe has not been granted a protected size). Gill nets proved selective mostly toward perch (as little as 4 t undersized by-catch) and fully selective toward juvenile roach (less than 0.5 t yr^{-1}).

Overall, all the catches obtained with fyke nets and perch-roach gill nets were estimated to have yielded 800 t (a rounded value) of undersized fish per year, roughly 150 t (⅓ of fyke net by-catch of cyprynids) surviving after they had been thrown overboard. Thus, an additional fishing mortality involved 650 t of juveniles and fry. About 300 t of undersized fish were used for direct consumption and/or were sold, the sales going mostly unrecorded. However, a part of those fish (an estimated 20-25%) were sold officially, albeit against regulations; this was primarily the case of pike-perch, perch, and eel. Most controversial are the records of eel purchases. Based on the opinions of ichthyologists and fisheries inspectors, Garbacik-Wesołowska et al. (1985) found the officially reported eel landings to have accounted for as little as 30% of true catches. Most of the unrecorded eel catches comprised undersized individuals.

The budget shows 300 t of dead fish to have been thrown overboard each year, whereas 50 t of dead fish were brought, as garbage, onshore or were disposed of close to the shore. Half of the undersized fish thrown overboard, i.e., about 150 t, could have been consumed by seagulls and other birds, or transported with the current away and out of the Lagoon. The shore of the Piastowski Canal and the Stara Świna are populated by hundreds of grey herons waiting for a food supply like this. In the case of the gill net catches brought on the shore with the gear, 50 t of undersized fish were dead. It is assumed that half of that offal was consumed by birds foraging mostly on the land. Those birds included primarily the very numerous European herring gull, the "garbage dump visitor", which consume relatively large fishes as well; less abundant were other seagull species. Cervinids were also important consumers of discarded fish. A sizable part of the by-catch was used on farms as livestock fodder or as fertiliser.

The major factor inflicting fishing mortality of undersized fishes was poor selectivity of the fishing gear. In addition, there were also certain random, and/ or purposeful, effects:

- during long periods of inclement weather, the fishing gear may be left in the water for a long time; the fishes caught in such gear die and disintegrate in the water;
- gill nets along with the fish trapped in the netting are frequently damaged and moved away by vessels, and are thus lost to the owners;
- the young fish protective sieves in fyke nets may almost completely lose their throughput capability due to the trapped fish (most often ruffe or juvenile pike-perch) which block them;
- occasionally, the fyke net cod-ends become rotated, as a result of which the protective sieves are disabled and, the catches, particularly those of juvenile fish and fry, are considerably increased;
- the fyke net cod-ends with protective sieves are occasionally turned around on purpose, for example to catch the undersized eel more efficiently.

Another problem related to the additional fishing mortality of all the fish is posed by angling and poaching. It is extremely difficult to estimate how big this problem is, particularly that some unknown quantity of angled and poached fish was officially purchased and hence registered in official regional fish catch statistics.

1.12.4 Migrations and Passive Transport of Fish into the Pomeranian Bay

Fish larvae are transported with the current from the Szczecin Lagoon into the Pomeranian Bay. This effect is evidenced by data on the occurrence of the larvae beyond the northern boundary of the Lagoon (Porębski and Szkudlarek-Pawełczyk, 1997; Szkudlarek-Pawełczyk, 2003). However, the quantitative dimension of the process remains unknown. In the Great Lagoon, the larval export is not particularly large because the time of occurrence of pro-larvae and small larvae, i.e., poorly mobile individuals, is relatively short and the spawning grounds are situated mostly close to the shore, outside of the main riverine water flow across the Lagoon. Those larvae transported to the Piastowski Canal or to the main Świna bed and branches of the delta may be returned to the Lagoon with the backflush. Those passive migrations are confirmed by the presence of considerable amounts of herring and gobiid larvae brought into the Great Lagoon with incursions of the brackish water from the Pomeranian Bay (Porębski and Szkudlarek, 1995).

Juveniles migrate mostly actively, occasionally in masses (Wysokiński et al., 1999). Juveniles of freshwater species have been frequently reported from the southern part of the Pomeranian Bay. In late July 1997, small-mesh trawl tows towards the shore line were conducted at three sites in the Bay, between

Świnoujście and Międzyzdroje (Wysokiński et al., 1999). The catches were found to contain, i.a., relatively abundant juveniles of ruffe, smelt, and perch measuring (*l.t.*) 4-7 cm. Also the data from special small-mesh trawl catches in the Pomeranian Bay (Garbacik-Wesołowska et al., 1998) provide evidence of the migrations. There are, however, no direct data on the quantities of migrating juveniles.

Back and forth migrations of young and adult freshwater fish between the Lagoon and the Pomeranian Bay are a constant phenomenon (fish shoals have been frequently observed on echosounder records from the Piastowski Canal and the Świna Strait). Those migrations are an important factor sustaining the annual freshwater fish catch potential of the Pomeranian Bay. The species concerned, however, mostly return to the Lagoon to spawn. There is hardly any basis for quantification of the Lagoon fish transport and migrations, including those taking place between the Great Lagoon and the Kleines Haff. As the Great Lagoon is an open, flow-through water body, a supply of fish from the upper part of the Odra River mouth system, i.e. from Lake Dąbie and from the River Odra channel system cannot be ruled out. It is therefore most plausible to assume that, generally, fish migrations in the Great Lagoon are a balanced process.

1.12.5 Fish Biomass and Production

1.12.5.1 Estimation of the Exploited Stock Biomass

Under conditions of rational, sustainable fisheries management, the commercial stock consists primarily of sexually mature fishes, which are larger than the protected size specified in appropriate regulations. Throughout the entire history of ichthyological research carried out by the former Sea Fisheries Institute (at present: the National Marine Fisheries Research Institute), biomass of commercially exploited stocks of certain fish species in the Szczecin Lagoon has been estimated on several occasions, including the stock of roach (Sołtysik, 1989) and bream (Kaczewiak, 1990). No such estimates were made in 1998-2002. However, conclusions concerning comparison of yields and exploited stock biomass estimates (i.e., Y/B) in the Lagoon and other water bodies, drawn from compilations of various data, repeatedly point out to a characteristic pattern:

- Sołtysik (1989) estimated the commercial roach stock biomass in the Great Lagoon in 1980-1986 at an average of 4930 t; at that time, the catches averaged 1285 t yr^{-1}, i.e. 26.1% of the biomass.

- According to Kaczewiak (1990), the mean commercial bream stock biomass in the Szczecin Lagoon was, in 1980-89, 1835 t; the mean annual catch was 400 t, i.e., 21.8% of the biomass.
- Based on their 25-year-long data series, Gröger et al. (2007) determined the pike-perch exploitation rate (Y/B) in one of the German Baltic coastal lagoons to average 24%.
- Penczak and Tatrai (1985) found the biomass of bream aged 3+ to 7+ in Lake Balaton to be 74.1 kg ha^{-1}, the catches amounting to 17.9 kg and producing Y/B of 24.2%.
- In his assessment report, Szczerbowski (1985) estimated the fish biomass in Polish lakes at 120-180 kg ha^{-1} (i.e., 150 kg ha^{-1} on the average). The yield from bream- and pike-perch- type lakes averaged 36 kg ha^{-1}, i.e., giving Y/B of 24% (30-20%).
- Calculations based on a pike-perch virtual population model for the Great Lagoon in 1994 (Annex, Section 1.17) showed the exploited stock biomass of pike-perch (i.e., the biomass of adult fish) in the Great Lagoon to amount to 1331 kg km^{-2}; the annual catch in 1993-1995 averaged 350 kg km^{-2} (according to the Maritime Office statistics), i.e., 26.3% of the biomass.

The data presented above allowed to assume that the biomass of the commercially exploited fish stock was approximately four times higher than the commercial catches.

1.12.5.2 Biomass of Major Age Groups as a Proportion of Total Population Biomass

The biomasses of major age groups as proportions of the total population biomass in each species were estimated based on the already mentioned pike-perch virtual population model (Annex, Section 1.17) as well as on corresponding calculations for bream based on modified data of Kakareko (2000). The resulting percent contributions of larvae and fry, juveniles and adults in total biomasses of these two fish species are illustrated in Table 1.32.

Table 1.32 Proportions of biomass of principal age categories of pike-perch in the Great Lagoon (as calculated in Annex, Section 1.17) and common bream in the Włocławek Dam Reservoir (after Kakareko, 2000).

Age group	Pike-perch in the Great Lagoon	Common bream in the Włocławek Dam Reservoir
Larvae and fry (0 and 0+)	10%	8%
Juveniles	28%	36%
Adults/commercial stock	62%	56%

Similar proportions of biomasses of three major age groups as determined by the pike-perch model and that developed for bream (cf. Kakareko, 2000) allowed to generalise the proportions and extrapolate them on all the fish species. It was thus assumed that larvae and fry (age groups 0 and 0+), juveniles, and adults (i.e., the exploited stock) accounted for 10, 30, and 60%, respectively, of the population biomass.

1.12.5.3 Production Rate (P/B)

Based on the published data (below), the P/B values of larvae and fry (age groups 0 and 0+) and juveniles were assumed to amount to 5 and 1, respectively; the P/B values of adults of non-predatory and predatory species were 0.6 and 0.4, respectively. The basic difficulty in applying the numerous published data on fish production rate to comparisons boiled down to the fact that many authors were not precise enough in defining what production rate they were discussing: that of the entire population or that of a certain age group only.

According to Sorokin (1972), P/B of fish aged 0 and 0+ in the Rybinsk Dam Reservoir was 5. Calculations of Wolnomiejski (unpubl.) showed P/B of perch in the Great Lagoon to be 4.8. The P/B value of 7.1 produced for the pike-perch of this age group by the pike-perch population model (cf. Annex, Section 1.17) cannot be extended to other fish species, because the larvae and fry of pike-perch grow extremely fast during the first year of life. In the Lagoon, the pike-perch accounted for as little as 4% of the total catch.

The P/B value of fish juveniles was assumed to be 1 based on the pike-perch population model (Annex, Section 17) and calculations for bream (Kakareko, 2000).

P/B of the adults (exploited stocks) of predatory species was assumed to be 0.4. Although the pike-perch population model (Annex, Section 1.17) produced P/B = 0.352, the P/B value of age groups 3-6 which make up 90% of the abundance and production of the exploited stock and 98% of the catch was 0.4. An identical value (0.4) was reported for pike-perch by Winberg et al. (1972). According to Nagięć (1964), the River Vistula pike-perch aged 3-9 years had P/B of 0.41. The general estimate given by Szczerbowski (1985) for P/B of predatory species was lower than that of the non-predatory ones and amounted to about 0.5; generally, he estimated the fish P/B at 0.3-0.7.

The P/B value of adults of non-predatory species was assumed to be 0.6 based on a compilation of data reported by Backiel et al. (1980), Pivnička (1982), Szczerbowski (1985), Penczak and Tatrai (1985), Wetzel and Likens (1995), Penczak (2000), and Kajak (2001). Particularly noteworthy and useful are the data provided by Penczak and Tatrai (1985), showing P/B of the Lake Balaton bream aged 3+ to 7+ to be 0.63.

According to the methodology adopted, the production (of benthivorous species as examples) was calculated in the following manner: the average annual catch from the exploited stock (adult fish) was 1776 t; the stock biomass was four times higher, i.e., 7104 t. As this is equal to 60% of the total biomass of benthivorous species, the total biomass was 11 840 t. The juvenile fish accounted for 30%, i.e., 3552 t, the larvae and fry (age groups 0 and 0+) contributing 10%, i.e., 1184 t. Using the P/B values adopted, the total production was estimated for the three trophic groups and the three age groups (Table 1.33).

Table 1.33 Average biomass and production of fish in the Great Lagoon in 1998-2002.

Age group	**Biomass [t]**	**P/B**	**Production [t yr^{-1}]**
Benthivorous: adults (commercial stock)	7104	0.6	4264
Benthivorous: juveniles	3552	1.0	3552
Benthivorous: larvae and fry (0 and 0+)	1184	5.0	5920
Benthivorous total	**11 840**	**1.16**	**13 736**
Piscivorous: adults (commercial stock)	2884	0.4	1154
Piscivorous: juveniles	1442	1.0	1442
Piscivorous: larvae and fry (0 and 0+)	481	5.0	2405
Piscivorous total	**4807**	**1.04**	**5001**
Planktivorous: adults (commercial stock)	1204	0.6	722
Planktivorous: juveniles	602	1.0	602
Planktivorous: larvae and fry (0 and 0+)	201	5.0	1005
Planktivorous total	**2007**	**1.16**	**2329**
Fish total	**18 654**	**1.13**	**21 066**

Generally, P/B of the Great Lagoon fish fauna was 1.13.

1.12.5.4 Caloric Value and Organic Carbon Content of Fish

The published data show relatively small differences in the caloric value of freshwater fish as a whole, the differences stemming mainly from different estimates of dry weight and differences in fish age and habitat type. The average caloric content of the entire fish fauna was found to range from 1.155 to 1.391 kcal $g_{w.w.}^{-1}$ (Westlake et al., 1972; Kozlova and Panasenko, 1977; Penczak, 1992; Penczak et al., 1982; 1984; Penczak and Tatrai, 1985; Prus, 1993). On the other hand, considerable differences are noticeable between species, e.g., between eel and other species.

The calculations made in this study were based on data reported by Schreckenbach et al. (2001) on fish of various species, aged 1-12 years, from 23 lakes in Brandenburg and environs of Berlin. Those data were used to estimate caloric content of the 9 species which accounted for about 90% of the total catch from the Great Lagoon. The average caloric content (weighted by the individual contribution to the total catch) amounted to 1.252 kcal $g_{w.w.}^{-1}$ and this was the value applied further on in this study.

The energy to carbon conversion factor of Vinogradov and Shushkina (1987) (1 kcal = 0.08 gC) produced the fish carbon content of 0.1 gC $g_{w.w.}^{-1}$. This is a value assumed to be typical of juvenile and adult fish.

The body of larvae and fry contains much more water than the body of adults, hence their caloric content is much lower and proportional to the body size. Based on data of Opaliński et al. (2008) and on the unpublished data kindly made available by Opaliński and Maciejewska, an average caloric content of larvae and fry (aged 0 and 0+) was 0.867 kcal $g_{w.w.}^{-1}$ and organic carbon content about 0,07 gC $g_{w.w.}^{-1}$.

1.12.5.5 Annual Production of Fish Fauna Per Great Lagoon Unit Surface

The production of fish fauna in the Great Lagoon, adjusted to the organic carbon content, is shown in Table 1.34.

The ratio between the total production and the production of predatory species was 4.2 (wet weight) and 4.3 (organic carbon-based). The corresponding ratio for adults only (i.e., the exploited stock) was 5.3.

Table 1.34 Fish production per standard 1 m² of the Great Lagoon. Cc, organic carbon conversion factor.

Trophic and age group	**Production [$g_{w.w.}$ m^{-2} yr^{-1}]**	**Cc [gC $g_{w.w.}^{-1}$]**	**Production [gC m^{-2} yr^{-1}]**
Benthivorous: adults	10.40	0.1	1.040
Benthivorous: juveniles	8.66	0.1	0.866
Benthivorous: larvae and fry (0 and 0+)	14.44	0.07	1.011
Piscivorous: adults	2.81	0.1	0.281
Piscivorous: juveniles	3.52	0.1	0.352
Piscivorous: larvae and fry (0 and 0+)	5.86	0.07	0.410
Planktivorous: adults	1.76	0.1	0.176
Planktivorous: juveniles	1.47	0.1	0.147
Planktivorous: larvae and fry (0 and 0+)	2.45	0.07	0.171
Fish total	**51.37**		**4.454**

1.12.5.6 Gonad Production in the Total Fish Production Budget

Estimation of gonad production was based on data illustrating the proportion of gonads in the total production of four species studied in the Vistula Lagoon (Wilkońska, 1996): roach, bream, pike-perch, and perch. In the Great Lagoon, the four species made up almost 90% of the total catch. The weighted mean gonad production in the adult fish was about 20% (12.7-26.8%) of their total annual production. In the adults and juveniles taken together, gonad production was – proportionally – about 10% of the total production. Similar data were reported by Backiel (1971), Penczak et al. (1977), Backiel et al. (1980) and Kakareko (2000).

1.12.6 Food of the Great Lagoon Fish

To determine food composition of the Great Lagoon fish and contributions of different food items to their diet, stomach contents of 2910 specimens with food-filled stomachs and intestines were analysed. The individuals examined in 1982-1992 (a total of 1258 individuals; Wolnomiejski and Grygiel 1994b, c, 1998a, 2002) represented perch, bream, and ruffe. In 1996-2004, food content was analysed in stomachs of 1652 individuals (Wolnomiejski, unpubl.). Stomach content was analysed in all age groups of 13 species. About 70% of the analyses involved four species: bream, roach, perch, and ruffe. The fish to be examined were yielded by 150 trawl tows performed for the purpose in 1984-2004 throughout the Great Lagoon, at the depth range of 1.5-8 m. Fishing operations were carried out mainly during the growing season, although several tows were conducted also in winter.

Benthic organisms found in stomach contents were identified to the lowest possible taxon. The food weight was determined using the consumption index, i.e., it was reconstructed from standard weights and size of food organisms. The standard weights used were developed by Wolnomiejski (unpubl.), based on faunistic studies in the Szczecin Lagoon. Conversion coefficients shown in Table 1.35 were applied to express the food wet weight in organic carbon units.

The percent contributions of individual food components to the total food weight, converted to organic carbon content, are shown in Table 1.36. The values are means weighted by contribution of each fish species to the total production of a given fish trophic group.

More detailed data on feeding of ruffe, perch, and bream in the Great Lagoon can be found in Wolnomiejski and Grygiel (1994b, c; 2003).

For the general characterisation of fish food, 12 basic dietary components were distinguished (Table 1.36):

Table 1.35 Carbon conversion factors applied to fish food items.

Diet component	gC $g_{w.w.}^{-1}$	Data source
Microalgae	0.060	Compiled from Bohr et al. (1975); Vinogradov and Shushkina (1987); Spodniewska (1988); Riccardi and Mangoni (1999)
Macrophytes (submersed and floating-leaved)	0.048	Based on Table 1.13, assuming 0.15 $g_{d.w.}\ g_{w.w.}^{-1}$
Mesozooplankton	0.064	Data from Table 1.10
Neomysis vulgaris	0.064	Data from Table 1.11
Chironomus sp.	0.098	Data from Table 1.28
Ostracoda	0.040	Data from Table 1.28
Oligochaeta	0.077	Data from Table 1.28
Amorphous remains, detritus and mud	0.050	According to opinion and data of Winberg (1979) and Job and Kannan (1980)
Dreissena polymorpha[a]	0.026	Data from Table 1.28
Other Mollusca[a]	0.026	Data from Table 1.28
Littoral bentos and macrozooperyphyton	0.075	Compiled from data in Table 1.28 and from Penczak et al. (1982)
Fish	0.085	From Table 1.31, assuming ratio of larvae and fry to juveniles and adult of 1:1

[a] with shells

Table 1.36 Percent contributions of diet components to fish food (estimates based on carbon units).

Diet component	Fish trophic group					
	Benthivorous		Planktivorous		Piscivorous	
	Larvae and fry	Juveniles and adults	Larvae and fry	Juveniles and adults	Larvae and fry	Juveniles and adults
Microalgae	2.8	1.6	1.8	0.2	0.6	0.0
Macrophytes	1.5	2.4	0.1	< 0.1	0.0	0.0
Mesozooplankton	66.5	11.0	86.5	34.1	61.5	11.6
Neomysis vulgaris	0.0	0.8	1.8	33.9	9.2	3.0
Chironomus sp.	10.8	58.1	4.3	20.5	19.6	12.1
Ostracoda	11.9	7.6	2.0	1.4	0.1	< 0.1
Oligochaeta	1.4	2.1	0.4	0.6	0.1	0.2
Unidentified remains and mud	1.3	1.4	0.1	0.1	0.0	0.0
Dreissena polymorpha[a]	< 0.1	4.1	< 0.1	0.3	0.0	< 0.1
Other Mollusca[a]	0.1	1.9	0.1	0.2	0.1	0.0
Other littoral benthos	3.7	9.0	2.8	2.7	2.1	3.3
Fish	0.0	0.0	0.1	5.9	6.7	69.7

[a] with shells

Microalgae. This group comprised phytoplankton as well as periphytic and phytobenthic algae. Most probably, those algae were taken up passively, when a fish was actively feeding on other, larger planktonic, benthic or periphytic organisms.

Aquatic macrophytes are seldom used as food by fish in Polish waters because, as argued by, i.a., Mann (1988) and Kornijów (1996), they contain a high proportion of indigestible cellulose, low level of nitrogen compounds, and certain protectants (repellents) such as alcaloids and phenols. Macrophytes were mainly consumed, as an auxiliary food, by roach the stomachs of which contained fragments of submerged and floating-leaved plants and macroalgae, particularly the filamentous ones; only rarely were macrophytes the only food item in the stomach.

Mesozooplankton was the basic, and occasionally the only, dietary item consumed by larvae and fry (age group 0) of all the fish species, and was the major auxiliary food component for juveniles. As pointed out by Kluchareva and Svetovidova (1968), all the fish are planktivorous when young. Most frequently, stomach contents of various individuals showed the domination of either copepods or cladocerans. *Leptodora kindti* was found in the stomachs in amounts much higher (and not infrequently in masses) than indicated by its contribution to the zooplankton abundance. Occasionally, those largest cladocerans dominated in the food of 50-cm long bream or eel. Mysids contributed highly to the diet of planktivores, whereby the planktivore feeding pressure on the mesozooplankton was reduced.

Macroplanktonic mysid *Neomysis integer* was a very important food item for predatory fish fry (particularly that of pike-perch) and the basic diet component of adult smelt. The major role of mysids in the diet of pike-perch juveniles in the Curonian Lagoon was emphasised also by Ložys (2003). In contrast, mysids were consumed by benthivorous fish to a minimal degree only.

Chironomids (*Chironomus* sp.) were consumed by fish as larvae and pupae. Larvae were the major diet component of juvenile and adult bream and ruffe, and – to a lesser extent of white bream and roach (they were occasionally eaten also by eel and large smelt). Lammens and Hoogenboezem (1991) contended that bream, carp, and white bream are particularly well adapted to pick chironomid larvae from muddy sediments. Bream used their filtration system of the palate and gills to separate larvae from the mud. A fish fills its mouth with mud and filters out chironomid larvae, the sediment being ejected "cloud"-like through the gills from below the *operculum*. White bream is much less efficient than bream. However, small benthivores are helpless when the chironomid larvae dwell deeper than 1 cm in the sediment. On the other hand, all the fish species, with a varying degree of preference, consumed large amounts of chironomid pupae from the water column, particularly during the two periods of mass emergence of chironomid imagines.

The high abundance and consumption of chironomid larvae is crucial for mass development and domination of benthivorous fish in the Szczecin Lagoon.

Semi-benthic, meio-, and macrobenthic ostracods, particularly abundant on the muddy bottom of the central part of the Lagoon, played an important role in the food of fry and juvenile benthivores, adult roach relying heavily on them as well. The importance of ostracods in fish nutrition was reported by many authors, e.g., Aristova (1965) and Brabrand (1984). Ostracods are a relatively easy prey as they stay mostly on the mud surface and above it.

Oligochaetes (almost exclusively tubificids), although present in all benthic habitats, are most abundant (about 85% of the total abundance and biomass) in muds and sandy muds. Although highly abundant, oligochaetes are poorly utilised as food by fish. This has been also noted in other water bodies. Some authors (e.g., Pliszka, 1956; Poddubnyy and Bakanov, 1980) are of the opinion that the role of oligochaetes is underestimated, because they are rapidly digested by fish; other authors, e.g., Pustelnikovas (1998), contend that the oligochaete role in fish feeding is unclear. Brabrand (1984) omitted oligochaetes from the list of diet components of benthivores such as bream and white bream. In this study, even when the stomach content was examined under a magnifying glass and a stereomicroscope, only few undigested oligochaete setae could be noticed. In the Włocławek dam reservoir, particularly rich in oligochaetes and *Chironomus* larvae (about 60 and 100 $g_{w.w.}$ m^{-2} muddy bottom, respectively) Kakareko (2001) found bream of all age groups to be minimally interested in consuming oligochaetes which were only an accessory diet component (a fraction of a percentage of the stomach content). It was only in the upper, rheolimnic, part of the reservoir, where mud was sparse and *Chironomus* larvae were scant, that he found oligochaetes to be an important food component, particularly for juvenile bream. They were retained in the stomach content as clearly identifiable specimens at various degree of digestion. Similarly, Wielgosz and Tadajewska (1988) found some amounts of oligochaetes in the food of bream in the Włocławek dam reservoir. Giles et al. (1990) reported a distinct effect of food supply on food consumption in perch, but only with respect to chironomid larvae. Wherever chironomids were abundant in the benthos, they featured prominently in the fish food. On the other hand, oligochaetes, despite their high contribution to the benthos, were only minimally utilised by the fish.

Mud, detritus, and unidentifiable remains. Mud lumps were occasionally found in stomach contents of some fishes shortly after a storm abated. Mud particles can also be accidentally ingested when a fish feeds on the bottom. Some amounts of mud or detritus may be supplied by intestines of the digested chironomid larvae. Mud (particularly its surficial layer containing microbiota *sensu* Johnson et al., 1989) – along with detritus, sedimented phytoplankton, faeces, and dead animals decomposed by microorganisms is a high-caloric diet component. Winberg (1979, and the references therein) emphasised the high nutritive value (high amounts of proteins, lipids, and easily assimilable minerals) of detritus and faeces decomposed by microorganisms, and provides examples of such food resources being utilised. The inshore zone sediment contains a plethora of large detrital particles which, when decomposed, can be a food resource also for fish (e.g., Mann, 1988). Michelsen et al. (1994) found detritus to feature prominently in fish food during periods where food fauna was deficient.

Dreissena polymorpha. The large bivalves living in aggregations were consumed mainly by roach; they were usually a basic dietary item for large roach individuals. The zebra mussel was much less exploited by white bream. For the remaining fish species, the bivalve was an accessory diet component, only small individuals (usually aged one year) being consumed. This finding is in agreement with evidence provided by other authors, e.g. Nagelkerke and Sibbing (1996). The largest zebra mussel individuals are consumed by roach because their mouth and pharyngeal gap is at the widest. The pharynx and pharyngeal teeth can exert pressure sufficient to crush the bivalve shell. On the other hand, the zebra mussel is of a marginal importance in bream diet, because shell crushing requires too much effort and energy on the part of bream.

Other molluscs. The two most common gastropods, i.e. *Valvata* and *Bithynia*, were the most common gastropods in the fish stomach contents. In addition, the fish food contained, particularly in some localities, small sphaerid bivalves.

Littoral macrobenthos and macrozooperiphyton. Littoral invertebrates were of a significant, but only supplementary, importance in the food of roach, ruffe, and bream and in juvenile perch. It is often impossible to tell if many ubiquitous forms have been consumed as a benthic or periphytic food. In addition to small bivalves and the zebra mussel, the littoral forms used as food included mainly large chironomid larvae of the genus *Glyptotendipes*, hirudineans, and benthic crustaceans, particularly *Asellus* and *Corophium*. According to Van de Bund and Groenendijk (1994), the chironomid larvae eaten by cyprinids (particularly the benthivorous bream) are usually larger than 4 mm. Of the fish species studied, white bream was found to feed most intensively in the littoral. The littoral, its shallow parts in particular, is inhabited by numerous small fish not captured during the study described. The true extent to which the littoral benthos was utilised as food was therefore higher than could be determined here. The degree to which the shallow-water benthos can be utilised by fish is determined, in addition to its abundance, by the accessibility of benthic invertebrates (Cobb and Watzin, 1998), as dense vegetation restrict access to benthic food. Some authors, e.g. Scheffer (2001) mention a sheltering role of macrophytes, particularly underwater meadows, which prevent fish from feeding on the benthos. The near-shore macrobenthos is also protected from predation by the presence of aggregations of live zebra mussel and mounds of its empty shells, by large deposits of plant remains, and macrophyte roots. Mastitsky and Samoilenko (2005) mentioned the protective role of *Dreissena* beds (the shells and byssus of the bivalve) with respect to the invertebrates which could have been fed upon by fish. On sandy bottoms, to avoid being consumed by fish, the abundant chironomid larvae penetrate the sediment down to the depth of several centimetres (Van de Bund and Groenendijk, 1994).

Fish. The fish were available as food in four forms: eggs, carcasses, easily available hatch and fry, and as larger individuals available to obligatory predators only. White bream was the main fish egg feeder, showing a particular preference towards the eggs of perch. During periods of perch spawning, white bream were observed to consume large masses of perch eggs. However, as the spawning period is short, the eggs eaten weighed little in the total amount of food consumed (average < 0.1%). The particularly strong attraction of fish eggs to white bream, compared to other fish species, was mentioned also by Braband (1984).

The fish carcasses were, to some small extent, consumed by white bream and roach. Fish larvae were fed upon mainly by fry and juveniles of predatory fish species. The adult (longer than 15 cm in total length) smelt fed periodically on fish larvae and small fry, particularly those of ruffe. Treasurer (1997) reported on strong cannibalism of perch with respect to the hatch, which was the reason of the low production of perch in Scottish lochs. In the Lagoon, mass-scale cannibalism was not observed in any of the predatory fish species. The most distinct signs of cannibalism were observed in pike-perch. Wysokiński (1998) and Winkler (1998) did not find any strong predation pressure of perch on the pike-perch fry. Wysokiński (1998) is of the opinion that the pike-perch larvae and fry are protected from perch predation by inhabiting different zones in the water body.

Obligatory predators fed most frequently on ruffe and smelt which jointly accounted for 75 and 60% of perch and pike-perch prey, respectively. Pike-perch preyed upon perch, too, and contribution of roach increased distinctly in the diet of the largest pike-perch. Beattie et al. (1972), too, listed smelt and ruffe as definitely the most important components of juvenile pike-perch food. A high importance of ruffe in the diet of predatory species was emphasised by Kozlova and Panasenko (1977) and by Ložys (2003).

1.13 Birds

1.13.1 Composition and Abundance of Avifauna

The Szczecin Lagoon supports a very abundant avifauna; the Lagoon's importance for birds increases particularly during autumn and winter migrations of waterfowl. Exceptionally abundant are flocks of anseriforms. Nineteen of the species recorded in the Lagoon are threatened by extinction and are listed in the Polish Red Book of Animals (Kalisiński et al., 2004). Racławski et al. (2004) described protection categories and degree of extinction threat faced by some waterfowl species. Kaliciuk (2004) wrote on Special Protection Areas (SPA) of Birds, established in the Province of Western Pomerania in the framework of the NATURA 2000 programme. In the Szczecin Lagoon, almost 45 thou. ha have been declared a NATURA 2000 SPA (PLB320018).

According to Czeraszkiewicz and Oleksiak (2003; 2004), the Lagoon supports, albeit only periodically, particularly abundant aggregations of waterfowl, i.a., cormorants, whooper swans, tufted ducks, lesser scaups, common goldeneyes, and common mergansers. Like in other parts of Poland, the abundance of black cormorants in the Odra mouth system has been observed to have greatly increased.

The composition and abundance of the Lagoon's avifauna in 1995-2003 (Table 1.37) was assessed based mainly on Kalisiński et al. (2004), on results of

Table 1.37 Waterfowl abundance in the Great Lagoon in 1998-2002, duration of occurrence and daily food requirement.

Species	**Abundance** [indiv. km^{-2}]	**Duration of occurrence** [d yr^{-1}]	**Weight** [g $indiv.^{-1}$]	**Daily food requirement** [g $indiv.^{-1}$ d^{-1}]
Resident birds:				
Mallard (*Anas platyrhynchos*)	17.8	365	1000	250
Coot (*Fulica atra*)	7.2	365	800	200
Cormorant (*Phalacrocorax carbo*)	7.1	365	1600	400
Great crested grebe (*Podiceps cristatus*)	3.6	365	800	200
Mute swan (*Cygnus olor*)	2.1	365	10000	2500
Grey heron (*Ardea cinerea*)	0.2	365	1400	350
Gulls and terns	0.1	365	400	100
Migratory birds:				
Tufted duck (*Aythya fuligula*)	36.5	150	800	200
Lesser scaup (*Aythya marila*)	14.2	150	1000	250
Black tern (*Chlidonias niger*)	13.4	120	80	20
Common merganser (*Mergus merganser*)	12.8	150	1600	400
Goldeneye (*Bucephala clangula*)	10.7	150	800	200
Smew (*Mergus albellus*)	8.8	150	800	200
Aythya indet.	8.0	150	1000	250
Pochard (*Aythya ferina*)	6.7	150	1000	250
Little gull (*Larus minutus*)	3.1	120	100	25
Widgeon (*Anas penelope*)	1.4	60	600	150
Whooper swan (*Cygnus cygnus*)	0.6	100	8000	2000
Red-throated and black-throated divers (*Gavia stellata, G. arctica*)	0.2	100	2000	500

the Western Pomeranian waterfowl census (Czeraszkiewicz and Oleksiak, 2003; 2004), and on personal information provided by ornithologists. The periods of local occurrence of individual species were assumed to be analogous to those in the Vistula Lagoon (as determined by Professor L. Stempniewicz of the University of Gdańsk).

Table 1.37 lists only those bird species which are obligatory inhabitants of aquatic areas, particularly those actively feeding directly in the water column or on the bottom.

1.13.2 Avian Food and Consumption

Data on body weight, daily food requirement (25% of the mean body mass), and diet composition were retrieved from Cramp and Simmons (1977) and Stempniewicz and Meissner (1999). For cormorants, the daily food ration was compiled from Gmitrzuk (2004), Krzywosz (2008), Cramp and Simmons (1977), and Bzoma (2008b). The data on the abundance of birds in the Great Lagoon and their food demand (Table 1.37) were used to assess the composition and estimate the weight of avian food by trophic groups, as illustrated in Table 1.38.

The following conversion factors were used to express avian food consumption in organic carbon units (Table 1.39):

- 0.046 gC $g_{w.w.}^{-1}$ for macrophytes (only the so-called soft plants, i.e. elodeids and nymphaeids);
- 0.064 gC $g_{w.w.}^{-1}$ for *Neomysis;*
- 0.026 gC $g_{w.w.}^{-1}$ for molluscs with shells;
- 0.075 gC $g_{w.w.}^{-1}$ for other macrobenthos (as an average of chironomids, oligochaetes, and other benthos);
- 0.08 gC $g_{w.w.}^{-1}$ for fish, assuming similar proportions of larvae, fry, juvenile, and adult fish in the food weight.

Among herbivorous birds, most important were mute swans and mallards, which consumed more than 95% of aquatic macrophytes (soft plants) eaten by all birds. Overall, birds consumed about 1500 $t_{w.w.}$ elodeids and nymphaeids, which amounted to about 1.8% of their annual production (as little as 0.6% of production of all the aquatic macrophytes).

Molluscs, mainly the zebra mussel, were the basic food of diving ducks and coots. According to Stempniewicz and Meissner (1999), diving ducks feed mostly on bivalves which account for almost 80% of their food. All the birds consumed a little more than 2% of the annual mollusc production. The molluscivores fed on other benthic invertebrates, too, which were consumed together with the zebra mussels the aggregations of which host numerous

Table 1.38 Food consumption of waterfowl in the Great Lagoon. Compiled data; see text for explanations.

Taxon	Consumption of food components [$kg_{w.w.}$ km^{-2} yr^{-1}]					
	Macrophytes (floating-leaved and submerged)	**Neo-mysis**	**Molluscs (with shells)**	**Other macro-benthos**	**Fish**	**Terrestial organisms collected from the water surface**
Herbivorous birds:						
Mute swan *(Cygnus olor)*	1916					
Whooper swan *(Cygnus cygnus)*	120					
Mallard *(Anas platyrhynchos)*	1137	243				243
Widgeon *(Anas penelope)*	13					
Herbivorous birds total	**3186**	**243**				**243**
Molluscivorous birds:						
Pochard *(Aythya ferina)*	25		293	50		
Tufted duck *(Aythya fuligula)*	109		1277	219		
Lesser scaup *(Aythya marila)*			754	80		
Aythya indet.	15		385	54		
Goldeneye *(Bucephala clangula)*		96	214	96		
Coot *(Fulica atra)*	259		432			
Molluscivorous birds total	**408**	**96**	**3355**	**499**		
Omnivorous birds:						
Gulls, mainly little gull *(Larus minutus)*		6			10	6
Black tern *(Chlidonias niger)*		13			7	13
Omnivorous birds total		**19**			**17**	**19**
Piscivorous birds:						
Cormorant *(Phalacrocorax carbo)*					1037	
Common merganser *(Merlus merganser)*					768	
Smew *(Merlus albellus)*					264	
Red-throated and black-throated divers *(Gavia stellata, G. arctica)*					10	
Great crested grebe *(Podiceps cristatus)*					263	
Grey heron *(Ardea cinerea)*					26	
Piscivorous birds total					**2368**	
Avifauna total	**3594**	**358**	**3355**	**499**	**2385**	**262**

Table 1.39 Food consumption by waterfowl trophic groups in the Great Lagoon, in organic carbon units [gC m^{-2} yr^{-1}].

Trophic group \ Food component	Macrophytes	Neomysis	Molluscs with shells	Other macrobenthos	Fish
Herbivorous birds	0.147	0.016			
Molluscivorous birds	0.019	0.006	0.087	0.037	
Omnivorous birds		0.001			0.001
Piscivorous birds					0.203
Avifauna total	0.166	0.023	0.087	0.037	0.204

hirudineans, oligochaetes, and insect larvae. Relatively large larvae and pupae of *Chironomus* could have been main predation targets.

The most important piscivores were cormorants, common mergansers, smews, and great crested grebes. Cormorants were most important due to their high abundance and persistence in the Lagoon throughout the year. When extrapolated to the entire area of the Great Lagoon, the birds consumed almost 980 t fish per year; cormorants consumed 425 t (43%) and common mergansers removed 315 t (32%). Generally, the actively feeding piscivorous predators consumed a mass of fish equivalent to more than 40% of the officially registered commercial landings. The piscivore predation pressure on the fish fauna is best described by the ratio between avian consumption and fish production (Table 1.40).

Table 1.40 Utilisation of fish production by piscivorous birds in the Great Lagoon. Birds were assumed to consume equal amounts of larvae and fry as well as juvenile and adult fish.

Fish age group	Fish production (P_F) [$t_{w.w.}$ km^{-2} yr^{-1}]	Fish consumption by birds (C_B) [$t_{w.w.}$ km^{-2} yr^{-1}]	C_B / P_F [%]
Larvae and fry	22.76	1.19	5.2
Juveniles and adults	28.62	1.19	4.2
Fish total	51.38	2.38	4.6

1.14 Hydrobiological Characteristics of the Great Lagoon: Comparison With Other Areas

1.14.1 The Biotope

On account of interactions between the riverine and Baltic waters occurring in the River Odra mouth system, some authors defined this area as an estuarine reservoir (e.g., Majewski 1972; 1980; Tadajewski et al., 1989; Poleszczuk, 2007). In such approach the Szczecin Lagoon, constituting an essential part of the system, was termed a second-order estuary. However, the physiography, geology, and hydrology of the Odra river mouth show that the area possesses many more characteristics typical of flow-through coastal lagoons than those of estuaries (Osadczuk, 2004; Osadczuk et al., 2007). Of key importance in this respect is the Szczecin Lagoon, an extensive, shallow water body separated from the open sea by a barrier intersected by three narrow and long straits. The lagoonal nature of the area is demonstrated also by its geological history (Osadczuk et al., 2007).

Therefore, when attempting to compare the Szczecin Lagoon with other areas, a most rational approach it to refer to other lagoons and/or coastal lakes of the southern Baltic dominated by the riverine regime, intrusions of the brackish Baltic water being of a secondary importance only. Such areas are mixo-oligohaline. The freshwater nature of the Szczecin Lagoon's flora and fauna permits also to carry out meaningful biocoenotic comparisons with inland water bodies, particularly with dam reservoirs and large natural lakes.

Morphology, hydrography and hydrochemistry of Baltic coastal lagoons and lakes have already been treated in numerous publications; those most noteworthy include Mikulski (1964), Majewski (1972; 1980), Lazarenko and Majewski (1975), Andrulewicz (1997), Chubarenko and Margoński (2002), Mudryk (2003), Paturej (2006), Aleksandrov et al. (2006), Schiewer (2008), Witek et al. (2010). Comparisons between the Great Lagoon and the Vistula and Curonian Lagoons and two Polish coastal lakes, the Gardno and the Łebsko, show that majority of hydrological and hydrochemical properties of the Great Lagoon are comprised within the range of variation of these parameters in other areas mentioned above (Table 1.41). The Lagoon is, however, distinct in being highly dynamic due to intensive mixing and flow. This is evidenced by a short freshwater residence time (FRT, about 52 days) and water exchange time (WET, about 44 days), which are the shortest among areas compared in Table 1.41, except very shallow Gardno lake. Owing to the relatively high Schindler's ratio and high riverine water input, the nutrient load brought with the river runoff per lagoon unit area is particularly high (Table 1.41). This was a basic argument used by some authors to classify the Lagoon as a hypertrophic water body.

Table 1.41 Basic characteristics of the coastal water bodies located along the southern coast of the Baltic Sea.

	Szczecin Lagoon	Vistula Lagoon	Curonian Lagoon	Lake Gardno	Lake Łebsko
Source of data (if not specified otherwise)	this study	Witek et al., 2010	Gasiunaite et al., 2008	Cyberski and Jędrasik, 2003	Cydzik and Soszka, 1988
Area [km^2]	687 [a]	838 [b]	1 584	25	71
Volume [km^3]	2.58 [a]	2.30 [b]	6.0	0.031	0.118
Mean depth [m]	3.8 [a]	2.7 [b]	3.8	1.3	1.6
Maximum depth [m]	8.5 [a] (10.5) [c]	5.2 [a] (12) [c]	5.8 (14) [c]	2.6	6. 3
Catchment area [km^2]	129 000 [a]	23871 [b]	100 458	897.6	1607.7
Salinity [PSU]	0.9 [d]	0.5 - 6.5; mean 3.2 (3.5 [e])	0 - 8 (Russian part 0.1) [f]	0.64 [g]	1.8 [h]
Annual riverine water input [km^3]	18.4 [i]	3.68 [b]	23.1 [j]	0.285	0.378 [k]
Mean seawater annual inflow [km^3]	3.6 [l] (11) [d]	3.4 [m] (17) [b]	5.1 [i]	0.018	no data
Freshwater residence time, FRT [days]	52	228	95	40	114
Water exchange time (based on salt budget), WET [days]	44	119	~ 90	~37	~100
Schindler's ratio (catchment and water body area to water body volume ratio, $m^2\ m^{-3}$)	50	11	17	30	14
P load from drainage basin [tonnes yr^{-1}]	3600 - 5600 [i] (1998 - 2002)	980 (1998 - 2000)	1200 - 4000 (no data)	33 - 51 [n] (1998 - 2000)	59 - 110 [n] (1998 - 2000)
N load from drainage basin [tonnes yr^{-1}]	56000-87000 [i] (1998 - 2002)	13600 (1998 - 2000)	33000 - 64000 (no data)	636 - 1096 [m] (1998 - 2000)	959 - 1610 [m] (1998 - 2000)
P load per water body surface area [tonnes $km^{-2}\ yr^{-1}$]	5.2 - 8.2	1.2	0.8 - 2.5	1.3 - 2.0	0.8 - 1.5
N load per water body surface area [tonnes $km^{-2}\ yr^{-1}$]	82 - 127	16	21 - 40	25 - 44	14 - 23
Secchi disc visibility [m]	annual range 0.6 – 1.5 (mean 1.0)	usually 0.3 - 0.8	0.45 - 0.6	mean 0.6 [g]	mean 0.7 [h]

[a] Majewski, 1980; [b] Lazarenko and Majewski, 1975; [c] in brackets - artificially deepened shipping channel; [d] Majewski, 1972; [e] Chubarenko and Margoński, 2008; [f] Aleksandrov et al., 2006; [g] Wielgat-Rychert et al., 2010; [h] Jarosiewicz and Ficek, 2008; [i] Anon., 1991 – 2003; [j] Pustelnikovas, 1998; Gasiunaite at al., 2008; [k] Bogdanowicz, 2004; [l] Wielgat and Witek, 2004 (acc. to salt budget); [m] based on salt budget; [n] Anon., 1998 – 2002.

However, due to the high water dynamics and high biological production at every level of the food web, high nutrient load per unit area doesn't hinder the efficiency of ecosystem functioning, including self-purification processes. This is particularly true with respect to denitrification, its rate being much higher in the Great Lagoon than in other Baltic coastal water bodies (Pastuszak et al., 2005), and twice as high as that in the Vistula Lagoon (Witek et al., 2010). In addition to denitrification, nutrients are ad- and absorbed by organisms and suspended particulates, the chemical load being periodically deposited in the sediment (Poleszczuk, 1998; Wielgat and Witek, 2004; Pastuszak et al., 2005). Those processes lead to horizontal variability in water chemistry along the major route of riverine runoff (Poleszczuk et al., 2005). Common in other systems symptoms and effects of over-eutrophication, like soaring phytoplankton biomass and excessive concentrations of organic seston, in the Great Lagoon are substantially reduced by intense biofiltration, particularly that of the zebra mussels. Similar phenomena have been reported also from other areas, including marine embayments supporting beds of suspension feeders, particularly large bivalves (e.g., Hily, 1991).

Thus, unlike in closed water bodies, the trophic status of the Lagoon does not increase via accumulation. It is mainly in the sediment accreting in the shipping channel that a substantial chemical load, including nutrients, is deposited (Minning, 2004). That sediment, however, is periodically removed by dredging. The nutrient budget, i.e., a comparison between the load introduced by the Odra and other sources and the amount removed into the Pomeranian Bay, developed by Pastuszak et al. (2005) showed half of the total nitrogen load and one-third of the phosphorus load to be retained and eliminated in the entire Odra river mouth system, the Great Lagoon playing a major role in the process. Despite the high water dynamics inducing sediment resuspension, riverine water inflow, and Baltic water incursions, the sediment-forming processes and bottom erosion remain in a specific equilibrium (cf. Section 1.2). The Great Lagoon doesn't shallow to the extent observed in the Curonian Lagoon or the coastal lakes as a result of intensified sedimentation of suspended particulates delivered with riverine runoff and deposition of autochthonous seston. Nor is its bottom eroded to the extent that observed in the Vistula Lagoon (Witek et al., 2010). In the latter, following the closure of the connection with the Vistula River, the removal of inorganic suspension is higher than the input from the catchment area and atmosphere.

A comparative discussion on the trophic status and pollution of river mouth systems *versus* inland water bodies is not meaningful, because ecological effects in river mouths are not commensurate with the chemical status. This is particularly well-visible in the Great Lagoon. The multi-year, seasonal or even short-term variability of the riverine runoff-borne load, the "cleansing" effects of Baltic water incursions, and random changes in the water chemistry regime under the influence of high water dynamics or mass development of aquatic organisms, as well as the export of seston to the sea (e.g., Gromisz et al., 1999) profoundly change the values of parameters serving as trophic status and pollution assessment criteria. No overly high amounts of organic matter, nutrients, and other substances are accumulated in the sediment because much of the sediment, resuspended and fluidised, is transported with the main water current, via the straits, into the sea. The opinion according to which all kinds of pollutants accumulate in the sediment was championed by, i.a., Wiktor (1976). However, the pollution status of the Great Lagoon sediment (outside of the shipping channel) is relatively low. According to the major conclusions of Protasowicki and Niedźwiecki (1991; 1995), the pollution status of the Odra mouth area sediment is much lower compared to other European rivers serving similar transport functions. In most cases, concentrations of various pollutants in the Great Lagoon sediments were at their natural background levels. Due to the continuous erosion of the bottom, it is in a few sites only (apart from the dredged shipping channel) that thicker, heavily organic matter-enriched mud deposits accumulate, the organic matter containing elevated pollutant concentrations, particularly those of heavy metals (Chudecki and Niedźwiecki, 1987). This is the reason why some sites show concentrations of certain pollutants above their critical levels. In his recent survey, Skowronek (2011) showed 16% of his samples to have slightly exceeded levels of Hg, whereas concentrations of As, Zn, Cd, and organic pollutants (including polyaromatic hydrocarbons, PAH) exceeded their critical levels only sporadically, and slightly at that. The open Lagoon has never shown longer-lasting symptoms of saproby (*sensu* Olszewski, 1971), i.e., a state when the biological production is hampered by excessive input of organic matter. Thus, the Szczecin Lagoon exhibits a high, albeit variable, trophic status, its hydrographic regime preventing nutrient load accumulation and counterbalancing adverse effects of hypereutrophication and input of pollutants.

When developing an aquatic area environmental status assessment, it is advisable that self-pollution processes be paid attention to. Those processes result from disturbances in the ecosystem functioning; their effects on aquatic communities are frequently stronger than those exerted by pollutants brought from outside. As already mentioned, in 1998-2002 there were no adverse chemical effects (substantial concentrations of hydrogen sulphide and ammonium or too high pH etc.) or clear-cut negative biocoenotic responses to excessive growth of planktonic algae, particularly under prolonged chemical stratification. Such effects, on the other hand, have been reported from the Curonian Lagoon, an area with a limited water exchange, with deleterious consequences for the aquatic fauna (Aleksandrov, 2003; 2009; Aleksandrov and Dmitreva, 2006). The importance of such responses in inland water bodies has been treated by, e.g., Barica (1990), Krzyżanek et al. (1993), and Błaszczyk et al. (2010).

1.14.2 Phytoplankton

As already mentioned in Section 1.3, the phytoplankton dominance structure was different in each year of the period covered by this study. A marked domination of *Microcystis aeruginosa* was recorded only in 1999. It is therefore difficult to draw unequivocal conclusions from comparisons with phytoplankton dominance structure in other water bodies, as the outcome of the comparison will depend on the year for which the comparison is conducted. On the average, about 85% of the Great Lagoon phytoplankton biomass should be attributed mainly to 4 species of cyanobacteria and 3 diatom taxa, listed in the order of decreasing contribution: *Microcystis aeruginosa, Actinocyclus normani, Cyclotella* sp. div., *Planktothrix agardhii, Pseudanabaena limnetica, Aulacoseira granulata,* and *Aphanizomenon flos-aquae.* In the Vistula and Curonian Lagoons (Dmitreva, 2005), the phytoplankton was dominated by *A. flos-aquae, M. aeruginosa* playing a secondary role.

Seasonal changes in the phytoplankton dominance structure, biomass, and density in the Great Lagoon were similar to those reported from most temperate eutrophic reservoirs: the spring peak was due to diatoms, cyanobacteria being responsible for the summer (or early autumn) one. In addition to the seasonal variability of nutrient concentrations, the variability pattern depended on seasonal changes in water temperature. A stronger development of diatom populations and/or their domination is enhanced, in addition to a sufficient supply of nutrients, including silica, by lower temperatures (< 15 °C). Zhang and Prepas (1995) contended that diatom domination in phytoplankton communities is strongly associated with water temperature < 15 °C. On the other hand, the mass development and domination of cyanobacteria require water temperature of 20 °C and above (Błaszczyk et al., 2010). Under late-summer and early-spring mineral nitrogen deficiency, only cyanobacteria – owing to their specific mechanisms of absorption, and even storage, of molecular nitrogen (Błaszczyk et al., 2010) – were capable of forming lush populations. The mass diatom development in spring was a major cause of a drastic reduction in silicate concentrations in May, the reduction impeding further diatom growth. Silicate concentration reductions to the minimum levels in spring were reported by Bangiel (2004) and Pastuszak et al. (2008). Following the summer reduction, the diatom biomass rebounded in autumn. Year-to-year differences in diatom biomass in the Great Lagoon were, however, minor only, much lower than those reported from most inland water bodies. To a large extent, this could have been due to the import of planktonic algae with the Odra runoff. Janukowicz (2006) quoted other authors as saying that the abundance of phytoplankton in lotic waters may remain almost unchanged throughout the year.

The average phytoplankton biomass over 1998-2002 was 15.5 $mg_{w.w.}$ dm^{-3}, the maximum biomass (60 $mg_{w.w.}$ dm^{-3}) being recorded in August 1999. Those values evidence moderate abundances and are within the range of average values reported from eutrophic water bodies (e.g. data compiled by Margoński et al., 2003). An almost identical value (15 mm^3 dm^{-3}) was reported as a mean for the Kleines Haff phytoplankton biomass (Lampe, 1999 in Radziejewska and Schernewski, 2008). The multi-year average and maximum phytoplankton biomasses in the central basin of the Great Lagoon were similar to those recorded in 2002 in the Vistula Lagoon (Dmitreva, 2005): in the growing season, the average and maximum (during the peak cyanobacteria growth) biomasses in the Vistula Lagoon were 14 and 47 g m^{-3}, respectively. According to Dmitreva (2005), the Curonian Lagoon supported much richer phytoplankton, with an average biomass in the 2002 growing season of 70 mg$_{w.w.}$ dm^{-3}. Such high phytoplankton biomasses were recorded in the Great Lagoon only in the 1980s. Winberg et al. (1972) reported a biomass of 10.9 mg dm^{-3} in Lake Batorin (characterised by advanced trophy). According to Chrobak (2004), the average biomass of the Szczecin Lagoon phytoplankton in the growing season of 2001 amounted to 3.14 gC m^{-3}, which – converted to wet weight– would render a value more than twice as high as that found in the present study. A similar value (3 gC m^{-3}) was quoted for Lake Gardno by Skurzak (2009). Those estimates, however, were based on the subjective choice of carbon to chlorophyll *a* ratios which may vary quite extensively (Cloern et al., 1995).

As shown by our own measurements, chlorophyll *a* concentrations averaged 52 mg m^{-3} and were thus higher than those reported by other authors for the same period. According to the Regional Inspectorate of Environmental Protection data (RIEP, 2007), the chlorophyll *a* concentration over 5 years (1998-2002) averaged 42 mg m^{-3}. Chrobak (2004) and Janukowicz (2006) reported still lower concentrations for the Great Lagoon in 2001 and 2002. The mean chlorophyll *a* concentration found in this study was, however, very similar to that (52.1 mg m^{-3}) recorded later, in 2004-2006 (Bucior, 2009). The chlorophyll *a* concentrations in the Szczecin Lagoon were within the range of concentrations in other coastal lagoons of southern Baltic (Table 1.42. In the coastal Lake Gardno in 2005 and 2006 the average chlorophyll *a* concentration was 75 mg m^{-3} (Wielgat et al., 2010).

Table 1.42 Chlorophyll *a* concentrations in the southern Baltic coastal lagoons in the years 1998-2002.

Area	range (mg m^{-3})	average (mg m^{-3})	source
Szczecin Lagoon (Great Lagoon)	14-86	52	this study
Vistula Lagoon (Russian part)	15-60	32	Aleksandrov (2005a, b)
Vistula Lagoon (Polish part)	29-84	43	Renk et al. (2001)
Curonian Lagoon (Russian part)	25-697	93	Aleksandrov (2005a, b)

According to Dmitreva (2005) and Dmitreva and Semionova (2009), the phytoplankton biomass denoting a harmful hyperbloom in the southern Baltic coastal lagoons exceeds 100 $mg_{w.w.}$ dm^{-3}. Biomass at this level poses a threat to other aquatic organisms due to toxins produced intravitally by the planktonic algae and/or those released from dead algal cells as well as due to excessive hydrogen sulphide or non-ionised ammonium produced during bacterial decomposition of dead algal masses (Błaszczyk et al., 2010). Barica (1990) considers the chlorophyll *a* concentration of 100 mg m^{-3} to be a critical level signifying a threat posed by a bloom.

Neither the algal biomass nor chlorophyll *a* concentrations in the Lagoon during the period of study were indicative of any hazard posed to the biota by the excessive phytoplankton development.

Compared to other water bodies, including the Baltic coastal reservoirs, in a similar period of time, the primary production in the Great Lagoon over 1998-2002 was high (Table 1.43).

Table 1.43 Phytoplankton net primary production in the Szczecin Lagoon and other water bodies. Net primary production was assumed to account for 80% of gross primary production; winter primary production was assumed to be equal to 17 % of the growing season production.

Water body and period of study	**Net primary production [gC m^{-2} yr^{-1}]**	**Data source**
Szczecin Lagoon (Great Lagoon) 1998-2002	400	This study; based on data of Westphal and Lenk (1998), modified
Polish part of the Vistula Lagoon 1999 (southern Baltic Sea)	243	Renk et al. (2001)
Russian part of the Vistula Lagoon 2001-2003 (southern Baltic Sea)	296-418	Aleksandrov (2005a)
Russian part of the Curonian Lagoon 2001-2003 (southern Baltic Sea)	337-580	Aleksandrov (2005a)
Lake Gardno (southern Baltic coast) 2006-2007	322-353	Wielgat-Rychert et al. (2010)
Lake Łebsko (southern Baltic coast) 2007	325	Wielgat-Rychert and Rychert (2008)
Estuaries	Average 255	Mann (2000)
Estuaries	Range 26-810 Average 287	McLusky and Elliott (2004)
Shallow European lakes (several tens of lakes)	Average 205 ± 37	Lacroix et al. (1999)
Deep European lakes (several tens of lakes)	Average 177 ± 13	Lacroix et al. (1999)
Pärnu Bay (eastern Baltic Sea)	156	Tomczak et al. (2009)
Gulf of Riga (eastern Baltic Sea)	125	Tomczak et al. (2009)
Puck Bay (southern Baltic Sea)	198	Tomczak et al. (2009)

The high level of primary production in the Great Lagoon resulted, in addition to the permanent (although variable in time) nutrient input, also from the absence of stagnation and vertical stratification of the water column. Under conditions of intensive and permanent mixing and water flow, live phytoplankton inhabits the entire water column and photosynthesis (alternating dark and light phases) proceeds throughout the water column as well. This has been confirmed by the research of Bucior (2009). In contrast, water exchange in the Curonian Lagoon, which shows a particularly high nutrient concentrations, is limited, and stagnation and water column stratification occur during the summer. In addition, a subsurface abundant cyanobacterial bloom limits light penetration (Aleksandrov, 2003; Aleksandrov and Dmitreva, 2006). Despite the much higher phytoplankton biomass in the Curonian Lagoon and the chlorophyll *a* concentration almost twice that found in the Szczecin Lagoon, the primary production level in the Curonian Lagoon is only slightly higher than that in the Great Lagoon.

The contribution of phytoplankton production to the total primary production in the Great Lagoon was similar to that recorded in other eutrophic water bodies (Table 1.44).

Table 1.44 Contributions of different primary producer categories to total primary production in some eutrophic water bodies.

Water body / Producer category:	Szczecin Lagoon (Great Lagoon)	Lake Mikołajskie[a]	Lake Tynwałd[b]	Lake Myastro[c]
Phytoplankton	82%	77%	74%	86.9%
Macrophytes	13.5%	14%	21%	7.5%
Phytoperiphyton	4.5%	8%	5%	5.5%
Microphytobenthos		1%		

[a] Hillbricht-Ilkowska et al.(1971); [b] Adamska et al. (1979); [c] Winberg et al. (1972)

Should the production of macrophytes (together with buried parts) and other plants be accounted for, the phytoplankton, macrophytes, and phytoperiphyton would contribute 79.1, 16.5, and 4.4% to the total primary production in the Great Lagoon.

As stated by McLusky and Elliott (2004), estuarine phytoplankton contributes, on the average, 85% to the total primary production. According to Winberg et al. (1972), the contribution of phytoplankton production increases with increasing trophic status of a water body; in some extreme cases such as the highly eutrophic Lake Batorin (Secchi depth of 0.3-0.6 m), phytoplankton contributed as much as 96% to the total primary production. On the other hand, the flow-through, mesotrophic Goczałkowice Dam Reservoir showed the phytoplankton to contribute 64% to the total primary production (Żurek, 2009).

1.14.3 Mesozooplankton

Generally, the zooplankton abundance in the Great Lagoon was high. In the Odra upstream of Szczecin, the average density of the mesozooplankton was 749 ind. dm^{-3} (Szlauer and Szlauer, 1994) and was similar to that in the Lagoon (765 ind. dm^{-3}), but the zooplankton biomass in the Lagoon was 3 times higher than that in the Odra. The higher biomass in the Lagoon resulted from a much higher contribution of crustaceans, and a lower share of rotifers, than found in the Odra.

The eutrophic lakes Myastro and Batorin (Winberg et al., 1972) supported zooplankton biomasses (2.5-5.7; 2.7-6.7 $mg_{w.w.}$ dm^{-3}, respectively) similar to the biomass estimate for the Lagoon (and 5.13 $mg_{w.w.}$ dm^{-3}).

Margoński et al. (2003) reported a clearly lower mesozooplankton biomass in the Vistula Lagoon. In 1998 and 1999, the zooplankton there was particularly impoverished in terms of the cladoceran presence. Since 1999, the Polish part of the Vistula Lagoon has been showing a mass invasion of and a rapid increase in the abundance of the alien cladoceran *Cercopagis pengoi*, for which reason the zooplankton biomass soared to values incomparable with areas not experiencing such an invasion.

The Great Lagoon mesozooplankton production was high and reached the upper limit (20.5 ± 4.9 gC m^{-2}) of a range inferred by Lacroix et al. (1999) from an empirical model incorporating data from 56 lakes of various trophic status and depth.

On the average, the total zooplankton P/B for the growing season amounted to 22 (wet weight) and 21 (organic carbon-based). These values are within the range of 15.7-28.4 reported for 56 lakes by Lacroix et al. (1999).

With regard to major zooplankton taxa, the following remarks have to be made:

- the P/B value of 60 for rotifers assumed following Vladimirova (1974) is very close to the mean of the range of 34-100 reported for warm-water rotifers in Lake Sevan by Winberg (1979);
- according to Winberg (1979), in the temperate crustacean zooplankton, daily production rates of cladocerans and copepods over the growing season are 0.050-0.300 and 0.020-0.080, respectively. In the Great Lagoon, the respective daily P/B values were 0.119 and 0.048, i.e., close to Winberg's (1979) mean values.

The Great Lagoon's mesozooplankton production was higher than that found in most Polish eutrophic lakes (Hillbricht-Ilkowska et al., 1966; Zawiślak, 1972). On the other hand, compared to the eutrophic Lake Batorin (Winberg et al., 1965), the crustacean plankton production was lower by the factor of 1.3.

Wetzel (2001) reported data on the contributions of major zooplankton taxa to the total production of six Russian eutrophic lakes. When averaged, the values are similar to those reported in this work (Table 1.45).

Table 1.45 Contribution of individual taxonomic groups to the total mesozooplankton production.

Taxonomic group	Data after Wetzel (2001) from eutrophic Russian lakes	This study, Great Lagoon	
		Contribution to wet weight-based production	Contribution to carbon units-based production
Rotifera	28.6%	28%	23%
Cladocera	45.5%	52%	56%
Copepoda	26.0%	20%	21%

According to Hillbricht-Ilkowska et al. (1971), production of the predatory zooplankton accounts for 3-70% of the non-predatory zooplankton production; the corresponding contribution in this study was 14%. In the polymictic eutrophic Lake Śniardwy, the ratio between the production of planktonic suspension feeders and the net phytoplankton production amounted to about 0.03 (Hillbricht-Ilkowska et al., 1971). In the present study, the ratio was 0.08 and 0.055 when based on wet weight and organic carbon, respectively. Thus, the data evidence a higher ecological efficiency of the Great Lagoon, compared to Lake Śniardwy.

However, general comparisons with lakes show the utilisation of primary production by zooplankton in the Lagoon to be rather low, the $P_{mesozooplankton}/P_{phytoplankton} = 0.0643$ being clearly lower than the levels reported by Lacroix et al. (1999) from various lakes (≈ 0.1). According to Kajak (2001), the zooplankton production accounts for 3-16% of that of the phytoplankton. In this study, it was 6.4%, i.e., clearly below the average. The reason could be sought in the fact that the estimate disregarded the production of macroplankton, i.e., that of *Neomysis integer* (there were, however, no suitable data). Another reason could be a high pressure on the zooplankton exerted by fish. The highest quantitative effect on the zooplankton is produced by the obligatory fish planktivores: smelt as well as larvae and young-of-the-year of all the fish species. A strong pressure is exerted by large concentrations of larvae and fry of perch (Treasurer, 1997; Scheffer, 2001).

Experimental studies led Ismann and Mehner (1977) to emphasise the feeding selectivity of perch with respect to large daphnids, particularly the ovigerous individuals. In the opinion of those authors, daphnids are strongly eliminated by the young-of-the-year perch biomass of 20 kg ha^{-1}; the biomass of 50-60 kg ha^{-1} is critical and leads to total elimination of daphnids from the plankton. In the Great Lagoon, the young-of-the-year perch biomass averaged 8.5 kg ha^{-1}, but the total biomass of planktivorous larvae and fry (aged 0 and 0+) of all the fish species was estimated at about 45 kg ha^{-1}. A large portion of the zooplankton biomass can be consumed also by maturing and adult fish, particularly bream. On account of specific adaptations in their filtration apparatus (a branchial filtering system), they are capable of much more efficient filtering of small plankters than smaller cyprinids of other species (Lammens and Hoogenboezem, 1991; Berg et al., 1992; 1993). When the chironomid biomass is too small to feed large (longer than 20 cm) bream, they switch to feeding on the zooplankton (Michelsen et al., 1994). In 1998-2002, the biomass of juvenile bream (total length of 10-35 cm) was about 30 kg ha^{-1}. The mesozooplankton contribution to the food of those fish averaged, in a growing season, about 15-20% (Wolnomiejski and Grygiel, 2002). Therefore the feeding pressure of those bream on the zooplankton could periodically be substantial. This was particularly the case in the second half of May and early June when, due to the emergence of chironomids and the absence of their large larvae, feeding conditions for the obligatory benthivores were poor.

1.14.4. Macroplankton

The mysids *Neomysis integer* (macroplanktonic crustaceans) are both common and abundant in other lagoons and brackish coastal lakes of the southern Baltic as well as in the inshore waters of the Baltic Sea itself (Węsławski, 1981; Kotta et al., 2007 and the references therein). *N. integer* are the most thermophilous mysids in the Baltic. They are abundant in the Great Lagoon as well. Large aggregations of those crustaceans enter the Lagoon or are brought in with seawater incursions from the Pomeranian Bay. Therefore, their highest abundances are recorded in the northern parts of the Lagoon. In addition, the area supports a stationary local population. The mysids are particularly abundant from June (when the young-of-the-year appear) through the end of September (Wiktor, 1980). As stated by Maciejewska and Opaliński (2002), *N. integer* are omnivorous, but may switch to predation. Their food composition is similar to that of larvae and fry of herring and smelt of a comparable size (the relevant food composition data were reported by, i.a., Węsławski, 1981; Monakov, 1998; and Ten, 2004). Kotta and Kotta (2001) reported data which show *N. integer* to feed on phyto- and zooplankton at night, close to the surface, and on detritus from the bottom sediment during the daytime. Similar observations were reported by Ten (2004). According to Scheffer (2001), predation of *N. integer* may greatly reduce the zooplankton abundance, particularly in summer.

The lack of reliable data on the abundance and biomass of mysids in the Szczecin Lagoon and in other shallow areas of the Baltic Sea, its coastal lagoon and lakes, stems mainly from sampling difficulties. As fast swimmers, *N. integer* are easily capable of escaping the slow-moving plankton nets or avoiding being captured by relatively small, standard water bottles. Other behavioural traits, too, contribute to difficulties with reliable sampling of *N. integer*. At their peak abundance, mainly in summer, they form large aggregations in shallow areas, particularly among macrophytes, where they are difficult to catch. In deeper areas, *N. integer* stay close to the bottom during the daytime, and swim up to the subsurface layer at night (Wiktor, 1961). Therefore, sampling conducted with horizontal tows of plankton nets is hardly effective. Diurnal vertical migrations of the mysids, particularly pronounced in late summer and autumn, were described by Kotta and Kotta (2001). It is only during heavy seawater incursions into the Lagoon that *N. integer* swarms are frequently visible close to the surface. Most probably, large amounts of the mysids are then consumed by fish, whereby their abundance drops markedly and few individuals only are capable of reaching the southern part of the Lagoon.

In view of the methodological difficulties discussed above, the estimates shown in Section 1.5 and based on mysid abundances reported by Wiktor (1961) cannot be treated as a reliable measure of the production potential of those crustaceans. However, there are no other data on the mysid abundance from the Lagoon. The published P/B data raise doubts as well. As shown by Chojnacki (1991), the local *N. integer* population in the Lagoon consists of two generations, the vernal and the autumnal. According to Wetzel (2001), P/B of bivoltine forms is high and amounts to about 5.5. In addition, at least half of the mysids present in the Lagoon originate in the Pomeranian Bay and thus provide an additional, imported production.

Therefore, a rational quantitative assessment of the mysid production can be expected to be produced by model calculations based on food requirements of fish, the most important consumers of *N. integer*.

1.14.5 Macrophytes

The composition and domination structure of macrophyte assemblages in the Great Lagoon were distinctly freshwater in their nature. The brackish components included only the small amounts of *Enteromorpha* found particularly frequently near the shores in the northern part of the area. Salinity effects on diversity of wetland and aquatic vegetation were investigated by Piotrowska (1966). Her study, however, focused on the Stara Świna delta only, and – to some extent – included Lake Wicko. The small salinity gradients and short periods of elevated salinity (particularly at the shores located away from the main flow of seawater incursions) did not affect the distribution of aquatic macrophytes in any noticeable manner.

At some sites, the water line showed the presence of wetland helophytes, particularly *Iris pseudoacorus, Oenanthe phellandrium, Sium latifolium, Cicuta virosa, Carex stricta, C. limosa, Volvulus sepium*, and *Solanum dulcamara*. In her study, Garbacik-Wesołowska (1973) did not measure the biomass of those species separately, but included it in the biomass of the dominant rush belt helophytes.

The Great Lagoon supported a particularly high biomass of helophytes, particularly that of reed, which accounted for about 60% of the total macrophyte biomass. Bernatowicz and Pieczyńska (1965) provided data evidencing macrophyte production in various natural habitats to reach 4-10 $t_{d.w.}$ ha^{-1} or even 14 $t_{d.w.}$ ha^{-1}. Gayevskaya (1966) showed the reed biomass in highly productive water bodies to be as high as 98 $t_{d.w.}$ ha^{-1}, eutrophic limans supporting reed biomass of 29 $t_{d.w.}$ ha^{-1}. In the Curonian Lagoon, the maximum helophyte biomass was 8.35 $kg_{w.w.}$ m^{-2} (Pustelnikovas, 1998), equivalent to about 40 $t_{d.w.}$ ha^{-1}, when applying conversion factors for reed and remaining helophytes of 0.5 and 0.2 $g_{d.w.}$ $g_{w.w.}^{-1}$, respectively (Gayevskaya, 1964; Bernatowicz and Pieczyńska, 1965; Rejewski, 1979). According to Aleksandrov (2003), the Curonian Lagoon reed biomass reached 6.8 $kg_{w.w.}$ m^{-2} (34 $t_{d.w.}$ ha^{-1}). In 1975-1976, the Vistula Lagoon was found to support the dry-weight based biomass of macrophytes averaging 19.3 t ha^{-1} phytolittoral, 93% of macrophytes being accounted for by reed (Pliński et al., 1978). The Great Lagoon supported the macrophyte biomass of 8.4 $t_{d.w.}$ ha^{-1}, reed contributing 59% only. The average biomass of reed, within a reed community, was by 25% higher in the Great Lagoon (3.4 $kg_{w.w.}$ m^{-2}) than in the Polish part of the Vistula Lagoon (2.7 $kg_{w.w.}$ m^{-2}; Ringer 1959). The biomasses of soft (submerged and floating-leaved) plants were similar to those reported from very fertile inland lakes (Bernatowicz and Pieczyńska, 1965; Mikulski et al., 1975; Rejewski, 1979; 1988; Kraska, 1990). A eutrophic lake studied for three years by Kraska (1990) supported the maximum biomass of *Myriophyllum* equal to 605 $g_{d.w.}$ m^{-2}.

The average biomass of soft macrophytes in the Great Lagoon, within the respective communities, amounted to (corresponding values for the Polish part of the Vistula Lagoon, quoted after Ringer, 1959, are given in parentheses) 0.416 (0.240), 0.156 (0.190), and 0.329 (0.320) $kg_{d.w.}$ m^{-2} for the floating-leaved, Potamogetonaceae, and *Ceratophyllum* + *Myriophyllum*, respectively.

Production of the above-bottom parts of macrophytes in the Great Lagoon totalled 66.1 gC m^{-2} yr^{-1}, i.e. 747 kcal m^{-2} yr^{-1} (as referred to the total surface area of the Lagoon). This production was higher than that in Lake Mikołajskie

(450 kcal m^{-2} yr^{-1}; Hillbricht-Ilkowska et al., 1971). However, submerged vegetation in the Lake Mikołajskie phytolittoral (19% of the lake surface area) covered a much larger area than that occupied by helophytes. Macrophyte production in the very fertile, eutrophic Lake Tynwałd (Adamska et al., 1979) was 393 kcal m^{-2} yr^{-1}, but the phytolittoral (reed contribution of 80%) occupied as little as 8.3% of the bottom surface area (15.5% in the Great Lagoon). The values are thus similar and comparable. According to Witek et al. (2010), the macrophyte production in the Vistula Lagoon was similar to that in the Great Lagoon (about 60 gC m^{-2} yr^{-1}).

Production of the above-bottom parts of the macrophytes in areas of their occurrence in the Great Lagoon was within the range of annual production values reported from other areas (Wetzel 2001):

	reported by Wetzel (2001) [g_{afdw} m^{-2} yr^{-1}]	in the Great Lagoon [g_{afdw} m^{-2} yr^{-1}]
Helophytes	1000-10000	3280
Nymphaeids	100-500	360
Elodeids	5-1500	160

The underwater surface created by the above-bottom parts of macrophytes was 175 km^2, i.e. almost 43% of the Great Lagoon bottom surface area. It is 2.7 times the surface of the littoral within the 0-2 m depth zone. Above 1 m^2 of the bottom, the halophytes created an average of 2 m^2, and the soft plants 3 m^2 (1.0; 0.3; and 7.7 m^2 formed by pondweed, nymphaeids, and *Ceratophyllum* plus *Myriophyllum*, respectively) of underwater surface amenable for colonisation by periphytic communities; for the entire phytolittoral, the colonisable surface amounted to 2.76 m^2. These values are comparable with those reported by other authors. Bohr et al. (1975) found reeds to form a 2 m^2 underwater surface above 1 m^2 of the bottom in the eutrophic Polish lakes of Jeziorak and Tynwałd. Pereyra-Ramos (1981) calculated the characean (*Chara* sp.) surface above 1 m^2 littoral bottom to exceed 7 m^2. In Lake Mikołajskie, Kajak et al. (1972) found 1 m^2 of the bottom overgrown by reed to offer 0.74 m^2 (2.08 m^2 at the maximum) for colonisation by periphyton; however, the Szczecin Lagoon's reed biomass was clearly higher than that in Lake Mikołajskie. In the submerged vegetation zone of Lake Mikołajskie, the surface area amenable for colonisation by periphyton averaged 4 m^2 per 1 m^2 phytolittoral surface area (Kowalczewski, 1975). However, the soft macrophytes in the Lagoon were less developed than in Lake Mikołajskie.

1.14.6 Phytoperiphyton

The periphytic algal assemblages in Baltic coastal lagoons and lakes are usually dominated by diatoms (Piesik, 1992; Wolańczyk, 1997; Piesik and Obolewski, 2000; Obolewski, 2002; Piesik et al., 2003) which account for up to 98% of all the algae. In the present study, diatoms constituted 88.6% of the periphytic algae. Unfortunately, there are no sound grounds for qualitative comparisons with other water bodies, as diatoms in the present study were identified to the genus level only. For example, the genus *Navicula*, accounting for more than ⅓ of the entire periphyton biomass, is represented by about 1000 species, mostly ubiquitous forms present in various ecological formations, mainly plankton and periphyton.

Due to various reasons, quantitative assessment of the phytoperiphyton in individual samples is very often difficult. The density of periphytic algae usually varies greatly, both between sampling dates and between years of study. The values recorded in different samples collected on various, even close, dates may be influenced by the abundance of periphytic consumers (particularly gastropods). Such effects, found in experiments, were described by, i.a., Lampert and Sommer (2001). Along the stretches of the shore exposed to prevailing winds, particularly important is the current or recent wave action. This problem was touched upon by, i.a., Pieczyńska (1964) and Kuczyńska-Kippen et al. (2004). Pieczyńska (1964) found the highest periphyton algal losses due to removal by wave action to occur in loose, semi-planktonic (sedimenting) forms, e.g. *Lingbya limnetica.* In contrast, the sessile forms, particularly those forming the base layer, were only slightly affected. Thus, the base-forming algae, e.g. *Cladophora* sp. protect other algae from being removed. Therefore the periphyton is usually dominated by diatoms, as many of them are attached to the substrate by strong styli and polysaccharide secretions (Kuczyńska-Kippen et al., 2004). The sites where the periphyton was sampled from reeds were moderately exposed to wave action. Differences between the phytoperiphyton biomass between individual samples were two-fold.

When converted to the caloric value, the phytoperiphyton biomass and production in the Great Lagoon averaged 41 and 578 kcal m^{-2} reed substrate in the growing season, respectively. The values were similar to those reported from the small eutrophic Lake Tynwałd (Bohr et al. 1975). When referred to the 210-day-long growing season, the values there averaged 50 and 750 kcal m^{-2}, respectively.

1.14.7 Mesozooperiphyton

The domination of rotifers and nematodes in the mesozooperiphyton of areas adjacent to the Great Lagoon and coastal Baltic areas was emphasised by

Piesik and Obolewski (2000) and by Piesik and Wawrzyniak-Wydrowska (2003). Compared to those areas, the Great Lagoon periphyton showed fewer copepod nauplii and adults of other zooplankton and crustacean meroplankton. This could have been an effect of predation by *Cordylophora caspia*, abundant in the periphyton. More detailed comparisons are rendered impossible by the absence of biomass data on smaller mesoperiphytic invertebrates in published studies, the quantitative data reported consisting of the abundances only.

1.14.8 Macrozooperiphyton

Apart from the studies of Czarnecka and Tymolewski (2003) and Czarnecka (2006), data on the macrozooperiphyton inhabiting natural substrates, i.e. mostly aquatic macrophytes of the Great Lagoon, are very scant. Some contributions were provided by Piesik and Wawrzyniak-Wydrowska (2003) whose studies were conducted in the southern part of the Odra mouth and in the adjacent areas. There is more information available on the zooperiphyton developing on artificial substrates placed in the Stara Świna (Piesik, 1992) and in Lake Wicko Małe, a specific part of the Lagoon (Wolańczyk, 1997; Miączyńska, 1997). For comparative purposes, reports on the zooperiphyton of some Baltic coastal lakes, i.a. the Bukowo and the Jamno (Piesik and Obolewski, 2001 and Piesik et al., 2003) may prove useful. However, quantitative data in those publications were expressed only as densities (abundances) and not as biomasses.

The taxonomic composition and dominance structure of the macrozooperiphyton, particularly its motile components inhabiting reed were, notwithstanding some exceptions, similar to those in other areas in the vicinity of the Great Lagoon and in the coastal lakes. In addition, a high similarity was observed between the macrozooperiphyton inhabiting reeds and artificial substrates (Wolańczyk, 1997; Piesik, 1992; Piesik and Wawrzyniak-Wydrowska, 2003). Moreover, comparative studies of Soszka (1975) on colonisation of natural (macrophytes) and artificial (plastic) substrates by epifauna hardly showed any differences, which he took as corroboration of an old theory of Shelford (1918) that macrophytes act by merely providing settlement surface for numerous invertebrates.

In a Dutch polder, Beattie et al. (1972) found, within 15 April - 17 November, a biomass of 257 $kg_{w.w.}$ ha^{-1} of motile macrofauna (mainly gammarids, chironomids, oligochaetes, and hirudineans) dwelling on reed and cattail stems. In the present study, 1 hectare of reed phytolittoral supported 262 $kg_{w.w.}$ of motile forms. Despite differences in the structure of major groups of the macroperiphyton (the polder fauna was dominated by *Gammarus tigrinus* and chironomids), the biomasses are very similar. The specification below shows a comparison between the biomass of motile epifauna (gammarids, chironomids, trichopterans, ephemeropterans, and hirudineans) dwelling on macrophytes in the eutrophic large Kiev Reservoir (Gak et al., 1972), referred to $kg_{w.w.}$ plants (corresponding data from the present study are given in parentheses):

on helophytes	4-6 $g_{w.w.}$ $kg_{w.w.}^{-1}$ (3.2 $g_{w.w.}$ $kg_{w.w.}^{-1}$),
on soft plants	12-27 $g_{w.w.}$ $kg_{w.w.}^{-1}$ (30.9 $g_{w.w.}$ $kg_{w.w.}^{-1}$).

These data, too, bear evidence of a high similarity between biomasses of epiphytic macrofauna in the two areas. The lower epifauna biomass on helophytes in the Great Lagoon could be a result of a less accurate sampling design, as the areas adjacent to roots, particularly in reeds, i.e., the zone of adventitious roots, were not sampled. It is there that the motile forms usually aggregate, particularly on shores exposed to wave action. In addition, there are no data on the fauna colonising dead helophyte stems remaining from the previous year.

The data shown above are averages of entire sets of samples. Macrozooperiphyton densities varied widely from sample to sample; between-samples differences in biomass at a single site could be even 10-fold. Pieczyńska (1964) found a high degree of removal of zooperiphytic oligochaetes and chironomids as well as small nematodes by wave action. Particularly prone to wave removal was the young periphyton. The removal was greatly reduced when the substrate was covered by filamentous algae, e.g., *Cladophora*, or by colonial hydrozoans (e.g., *Cordylophora caspia*). The removal and transport of periphytic organisms is the cause of the high similarity between periphytic assemblages on different substrates. In the Szczecin Lagoon, the wave action is frequently very strong, particularly at the eastern shore. The absence, or low densities, of the epifauna found at a certain time may be an inadequate measure to be used to estimate the production.

The majority of studies published so far provide quantitative data as abundances rather than biomass. Comparisons between the abundance data obtained in this study, after appropriate conversions, were possible for some taxa only, because – in this study – individuals in colonies of hydrozoans or bryozoans were not counted, whereas Piesik and Wawrzyniak-Wydrowska (2003) reported up to 312 thou. zooids of *Cordylophora caspia* per m^{-2} substrate. A relevant comparison for major zooperiphyton components is presented in Table 1.46.

The comparisons show the abundance of macrozooperiphyton on reeds in the Great Lagoon to be generally high. Particularly high was the abundance of amphipods (represented primarily by gammarids). It might have been due to the high abundance of *Cordylophora caspia* which not only offered refuges for invertebrates, but was a food source for some predators. Piesik and Wawrzyniak-Wydrowska (2003) found gammarids to feed on those predatory hydrozoans. As shown by Czarnecka (2006), the abundance of motile epifauna on pondweed

Table 1.46 Comparison of macrozooperiphyton abundances [ind. m^{-2} substrate].

Taxon \ Water body (substrate)	Great Lagoon (reed)[a]	River Odra mouth and Roztoka Odrzańska (reed)[b]	Old Świna Canal (nylon nets)[c]	Coastal Baltic lakes (reed)[d]
Oligochaeta	5440	1650	19400	490-39530
Chironomidae	3860	4587	5940	205-1545
Amphipoda	5160	71	133	0
Gastropoda	78	46	1	0-194

[a] this study; [b] Piesik and Wawrzyniak-Wydrowska (2003); [c] Piesik (1992);
[d] Piesik and Obolewski (2000); Piesik et al. (2002/2003)

in the Great Lagoon was high, compared to some other areas. The richness of epifauna dwelling on elodeids has been demonstrated in many other water bodies, e.g., by Dukowska (2007).

1.14.9 Macrozoobenthos

The Great Lagoon macrozoobenthos consists almost exclusively of freshwater forms. The few Baltic forms occur, in low abundances, in the northern part of the area only, in the vicinity of the Piastowski Canal and Świna inlets. Many more brackish forms are found in the Vistula Lagoon (Ezhova et al., 2005). The species composition of the Great Lagoon macrobenthos has been discussed in depth by Wiktorowie (1954), Kolasa (1972), Masłowski (1992), Wolnomiejski and Grygiel (1994a), and Wawrzyniak-Wydrowska (1998). Masłowski (1992) published a list of 97 benthic taxa of various rank occurring in the Great Lagoon. Rödiger (2004), based on studies conducted in 2001 and 2002, identified 63 taxa in the Kleines Haff and in the Great Lagoon. A list of macrobenthic taxa (mostly species) was published by Wolnomiejski (1998) based on his own data and those published in the literature by that time; the list featured 152 taxa of various rank. The taxon-richest groups were the insect larvae (51 taxa), molluscs (45), crustaceans (19), oligochaetes (11), and hirudineans (10). Considering that numerous taxa were not identified to the species level, the taxonomic richness presented is considerable. A 2004 study in the Curonian Lagoon (Zettler and Daunys, 2007) revealed 174 different macrobenthic taxa.

In the final decades of the 20th century, invasions of alien benthic species intensified in the Baltic Sea coastal water bodies. In the Szczecin Lagoon, most alien species were amphipods (Gruszka, 1999; Jażdżewski et al., 2005) and the polychaetes *Marenzelleria sp.* (Żmudziński et al., 1997). The invasions increased the taxonomic richness; for example, 10% of the fauna in the Curonian Lagoon are alien species (Zettler and Daunys, 2007).

Most of the benthic studies so far have revealed a considerable abundance of the macrobenthos in the Great Lagoon, compared to coastal lakes and other Baltic lagoons. The macrobenthic biomass in the Great Lagoon averaged 57 $g_{w.w.}$ m^{-2} without *Dreissena polymorpha* and 42 $g_{w.w.}$ m^{-2} without *D. polymorpha* and unionids. The corresponding average biomass, found in the same period in the Vistula Lagoon, was 61 $g_{w.w.}$ m^{-2} (Ezhova et al., 2005), 80% of which was attributed to *Marenzelleria sp.* The polychaete has been observed to increase in abundance there since 1990, which generally increased the total benthic biomass. In 1988 prior to the *Marenzelleria* invasion, the average benthic biomass over the entire Vistula Lagoon amounted to 27 $g_{w.w.}$ m^{-2} (Żmudziński, 2000). In the Great Lagoon, no adults of the species have been found, the invasion involving juvenile forms, mainly in 1991, with average abundance and biomass of 73 ind. m^{-2} and 0.7 $g_{w.w.}$ m^{-2}, respectively (Wolnomiejski, 1994; Żmudziński et al., 1997). In later years, the species was not observed in the open basin of the Lagoon and was sporadically recorded in the littoral.

The muddy bottom benthos biomass in the Great Lagoon averaged 32.5 $g_{w.w.}$ m^{-2}, whereas the biomasses recorded in the coastal Lake Gardno and in the Vistula Lagoon in 1998 were 18.8 $g_{w.w.}$ m^{-2} (Dobrowolski, 2001) and 22 $g_{w.w.}$ m^{-2} (Żmudziński, 2000), respectively.

It is rational to consider the biomass of large bivalves, unionids and *D. polymorpha*, separately from the biomass of the non-shelled benthos and small molluscs. Biomass of large bivalves, particularly that of the bed-forming zebra mussel, may locally be an order (and more) of magnitude higher than the biomass of small benthic invertebrates which almost entirely are a potential fish food.

The unionid biomass in the Lagoon, averaging 12.2 $g_{w.w.}$ m^{-2}, was similar to that reported for the Goczałkowice Dam Reservoir in the 1990s (Kasza, 1995; Żurek, 2007). In the early 1970s, the unionid biomass in that reservoir was still higher (Wróbel, 1975). Lewandowski and Stańczykowska (1975) found the unionid biomass in Lake Mikołajskie to average 6.56 $g_{w.w.}$ m^{-2}, which they considered as a relatively low value. They referred to the published data evidencing average biomasses of 16.2 and 12.1 $g_{w.w.}$ m^{-2} in a Romanian lake and in the Thames, respectively.

The particularly high abundance of the zebra mussel (*Dreissena polymorpha*) is a special biocoenotic aspect of the Szczecin Lagoon. The bivalve's resources in the Great Lagoon alone were estimated at almost 60 thou. $t_{w.w.}$ (with shells). Fenske (2002) found the zebra mussel to be abundant in the Kleines Haff, too. Similar is the case in several other southern Baltic coastal water bodies, e.g., in the Curonian Lagoon (Gasiûnaté et al., 2008; Zaiko et al., 2010) and in the Neva River mouth area in the Gulf of Finland (Orlova et al., 2004). Outside the Great

Lagoon, in the remaining areas of the Polish part of the Odra mouth system (about 220 km^2) the zebra mussel resources are estimated at 50 thou. $t_{w.w.}$ (as shown by data published by Woźniczka and Wolnomiejski, 2005, and Wolnomiejski and Woźniczka, 2008). Combined with the resources in the Great Lagoon, the system supports a total of about 110 thou. t, i.e., close to the value reported in the late 1960s by Wiktor (1969). According to that author, the zebra mussel occurring in such an abundance is capable of filtering the entire water volume of the Lagoon during some thirty days. In addition, the mussel's environmental effects include habitat formation and contribution to sediment accumulation (due to the accumulation of faeces, pseudofaeces, and shells in the sediment) (e.g. Wiktor, 1969; Stańczykowska and Planter, 1985; Orlova et al., 2004; Zaiko et al., 2009; 2010). The zebra mussel is regarded as a habitat engineer (e.g., Gasiûnaté et al., 2008). Both the immense filtering potential of the bivalve (Wiktor 1969, and Woźniczka et al., 2006), of a major importance for chemical and biological status of the water, and its heavy biocoenotic pressure on the biotope and remaining aquatic organisms, including effects on fish feeding, are extremely important for the functioning of the Great Lagoon.

Beds of *D. polymorpha* create a convenient spatial (and most probably feeding) niche for numerous benthic animals, particularly hirudineans (mainly *Herpobdella* and *Helobdella*), gastropods, crustaceans (particularly *Asellus aquaticus* and *Corophium* sp.), and chironomid larvae. The macrofauna associated with zebra mussel beds was particularly abundant, which was the case in the Kleines Haff as well (Radziejewska et al., 2009). In the Curonian Lagoon, the macrobenthic biomass on the bottom supporting abundant zebra mussel aggregations (about 1500 ind. m^{-2}) was more than three times higher than that on the bottom without *D. polymorpha* (Zaiko et al., 2010).

That the *D. polymorpha* beds offer a special feeding niche has been convincingly demonstrated for the chironomid *Dicrotendipes* gr. *nervosus* the mass occurrence of which was observed only among a densely packed zebra mussel bed. A similar situation was reported from other water bodies by Wolnomiejski (1970) and Mastitsky and Samoilenko (2005). According to Monakov (1998), larvae of *Limnochironomus* (syn. *Dicrotendipes)* are commensals feeding on agglutinates formed in the water column under the influence of flocculants produced by the zebra mussel. On numerous occasions, Mastitsky and Samoilenko (2005) found *Dicrotendipes* gr. *nervosus* to occur even in the zebra mussel mantle cavity. Most probably, commensalism occurs also in other macrobenthic organisms present in association with zebra mussel beds. High amounts of faeces and pseudofaeces produced by the bivalve enhance detritophagy. Dombrovskiy (2009) reported on zebra mussel bed-associated macrobenthic guilds reaping habitat benefits which would not have been possible without the bivalve.

Biomass estimates for the macrobenthos of soft sediments of the Great Lagoon, which cover most of the bottom surface, do not raise any doubts.

The heavy Ekman-Birge grab collected representative quantitative samples. Similarly, estimates of the zebra mussel biomass are reliable. On the other hand, samples collected from the bottom covered by abundant organic remains and overgrown by macrophytes have to be treated with caution. The corer used allowed to collect representative quantitative samples of the infauna and molluscs. However, the highly mobile forms were captured only sporadically, particularly the predatory coleopterans, gammarids or ephemeropterans which efficiently escaped the capture with a small-opening gear. Such samples contained no large larvae and imagines of the Coleoptera, larval Rhynchota (Hemiptera), and Odonata. That was particularly the case on the phytolittoral bottom. According to Wawrzyniak-Wydrowska and Gruszka (2005), locally high aggregations of alien, invasive crustaceans *Gammarus tigrinus* and *Pontogammarus robustoides* dwelled close to the bottom, at the depth of about 0.5 m, on macrophytes and in the water column. It should be then assumed that the present study underestimated the littoral benthos resources. Grzybkowska et al. (1996) and Grzybkowska (2002) pointed out to the very high P/B values of some chironomid taxa of small size. This is particularly true with respect to the very small chironomids of the tribe Tanytarsini or the subfamily Orthocladiinae which produce numerous generations a year. Here, though, there is no fear of any significant underestimation of the production of littoral chironomids. As small chironomids contributed very little to the macrobenthic biomass in the Great Lagoon, their role in the total production budget was small, too.

The land-water ecotone, an extremely important upper littoral habitat, still remains completely unknown. Usually, the habitat supports a high benthic diversity and abundance (e.g., Poznańska et al. 2010). The spatial extent of this habitat depends on changes in water level and shore profile. In the Great Lagoon, with its substantial and frequent changes of water level, the ecotone may locally occupy vast areas.

Generally, the benthos production was high, but there are no corresponding comparative data from other coastal water bodies. The total macrobenthos production in the Great Lagoon, averaging 22.4 gC m^{-2} yr^{-1}, was by an order of magnitude higher than production of the macrobenthos in the Baltic (e.g., 1.6 gC m^{-2} yr^{-1} in the Gulf of Gdańsk; Witek, 1995; and 3.0 gC m^{-2} yr^{-1} in the Central Baltic; Elmgren, 1984). It should be remembered, however, that considerable parts of the bottom in the two Baltic areas mentioned above are affected by oxygen deficiency. The Pomeranian Bay, which is occasionally considered the first-order Odra estuary, has good oxygen conditions throughout, due to small depths. Nevertheless, the macrobenthos production in the Bay was much lower than that in the Great Lagoon. It was only in a relatively small (5.5 km^2) area at the forefront of the Świna mouth, with bottom sediment enriched with silt-clay, that the macrofauna biomass was at its highest and the production was usually high. Based on the 1976-2005 data published by Masłowski (2010), the macrobenthic production in this small area in the vicinity of the river mouth was estimated to range within 1.5-13 gC m^{-2} yr^{-1} and to average 7 gC m^{-2} yr^{-1}.

A particularly important characteristic of the Great Lagoon which greatly affects the food web in the area is the high biomass and production of macrobenthos in the central, muddy bottoms occupying about 50% of the total bottom area (Tab 1.23). The benthos production there averaged 228 $g_{w.w.}$ m^{-2} yr^{-1} or 21.4 gC m^{-2} yr^{-1}, 98% being contributed by *Chironomus* sp. and oligochaetes. Conditions for the life of macrobenthos in eutrophic inland lakes in which, due to stratification, the near-bottom water layer in the central basin is usually separated from the rest of the water column, are not favourable. For example, production of *Chironomus* sp. inhabiting muddy bottoms of two large and shallow eutrophic lakes, the Jeziorak and the Gopło, was 18.7 $g_{w.w.}$ m^{-2} yr^{-1} (Giziński and Wiśniewski, 1971) and 31.5 $g_{w.w.}$ m^{-2} yr^{-1} (Wiśniewski, 1976), respectively. In contrast, as shown by the present study, the Great Lagoon *Chironomus* sp. production was 183 $g_{w.w.}$ m^{-2} yr^{-1}. The P/B values were similar in the three areas.

It is very difficult to compare benthos production in different water bodies on account of differences in P/B values adopted, different seasons of sampling, different developmental stages considered in the analysis, etc. Those problems were pointed out also by Witek (1995). Most frequently, like in this study, the annual production is assumed to be equal to production in the growing season. In their empirical study on production of *Chironomus* sp., Wolnomiejski and Grygiel (1989) found no weight increment within December-February. In winter, weight increments in the benthic organisms are minimal; some forms disappear, leaving only resting stages, or die out altogether. For example, Andronikova et al. (1972) reported the chironomid growth to be arrested at 5 °C.

Ecological efficiency of the macrobenthos relative to the primary production in the Great Lagoon was high and amounted to 5.6% for the total macrobenthos or 4.7% for the macrobenthos without large bivalves. According to Kajak (2001), ecological efficiency of the macrobenthos in inland water bodies ranges within 0.2-6.0%.

1.14.10 Meiobenthos

The Great Lagoon meiobenthos was dominated by ostracods. Radziejewska et al. (2009) showed the meiobenthos of muddy areas in the Kleines Haff to be dominated by ostracods as well, the domination being, however, less distinct.

According to Kajak (1988), the meiobenthos in inland water bodies may reach very high abundances, up to 1 million ind. m^{-2}. In a shallow part of a Bothnian Bay, Merilaeinen (1988) recorded meiobenthic abundance of up to 2.4 million ind. m^{2}.

The average meiobenthic biomass in the Great Lagoon, 42 $g_{w.w.}$ m^{-2}, should be regarded as high. According to Kajak et al. (1980), biomass of meiobenthos may exceed 100 $g_{w.w.}$ m^{-2} and account for 10-60% of the total benthic biomass, regardless of the trophic status of a water body. In the present study, the meiobenthic biomass was 42 $g_{w.w.}$ m^{-2}, the total macrobenthos and macrobenthos without large bivalves showing biomass of 197 and 42.5 $g_{w.w.}$ m^{-2}, respectively (Table 1.27). Thus, the meiobenthos biomass accounted for 18 and 50% of the total benthic biomass and benthic biomass without large bivalves, respectively. Both values are within the range reported by Kajak et al. (1980).

The meiobenthos production of 16.3 gC m^{-2} yr^{-1} was only slightly lower than the total production of the macrobenthos without bivalves (18.1 gC m^{-2} yr^{-1}). McLusky and Elliott (2004) reported 20.4 gC m^{-2} as the annual meiobenthic production in an English estuary.

In the wet weight biomass-based production budget of the meio- and macrobenthos combined, the meiobenthos production accounted for 47.5% (42% when based on organic carbon). When the large bivalves were excluded, the meiobenthos contributed 60% (47% when based on organic carbon) to the combined benthos production. These values are similar to findings of Stead et al. (2005) from streams showing the meiobenthos production to be about 50% of the total benthic production (meio- and macrofauna combined).

As revealed by the analyses of fish diets, ostracods were the only meiobenthic taxon utilised, to a noticeable degree, by fish as food. Ostracods are largely semi-benthic and hence more amenable to predation, compared to the small infauna. A relatively high proportion of ostracods in fish food had been earlier reported in other studies, e.g. by Silbert (1979). On the other hand, opinions on effects of macrobenthic predation on the meiobenthos differ. For example, McIntyre (1969) found macrobenthic predators to affect the meiofaunal abundance only slightly.

1.14.11 Fish Fauna

The Szczecin Lagoon is a flow-through, highly productive water body supporting all food web components. The ecosystem is very efficient and shows a high homeostatic potential (Wolnomiejski, 1994). The fish fauna, in addition to benefitting from abundant food resources in the water column and on the entire bottom, is enhanced by the specific hydrography of the whole Odra River mouth area, which facilitates extensive fish migrations to the Lake Dąbie and upstream reaches of the river mouth system (the so-called Lower Odra and Międzyodrze), into the straits and canals of the Świna storm delta, and to the Pomeranian Bay. The fish can thus avoid deterioration of local environmental conditions (pollution, sediment sulphide contamination) or an episodic, usually short-lasting food deficiency. The fish growth rate in the Lagoon is good or very good. In winter, the Lagoon fishes continue feeding, although the feeding activity then is less intense than during the growing season. This has been reported from, i.a. ruffe and perch (Wolnomiejski and Grygiel, 1994b, c), bream (Wolnomiejski and Grygiel, 2002), and some other species (Wolnomiejski, unpubl. data). Winter feeding of fishes in the Lagoon was also reported by Pęczalska (1973).

The Szczecin Lagoon fish fauna is highly diverse and abundant (e.g. Pęczalska, 1973; Wiktor and Garbacik-Wesołowska, 1993; Skóra, 1996; Bartel et al., 1998). The Great Lagoon was reported to support from 31 (Mutko et al., 1994) to 55 fish species (Wysokiński, 1996; 2000). According to Wysokiński (1996; 2000), the Lagoon is inhabited by 25 freshwater, 6 migratory, 14 marine, and 10 accessory fish species. The data of Wolnomiejski (1998) and Skóra (1996) evidence the fully documented presence of 49 fish species in the Great Lagoon. The species richness is thus very similar to that reported from other, well-studied coastal Baltic water bodies. In the Darss-Zingst Bodden, a chain of coastal Baltic lagoons, Winkler (1996) reported 52 species (23 freshwater, 7 migratory, and 22 marine). In the Curonian Lagoon, Repechka et al. (1998) and Gasiûnaité et al. (2008) recorded 53 and 57 species, respectively.

As held by numerous opinions as to the fishery type of such water bodies (e.g. Stallone, 1979), these shallow, turbid areas are particularly favourable for benthivorous fish species. Commercial catches in the Great Lagoon were dominated by benthos-feeding fish (mainly roach and bream) accounting for 64% of the total annual landings. Commercial catches in other Baltic coastal water bodies – in the Polish part of the Vistula Lagoon, Curonian Lagoon, and Lake Gardno – featured benthivorous fish contribution amounting to 52, 68, and 90%, respectively (Borowski, 2000; Aleksandrov, 2003; Psuty, 2009; Skurzak, 2009).

A comparison of fish yields in the largest Polish estuaries and lagoons (Skóra, 1996): the Szczecin Lagoon, Puck Bay, and the Polish part of the Vistula Lagoon, showed the highest yields to be typical of the first area. The landing records in 1998-2002 revealed the yield to have averaged 56 kg ha^{-1} yr^{-1}. In contrast, the yields in the Polish part of the Vistula Lagoon in the same years (Borowski, 2000; Psuty, 2009) were much lower, 20 kg ha^{-1} yr^{-1}, not accounting for high catches of herring entering the Lagoon for a short time to spawn. In the Curonian Lagoon, commercial yields ranged within 20-30 kg ha^{-1} yr^{-1} (Aleksandrov, 2003; Repechka et al., 1998). According to Gasiûnaité et al. (2008), yields from the Lithuanian part of the Curonian Lagoon averaged as little as 18.3 kg ha^{-1} yr^{-1}. Yields in some brackishwater coastal lakes were somewhat higher; for example, Lake Gardno produced 33 kg ha^{-1} yr^{-1} over 1998-2002 (Skurzak, 2009).

Healthwise, the Great Lagoon fish are safe for human consumption. According to the 1973-1979 data cited by Mutko et al. (1994), mean concentrations of Hg, Cd, Pb, Cu, and Zn in the major commercial species caught from the Odra mouth area were clearly lower than the allowed (critical) levels. However, the Hg and Cd concentrations were occasionally and locally exceeded (particularly in the eel). According to Protasowicki (1987; 1991), the bream caught in 1980-1988 showed mean concentrations of Hg, Cd, and Pb of 0.053, 0.051, and 0.45 $\mu g\ g_{w.w.}^{-1}$, respectively, the respective critical levels being 0.300, 0.050, and 0.60 $\mu g\ g_{w.w.}^{-1}$. The Cu and Zn concentrations were way below the respective critical levels. As pointed out by Szefer et al. (2003) and Falandysz et al. (2004) who studied the perch caught in 1996 and 1997 in the Great Lagoon, "The maximum levels of muscle Pb and Cd are significantly lower than the PTWIs (permissible tolerable weekly intakes) for these elements (FAO/WHO, 1989; WHO, 1993) and do not constitute any threat for fish consumers." "The concentrations of muscle Hg (corresponding to 70-105 µg CH_3Hg eaten weekly) are comparable to the PTWI recommended by WHO (200 µg CH_3Hg)." And "All samples of muscle tissue examined contained detectable amounts of both organochlorine pesticides and PCBs, and concentrations were lower (or within the range) of those noted in perch elsewhere in the Baltic Sea in early 1990s." They referred to studies by various authors, conducted in the Bothnian Bay and Bothnian Sea, the Gulf of Gdańsk, the Swedish waters, and the coastal waters of Latvia. In a more recent study, Szlinder-Richert et al. (2010a) found that "The levels of OCPs, and PCDD/F/dl-PCBs in the muscle tissues of eels captured in the Vistula and Szczecin Lagoon were compliant with European regulations. Taking into account the average fish consumption in Poland (125 g per week), one can conclude that the consumption of eel tested in the current study does not pose a health risk for Polish consumers, however, the PTWI value estabilished for PCDD/F/dl-PCBs might be exceeded by populations that have higher fish consumption." In addition, with respect to the Pomeranian Bay, they wrote that "The intake of PBDEs from the consumption of fish from the southern Baltic seems to be low according to the PTWI (Provisional Tolerable Weekly Intake) proposed by EFSA (European Food Safety Authority) for these contaminats."

The quantitative assessment of resources and structure of the fish community in the Great Lagoon was based on official commercial landing and purchase records. The principles of virtual correction of the catch weight and composition, necessary for obtaining a more comprehensive picture of the fishery potential of the Lagoon, were presented and discussed in Section 1.12.2.

Commercial fisheries produced an additional fishing mortality resulting from poor gear selectivity, particularly the fyke nets which destroy substantial amounts of undersized fish. An estimate of this mortality in mass units was presented in Section 1.12.3. Such estimates have not been published so far for the coastal Baltic water bodies, because both gear selectivity and by-catch used to be determined in terms of the number of fish of different size that were captured. This approach greatly limited a meaningful discussion and comparisons with results of the present study. However, when the numbers of undersized individuals caught were compared to the numbers of fish of legal length, the data produced by this study proved consistent with earlier estimates published by Wysokiński and Garbacik-Wesołowska (1995b) with respect to, i.a. the by-catch of undersized bream in fyke nets. Similar data were reported also by Borowski et al. (1998) for the bream from the Vistula Lagoon. For the by-catch of undersized pike-perch, similar data were obtained by Borowski and Dąbrowski (1996) and Psuty-Lipska et al. (2004) for the Vistula Lagoon.

Critical remarks on inadequate selectivity of fishing gear used in the Szczecin Lagoon can be found in several papers, e.g. Wysokiński and Garbacik-Wesołowska (1995a, b), Garbacik-Wesołowska et al. (1998), and Psuty-Lipska et al. (2006). Most of those negative comments were applicable also to fishing gear used in the Vistula Lagoon (e.g. Borowski et al., 1998; Borowski and Dąbrowski, 1996; Psuty-Lipska et al., 2004).

Catch and landing statistics provide thus incomplete evidence of fishing pressure on the fish resources. The differences, however, cannot change the overall assessment of the Great Lagoon fish resources based on the magnitude and composition of catches effected each year. In a similar manner, differences in the magnitude and structure of catches could have been produced by weather conditions which affect catch per unit effort and fishing efficiency in different seasons of the year.

The magnitude of fish resources in the Great Lagoon may be, to some extent, affected by fish migrations. Žiliukienė and Žiliukas (2000) found a significant effect of fish larvae (mainly of smelt) transport to the Baltic on their abundance in the Curonian Lagoon. Ložys (2003) found also that migrations of adult pike-perch between the Curonian Lagoon and the Baltic inshore waters are of such mass dimensions that they become the main controlling factor of catch magnitude in the area. However, as already pointed out in Section 1.12.4, this process in the Great Lagoon cannot be assessed quantitatively due to the lack of relevant data. As the fish can move periodically throughout the entire Odra mouth system, it seems most rational to assume that emigration and immigration of fish are balanced in the Lagoon and do not affect the annual average. This conclusion is confirmed by annual catches in 1998-2002, with both the structure and catch magnitude showing only slight differences from year to year.

Estimates of fish biomass and production in this work stemmed the authors' concepts, based on compilation of published data and model calculations (Annex, Section 1.17). The published and unpublished data pertaining to the Szczecin Lagoon fish fauna were discussed in Section 1.12.5.

At an average Great Lagoon fish biomass of 45.5 t km^{-2} (4.41 gC m^{-2}), the production was 51.4 $t_{w.w.}$ km^{-2} yr^{-1} (4.45 gC m^{-2} yr^{-1}). Thus, P/B averaged 1.13 (1.01 when organic carbon-based), a value within the range reported in the literature for whole fish populations. According to Kajak (2001), the annual fish P/B averages 1.0-1.5. Morgan et al. (1980) published a table summarising results of numerous studies which show P/B to range within 0.5-2.3. Sorokin (1972) determined P/B of the Rybinski Dam Reservoir fish at 0.85, while Penczak et al. (1982) arrived at the value of 0.94 for a total of 11 fish species inhabiting a small lowland river. Wetzel and Likens (1995) found the percid P/B to range within 0.35-2.4, whereas the Włocławek Dam Reservoir bream population showed P/B of 0.93 (Kakareko, 2000). According to Szczerbowski (1993), the cyprinids living under natural conditions have P/B ≈ 1, a value similar to that reported by Pivnička (1982) for different fish species from Czech dam reservoirs.

The ratio between the total production of all the fish and that of predatory species was 4.2 and 4.3 (wet weight- and organic carbon-based, respectively). For adult fish, i.e., the exploited population, the ratio was 5.3. According to Morgan et al. (1980), the ratio ranges within 3.9-6.5 in dam reservoirs in the European part of Russia.

The contribution of larvae and fry (aged 0 and 0+) to the total annual production of all fish age groups was 44.3% (wet weight-based calculations). This is consistent with findings of Morgan et al. (1980) that fry production accounts for 30-80%, depending on the species, of production of the whole fish population. When salmonids are excluded, the range is 40-70%, ≈ 50% being the accepted average.

1.14.12 Birds

Kaliciak et al. (2004) regard the Szczecin Lagoon as the major aquatic area for the migratory waterfowl in the region. According to Czeraszkiewicz and Oleksiak (2003; 2004), the area supports a staggering – on the Polish and European scale – abundance of waterfowl, including migratory cormorants, whooper swans, widgeons, and common mergansers. The authors quoted reported observing periodically large flocks of tufted ducks (9 800 individuals), lesser scaups (10 500), and common mergansers (3 300). Thus, in addition to the local avifauna, the Lagoon's trophic web structure is considerably affected by migratory birds as well.

Until Bzoma's (2008a, b) research on cormorant feeding, food of birds in the Szczecin Lagoon had not been studied. Estimates of the food consumption compiled for the purpose of this work should be then treated as tentative only. However, it is worthwhile to provide some comments based on the authors' own observations.

The herbivorous waterfowl include primarily swans and mallard which, however, have become – to a large extent – synanthropic and, having bred and finished caring for their young, move away from the Lagoon to sea beaches where they rely on additional food provided by humans.

The mollusc resources, mainly those of the zebra mussel, are relatively poorly utilised (2%) by birds in the Szczecin Lagoon, compared with, e.g., the Gulf of Gdańsk where the mollusc-feeding birds consume about 13% of the total mollusc production (Stempniewicz and Meissner, 1999). Meire (1993) provided examples of substantial effects of molluscivorous birds on bivalve populations in coastal waters of seas and highly saline estuaries (e.g., those of the North Sea). However, those areas supported more abundant large molluscivores, primarily the eider ducks. In open seas and in many estuaries, intensive consumption of bivalves is related also to tides: during the ebb tide, vast areas of the bottom are exposed and prone to intensive foraging of seagulls and oystercatchers. Changes of the water level in the Lagoon, unrelated to tides, reveal relatively small surfaces of the upper littoral where the bivalve biomass, particularly that of the zebra mussel, is low. The heavy ships' traffic and, in winter, the ice cover are important obstacles to feeding by migratory birds. Consequently, they frequently move away to feed in the shallow expanses of the nearby Pomeranian Bay.

As the black cormorants, both those nesting in the Lagoon and those appearing in masses in cold seasons, have dramatically increased in abundance, water users focused their attention on those obligatory piscivores. The cormorant abundance has been increasing throughout Europe, a particularly rapid increase occurring since the 1980s (Krzywosz, 2008). According to Gromadzki et al. (1994), the Great Lagoon supported 1200 black cormorants as early as in 1994, additional 1500 migrants appearing in autumn in the Świna delta. The 1998-2002 cormorant abundance estimate of 7.1 ind. per km^2, i.e., about 2900 individuals in the entire Great Lagoon, is plausible, particularly that the 2006 data (Bzoma 2008 b) provide evidence of a further increase in the population size. The major colony, located on one of the islets on the Świna, supported an average of about 5000 individuals throughout 2006. According to tentative estimates provided by Tyluś (2006), the Great Lagoon together with the Świna delta supported then as many as 8–10 thousand cormorants!

The black cormorants feed exclusively on fish. Their feeding has three major aspects: selectivity, amount (weight) of food consumed, and maiming of fish during cormorant hunts.

The cormorant diet composition usually depends on what a given water body has to offer. According to opinions expressed in the literature and referred to by Gmitrzuk (2004), cormorants feed mainly and non-preferentially on the most accessible fish in the reservoir. For example, in the Wigry National Park lakes, cormorants feed mainly on roach and perch; similar food resources are relied on in the Koronowo Dam Reservoir (Wziątek and Martyniak, 2009). According to Bzoma (2008a, b), the ruffe accounted for slightly more that 50% of the total number of prey consumed by cormorants in the Szczecin Lagoon near the Odra river mouth, followed by three-spined sticklebacks, perch, and roach.

In the Vistula Lagoon (Martyniak et al., 1998), the domination of ruffe in the cormorant diet was still higher (75% of the food weight in 1997). The prey size is extremely important in food selection. According to Krzywosz (2008), the prey size averages about 9 cm. In the Vistula Lagoon, the average prey size *(l.c.)* was 7-8 cm in ruffe, 9-11 cm in roach, and 6-10 cm in perch (Martyniak et al., 1998). Similar values were reported from the Koronowo Dam Reservoir (Wziątek and Martyniak, 2009). In the Szczecin Lagoon, such body length ranges are typical of fry aged 0 and 0+ and of immature fish (age group 1). Considering the maximum prey size, individuals representing age group 2 and mature individuals of, e.g., ruffe or roach, are much less common. With respect to the individual weight at different age groups, it can be tentatively assumed that half of the piscivorous birds' food weight is accounted for by immature fish (together with a small proportion of mature fish of some species), the other half being supplied by fry aged 0 and 0+.

In this work, the daily cormorant food ration was assumed to be 400 g fish weight. The value is intermediate and based on the following published data: 207 g (Gmitrzuk, 2004), 350 g (Krzywosz, 2008), and 200-450 g (Cramp, 1977), while Bzoma (2008b) reported 539 g consumed daily by birds not nourishing their young. Correcting for nourishing the young (Krzywosz, 2008), a daily ration averaged over 365 days would amount to 400 g.

The cormorant feeding pressure on the fish fauna in various water bodies is being increasingly frequently treated in the literature. The estimates differ. Gmitrzuk (2004) found no significant effects of cormorant feeding on the status of the ichthyofauna in the Wigry National Park where cormorants feed mainly on roach and perch. On the other hand, Tyluś (2006) reported alarming data showing that, in the Great Lagoon and the Swina storm delta, the annual cormorant fish consumption was 5400 t, i.e., twice the magnitude of commercial landings! However, this result had been based on incorrect assumptions. A comparison between the cormorant fish consumption and the commercial catches is not meaningful, because the two fish pools differ in terms of both species composition and size. As shown by the data reported in this work, all the piscivorous birds which actively capture their fish prey consume little less than 5% of the total ichthyofauna production, cormorants alone consuming 2%. When referred to the official commercial fish landing/purchase data, the corresponding percentages would amount to 43 and 18.5%, respectively.

From the ecological standpoint, conclusions on the positive ecosystem effects of cormorant feeding brought about by the removal of "pests", i.e., the fish of low commercial value (Martyniak et al., 1998; Stempniewicz et al., 1998) are erroneous as well. Those fish are food for predators (mainly pike-perch) and therefore may indirectly control the predator stocks. Besides, part of the cormorant prey consists of juveniles of the commercially valuable species.

To complement the picture and scale of cormorant feeding effects on the local fish fauna, it is important to recall Bzoma's (2000b) description of the cormorant food composition which, generally in the entire Swina mouth area, is strongly dominated by freshwater fish. The freshwater fish domination in the food was also observed in those months when the Lagoon was totally ice-bound and cormorants were feeding mainly in the Pomeranian Bay. The Świna bed and delta canals as well as the Piastowski Canal were ice free; cormorants were observed to feed very voraciously there (Wolnomiejski, pers. obs.). Thus, in winter, cormorants aim at catching mainly the freshwater fish migrating between the Lagoon and the sea. This evidences that the cormorant feeding pressure is stronger than presented so far, as the migrating fish are being fed upon also by the thousands of cormorants from colonies located outside the Great Lagoon.

Unfortunately, there are no data on the local common merganser's food composition and size structure of the food items. It is estimated that in 1998-2002, common mergansers consumed about one-third of the total food taken up by piscivorous birds. According to Bzoma (2008 a), the common merganser abundance in the Vistula Lagoon in February and March was several times that of black cormorants and reached 17 thou. individuals. Throughout the year (in 2005-2006), the ratio between the total number of feeding days of the common merganser and cormorant amounted to 0.47. In the Szczecin Lagoon, with its lower cormorant abundance, the ratio was 0.74.

Negative effects of black cormorants and other piscivorous birds on the fish fauna include also maiming the potential victims that are difficult to capture or too large to swallow. The fishing yield is also affected as the birds attack and maim the fish stuck in the nets. In view of the continuous and dynamic increase of the black cormorant population within the Szczecin Lagoon, the negative effects of cormorant feeding on the local fish stocks are expected to escalate.

The Lagoon and areas adjacent to it support substantial flocks of seagulls, facultative piscivores feeding on dead fish thrown away by fishermen into the water or on the shore. Those seagulls are synanthropic; they congregate in human settlements on the Lagoon, particularly in the vicinity of fishing harbours and waste dumps. This problem has been touched upon in Section 1.13.2.

The avian fauna is regarded as top predators in the Szczecin Lagoon food web, because avian production remains virtually unused by other components of the ecosystem.

1.15 The Multi-year Variability of the Great Lagoon Ecosystem

1.15.1 Variability in 1998-2002

Descriptions of the biotope and biota of the Szczecin Lagoon published by different authors have frequently reflected their differing opinions as to the status of the Lagoon's environment and communities. The divergent opinions stem from different methods of study and short periods of sampling, occasionally shorter than a single growing season. As the dynamics of environmental factors is at present very high as a result of anthropogenic pressure and climate change, resulting in fluctuations in the abundance and biomass of aquatic biota, data collected in different years may be very (even drastically) different. A reliable assessment of the biological status of a water body, particularly a representative estimation of abundance, biomass, and production of individual community components, calls for averaging the results produced by simultaneous multi-disciplinary studies carried out over several (preferably consecutive) years and employing identical methods. Based on their multiyear studies of different lakes of northern Poland, Giziński et al. (1992) stated that "On the basis of one year investigations it is not possible to characterize a reservoir, assign it to a specific type, and classify it. The evaluation of the trophic level (or pollution) of the lake on the basis of one time investigations is pointless as we cannot assume a priori which season is representative for a given reservoir." With respect to certain ecological categories, e.g., the macrobenthos, those authors concluded that qualitative and quantitative changes occurred as a response to changing environmental conditions with a time lag of at least 2-3 years.

During author's own multi-year research in the Great Lagoon, the relatively largest set of data – concerning both the number of study objects and seasons – was assembled during the 5-year period of 1998-2002. Although the plea for at least several years of study was answered in part only, data concerning the most important biotic components of the ecosystem were collected over 5 (phytoplankton, macrozoobenthos, ichthyofauna) or 3 years (mesozooplankton, macrozooperiphyton). The net phytoplankton production accounted for more than 75% of the total primary production; the total production of mesozooplankton, macrozooperiphyton, macrozoobenthos, and ichthyofauna made up about 75% of the total production of metazoan consumers. Between-years differences in abundance and/or biomass of certain groups of organisms within 1998-2002 were considerable. Biomasses of the pelophilous macrofauna, macrozooperiphyton, and phytoplankton were found to show 5.5-, 4.5-, and 2.5-fold differences. In contrast, small differences only were recorded in the fish fauna (1.2-fold, as assessed from commercial catches) and mesozooplankton (1.3-fold). In the case of large differences, it seemed rational to average the data. In contrast, the hydrochemical parameters showed only small between-years differences within the 5-year period of study. The largest (2-fold) differences were observed in average concentrations of nitrates and orthophosphates. Salinity values showed 1.7-fold differences, whereas the remaining chemical parameters and the annual nutrient load in the riverine runoff and the magnitude of runoff itself differed by a factor of about 1.5.

1.15.2 Variability in the Second Half of the 20th Century

Another aspect of the assessment of habitat conditions and metrics associated with the biotic components was the analysis of long-term trends, as these could reflect changes driven by multi-year cycles (including succession) and climatic variability. For a flow-through lagoon type system which receives riverine discharge and loads brought from far inland, important may be all the supra-regional changes in hydrographic (e.g., floods) or climatic conditions as well as effects of economic activities within the catchment. Particularly important in this respect is the input of nitrogen and phosphorus compounds to rivers, the input changing dynamically and affecting the trophic status of the Szczecin Lagoon in the second half of the 20th century.

The major objective of the analysis was to find out what phase of the anthropogenic influence in the Lagoon was reflected in the data collected during the period of study (1998-2002), and to determine whether the status of the Lagoon's environment as observed during the 5-year period in question differed from that regarded as the "normal" functioning of the water body so that living conditions of organisms were, at least temporarily, changed.

1.15.2.1 The Lagoon as a Habitat

The Szczecin Lagoon water chemistry is shaped mainly by the riverine discharge controlling the chemical load brought in and the actual concentrations of various compounds. The Odra discharge was observed to vary greatly during the multi-year period examined (Fig. 1.4). The annual riverine discharges varied extensively.

The discharge varied from 9.5 to 26 $km^3 yr^{-1}$ (Mikulski, 1966 in Majewski, 1980; Tomaszewski, 1990; Grelowski and Pastuszak, 1996; Radziejewska and Schernewski, 2008). The annual discharge averaged over 1998-2002 amounted to 18.5 $km^3 yr^{-1}$ (Anon.,1998-2003), a value higher by 9% than the 1951-1990 average (16.87 $km^3 yr^{-1}$; Fal et al., 1997) and by 8% higher than the average for the previous 5-year period of 1993-1997 (17.1 $km^3 yr^{-1}$; State's Water Resources, 1993-1998; Table 1.47).

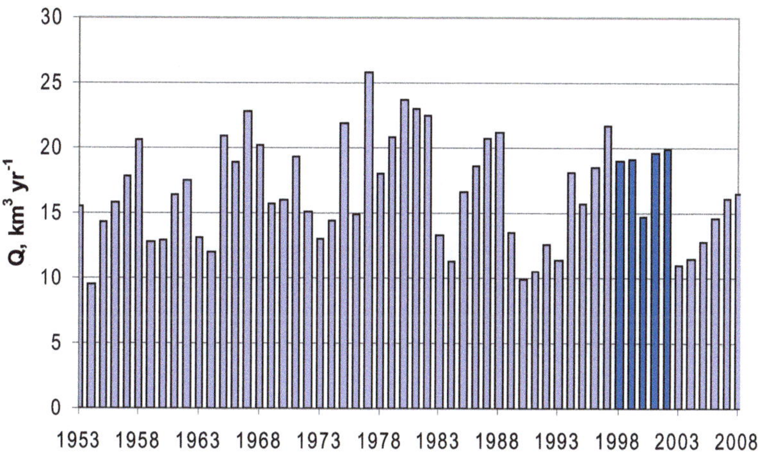

Figure 1.4 The River Odra annual discharge (based on Anon.,1991-2003; Fal et al.,1997; and Anon.,2003-2008).

Table 1.47 Comparison of River Odra discharge and nutrient loads in two periods (based on Anon, 1991-2003).

Parameter / Period	1993-97	1998-2002
Mean flow [$m^3 s^{-1}$]	542	585
Annual water discharge [km^{-3}]	17.1	18.5
Total phosphorus load [$t y^{-1}$]	5901	4423
Total nitrogen load [$t y^{-1}$]	84882	73556

The chemical and suspended particulate loads in the Odra discharge, mainly in the 1980s and in the first half of the 1990s, were summarised by, i.a., Poleszczuk et al. (1995) and Bangiel et al. (2004). The nutrient loads discharged by the Odra in a long-term perspective were discussed by Wielgat (2003), Wielgat and Witek (2004), and Pastuszak and Witek (2009a, b). An ample set of data on the 1988-1994 Odra discharge and loads of a wide spectrum of compounds (including nutrients, heavy metals, organic matter, and organic toxic compounds) was presented by Grelowski and Pastuszak (1996).

Based on an insightful analysis of monitoring data and published information, Wielgat (2003) reconstructed nutrient loads brought into the Szczecin Lagoon by the Odra in the second half of the 20th century (1950-1999). Results of the reconstruction with respect to mineral forms of phosphorus and nitrogen, as a sequence of 5-year averages supplemented by data for 2000-2007, are shown in Fig. 1.5.

The reconstructions illustrate profound changes which occurred in the nutrient supply to the Lagoon with intensifying urbanisation, industrial development, and use of chemicals in agriculture in the Odra catchment. In the 1960s, 1970s, and 1980s, DIP loads increased from a few hundred t to a few thousand t a year, DIN loads increasing from a level of about several thousand t to about 50 thousand t a year. The high phosphate input continued until the end of the 1980s, the elevated nutrient supply persisting until the end of the 20th century. In the 1990s, during transformation and modernisation of the Polish

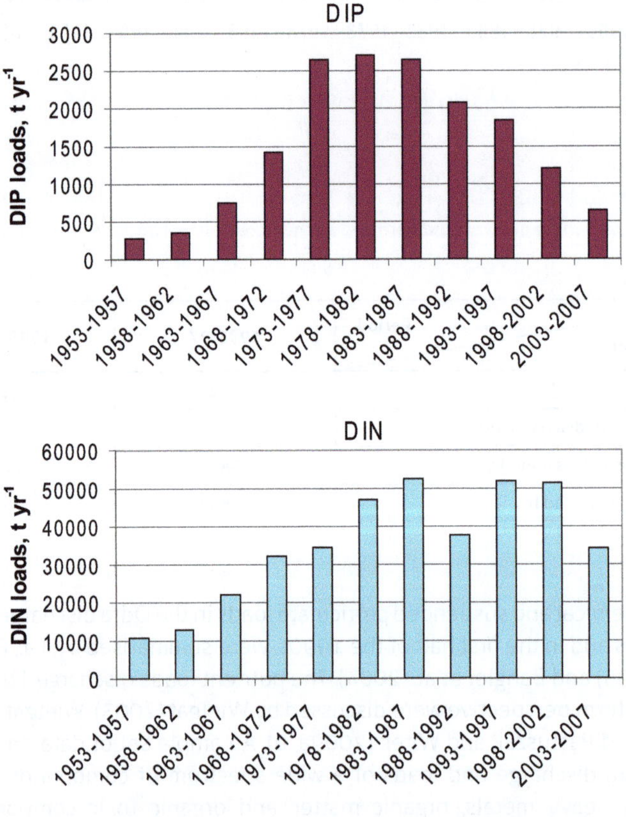

Figure 1.5 Five-year average loads of dissolved inorganic phosphorus (DIP) and dissolved inorganic nitrogen (DIN) discharged by the River Odra to the Szczecin Lagoon in 1953-2007 (based on Wielgat, 2003; and Pastuszak and Witek, 2009a).

economy, phosphate loads began to decrease down to less than a thousand t a year in the first decade of the 21st century. A reduction in the nitrate load during that time was less distinct, but in the mid-2010s, the loads did not exceed 30-40 thou. t N a year.

The heaviest eutrophication of the Szczecin Lagoon occurred in the 1980s. Phosphate concentrations during the growing season averaged then 6-8 μmol dm^{-3}, i.e., an order of magnitude higher than in the 1950s; nitrate concentrations averaged 55-60 μmol dm^{-3}, i.e., 30-50% higher than those at the turn of 1950s and 1960s (Fig. 1.6). The 5-year period of the present study (1998-2002) occurred during the phase of receding eutrophication of the Lagoon. Phosphate concentrations were by one-third lower than those recorded at peak eutrophication. Moreover, although the average concentration of nitrate

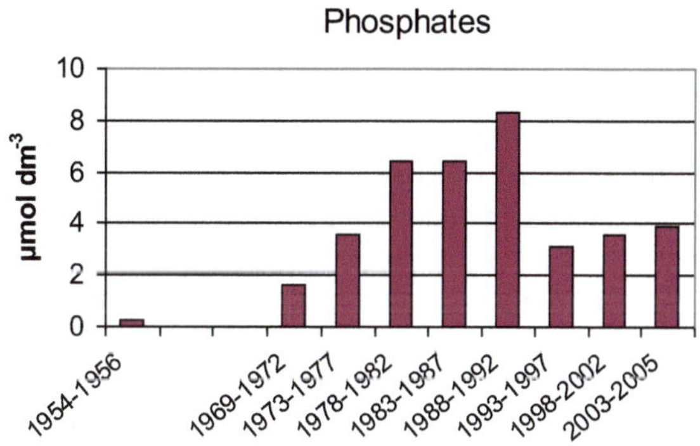

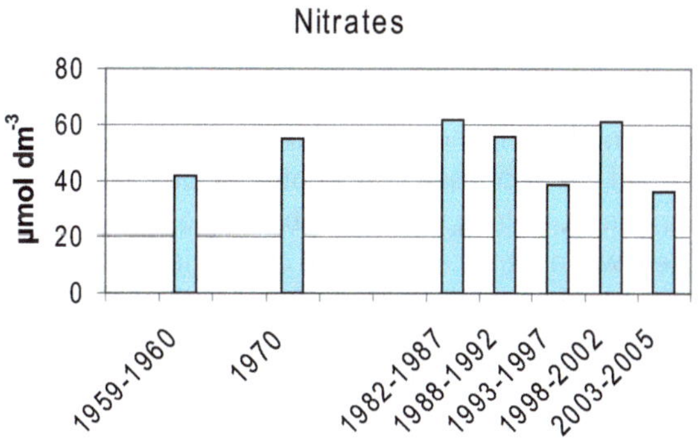

Figure 1.6 Five-year averages of phosphate (P-PO_4) and nitrate (N-NO_3) concentrations in the Great Lagoon in the growing seasons of 1954-2005 (based on author's own data, supplemented by evidence published by Wiktor and Zembrzuska, 1959; Majewski, 1974; and Młodzińska, 1980). Nitrate concentrations in 1959-1960 were measured in the Roztoka Odrzańska only (Majewski, 1974).

nitrogen in the growing seasons of 1998-2002 was similar to that recorded in the 1980s, the increase was episodic only (perhaps related to a higher-than-average riverine discharge) and the general trend was decreasing, which was confirmed by the 2003-2005 data and the evidence provided by RIEP (2009).

According to Bangiel et al. (2004), the nitrogen load in a riverine discharge depends, in addition to the emission in the catchment, also on whether the year was wet or dry. A similar idea of relating the load to a set of climatic conditions was presented by, i.a., Schernewski (2006). However, this is not enough to explain all changes in nutrient concentrations in the Lagoon. Wolnomiejski (1994) is of the opinion that in years with stronger Baltic water incursions into the Lagoon, the inflowing Baltic water, with its low mineral nitrogen contents, dilutes the Lagoon water, whereby the nitrogen concentration in the Lagoon decreases. The data collected over 21 years (1982-2002) (Wolnomiejski, unpubl.) evidence a significant inverse correlation between the average salinity and nitrate nitrogen concentration (R = -0.714; the critical R value at k = 19 and p = 0.05 is 0.433). In addition, intensive influxes of the Baltic water enhance reducing processes in the sediment, whereby nitrates are reduced to labile nitrites and ammonia.

In 1998-2002, average concentrations of dissolved orthophosphates, although lower than those recorded during the peak eutrophication in the 1980s, were still much higher than those in the 1950s. According to Mutko (1986) and Mutko et al. (1994), since 1970 the phosphorus concentrations in the Lagoon were clearly increasing, the increase being particularly pronounced after the Police Chemical Works (in the Odra mouth), manufacturing mainly phosphate-based fertilisers, began operating. On the other hand, from the 1990s onward, a substantial reduction in phosphate concentration was being recorded. However, any assessment of the nutrient concentration dynamics in individual years is difficult due to phenomena and processes such as floods, incursions of Baltic waters, atmospheric transport of chemicals, sediment nutrient retention, or nutrient recycling. Orthophosphate recycling from muddy sediments is enhanced by the increased ionic strength of saline water and the decreased redox potential observed during the Baltic water incursions. In the 21-yr-long period of observations carried out by Wolnomiejski (unpubl.), there was a significant positive correlation between the average salinity and the orthophosphate concentration (Pearson correlation coefficient R = 0.551; the critical R value at k = 19 and p = 0.05 is 0.433). This is one of the major aspects of the so-called trophic trap (Odum, 1971).

Data on the variability of other chemical parameters of the Lagoon water column in the two final decades of the 20th century were reported by, i.a., Wolnomiejski (1994) and Poleszczuk (1998). The latter author provided also an in-depth expert discussion of the data. The Lagoon chemistry in the earlier periods (since 1970) was described by Majewski (1980) and Mutko et al. (1994). Table 1.48 compares 1998-2002 averages of selected chemical parameters with those calculated for the multiyear period of 1982-2002.

Table 1.48 Average values of selected chemical parameters of the Great Lagoon water column in the growing seasons of 1998-2002 compared with multi-year averages of 1982-2002 (based on author's own data).

Parameter	1998-2002 average	Range of averages for growing seasons of 1982-2002	1982-2002 mean and 95% confidence interval
Salinity [PSU]	0.9	0.5-1.9	1.0 ± 0.4
BOD_5 [mgO_2 dm^{-3}]	2.5	1.9-4.4	2.8 ± 0.3
Dissolved oxygen content [mgO_2 dm^{-3}]	10.1	9.3-12.1	10.3 ± 0.3
Oxygen saturation [%]	101	88-122	102 ± 3

The 1998-2002 salinity average was only slightly different from the multi-year annual average. In the previous years, there were periods with salinity deviating substantially from the average, e.g., 0.6 PSU in 1985 and 1.6 PSU in 1989-1993 (Wolnomiejski, 1994; and unpubl. data). During 1998-2002, no short high salinity episodes were recorded, the maximum salinity amounting to as little as 2.1 PSU. In the preceding period, with strong Baltic water incursions, the salinity was observed to exceed 5 PSU. It can be thus concluded that the salinity prevailing in 1998-2002 was not a factor which could substantially affect the functioning of the Great Lagoon biota. According to Wiktor (1980), the Lagoon's salinity is not a limiting factor for the majority of freshwater organisms in the area, but is much below the lower salinity tolerance limit of most species from the Baltic Sea.

Dissolved oxygen contents in and oxygen saturation of the entire water column remained within the range of multi-year averages reported by, i.a., Mutko (1986), Mutko et al. (1994), Wolnomiejski (1994), Poleszczuk (1998), and Bangel et al. (2004). In 1998-2002, the difference between the minimum and maximum values in individual sampling campaigns was relatively small (oxygen saturation of 67-159%). For the 1970s and 1980s, Mutko (1986) reported the most severe oxygen deficiency (20% saturation) in 1984. Wolnomiejski (1994) recorded a single, local 49% saturation in 1989. Such deficiencies usually accompany short-lived stagnation emerging during strong Baltic water incursions which hamper the riverine flow and intensify sediment metabolism, particularly in summer. On the other hand, water oversaturation with oxygen is caused by intensive oxygen release during photosynthesis by large phytoplankton masses. In the previous periods, the maximum values recorded were very high: 206% in 1975 (Mutko, 1986) and 256% in 1983 (Wolnomiejski, 1994). During the period covered by this study (1998-2002), both oxygen deficiencies and oversaturation were slight and posed no threat to aquatic organisms.

Values of the remaining chemical parameters and water transparency remained within the range of multi-year averages reported by other authors (e.g., Mutko et al., 1994; Nowak, 1980; Wolnomiejski, 1994; Poleszczuk, 1997; 1998).

Since the 1970s, the Szczecin Lagoon has been regarded (e.g., Wiktor, 1976; Mutko, 1986; Drzycimski, 1989; Knasiak et al., 1990) as polytrophic and substantially polluted. Applying the criteria of Vollenweider (1968), Poleszczuk and Sitek (1993) and Poleszczuk (1998) defined the Lagoon as eu-/polytrophic. Carlson's Trophic State Index (Carlson, 1977) calculated by Wolnomiejski and Woźniczka (2008) based on the 2001 and 2002 data for the central part of the Great Lagoon was 69, signifying the high (but not overly so) trophic status. That the eutrophication level of the Lagoon decreased was corroborated also by data referred to by Wielgat and Schernewski (2002). According to the German system of trophic state classification and conclusions drawn from model studies (Wielgat and Schernewski, 2002), the Lagoon was polytrophic until late 1980s, but since the 1990s it has been "strongly eutrophic", i.e., the trophic state descended one step down in the German classification system. This status will most likely remain a characteristic of the Lagoon for many years to come. A water body receiving and retaining high riverine discharges cannot be expected to rapidly improve in terms of all water quality parameters. For example, this applies to nitrogen compounds which originate from diffuse sources. Effects of limitations, even drastic, of the use of artificial fertilisers in agriculture may be visible with a time lag of several years, because the heavy chemical inputs supplied to the entire catchment will be diminishing very slowly. Such conclusions can be drawn from the very slow reduction of chemical contamination of numerous European rivers (Stälnacke et al., 2004). This is also evident from data collected in the Great Lagoon. The reduction in orthophosphate concentrations has not been accompanied by a commensurate reduction in concentrations of nitrates.

The Szczecin Lagoon, as a river mouth area, is almost "by definition" regarded as heavily polluted, a conviction that may triggered a false alarm in the past (e.g. Pawlik, 1976; Wiktor, 1976). The belief in the irreversibility of anthropogenic pollution in the Lagoon is regrettably deeply rooted even today, although the 1970 and 1980 data on pollution loads in the Lagoon catchment (e.g., Bosy, 1990; Anon., 1992; Rutkowski, 1993; Mutko et al., 1994; Poleszczuk, 1995) have become obsolete. As early as in the 1990, the wastewater management in Poland greatly improved (owing to, i.a., a more effective implementation of river quality standards, rapidly advancing construction of sewage treatment plants, etc.). The pollution load transported by the Odra diminished already in the late 1980s, the improvement – compared to the earlier part of the decade – being distinct, but mentioned by few authors only (e.g., Tadajewski et al., 1989; Anon., 1992). For a balanced assessment of the pollution status, it is very important to refer to Grelowski and Pastuszak (1996) who remarked on a substantial reduction, in the 1990s, of the discharges of heavy metals (particularly those most noxious) and toxic organics.

Processes leading to self-deterioration, e.g., formation of excessive amounts of hydrogen sulphide in muddy sediment, a too high pH, and hyperoxygenation of water as a result of mass cyanobacteria blooms (and the associated effects of their toxins), have been substantially restricted as well. Such processes are considered in the Standard Statistical Classification of Surface Freshwater Quality for the Maintenance of Aquatic Life (ECE, 1992) proposed by a group of experts of the United Nations' Economic Commission for Europe (UNECE) (Soszka, 2002). Formerly, deleterious effects of self-deterioration had been periodically observed in the invertebrate fauna and fish of the Lagoon (e.g., Piesik, 1992; Świerczyński and Czerniawska, 1992; Wolnomiejski and Grygiel, 1992a).

The Lagoon's homeostatic potential was put to a specific, practical test during the large, catastrophic flood in 1997, termed the "flood of the century". The greatly augmented amount of flood water carried masses of suspended particulates consisting of soil with its fertiliser load as well as of various substances washed away from flooded towns and villages in Poland and Germany, from septic tanks, garbage dumps, factories (including chemical plants), petrol stations etc. However, the nutrient concentrations in the Lagoon water entering the Baltic Sea did not change markedly (Mohrholz et al., 1998; Pastuszak et al., 1998). At the beginning of the flood event in the Lagoon, the flood water showed high concentrations of nitrites and ammonia, higher than those reported at any time previously. However, those concentrations were not lethal for the aquatic fauna. Concentrations of chloroorganic pesticies and other organic substances, including polycyclic aromatic hydrocarbons (PAHs), although elevated as well, were not alarming (Trzosińska and Andrulewicz, 1997). No reduction in the abundance of the macrobenthos was recorded, the abundance remaining at the high level observed during the 3 preceding years (Wolnomiejski and Grygiel, 1998b). Such conditions were maintained also in autumn, when the flood abated, although the Odra had received enormous amounts of detergents and, frequently corrosive, chemicals used for cleaning and disinfection operations over the 600 thou. ha of completely flooded areas adjacent to the Odra, with about 1300 towns and villages. The commercial fish catches in the two years following the flood exceeded 3 thou. t yr^{-1}, which means that the fishing yield was still high and exceeded 60 t ha^{-1}.

1.15.2.2 Phytoplankton

The number of phytoplanktonic taxa reported by various authors (Zembrzuska, 1962; Pełszyńska, 1973; Pliński, 1979; Chojnacki, 1989; Drzycimski, 1989;

Poleszczuk, 1994; Mutko et al., 1994; Luścińska, 2006) at different times of the multi-year period ranged from 98 to 388. The authors referred to differed not only in their respective expertise in algal taxonomy and sampling methods, but also in premises and objectives of their research. Thus, the data on taxonomic richness reported by various authors, viewed from temporal perspective, are not amenable to a meaningful analysis of changes in biodiversity in the multi-year period considered and to concluding on changes in the trophic status. On the other hand, the trophic status increase taking place in the second half of the 20th century was accompanied by particularly substantial changes in the phytoplankton abundance and dominance structure. As reported by Pliński (1972a, b) and Wiktor (1976; 1980), the permanent and distinct domination of diatoms was replaced, in the late 1960s, by a pronounced domination of cyanobacteria. This was particularly true with respect to *Microcystis aeruginosa*, a species accounting for 98% of the total phytoplankton abundance in periods of extreme blooms. During a bloom, the cyanobacterial biomass (as determined volumetrically by sedimentation) reached up to 6.56 ml dm^{-3} (Piesik, 1992; Świerczyński and Czerniawska, 1992). Massive blooms of *M. aeruginosa* persisted in the Lagoon until the late 1980s, which even triggered fears of an ecological catastrophe in the area (Tomaszewski, 1990). In 1998-2002, no major blooms similar to those occurring up to the late 1980s were observed in the main basin of the Lagoon. In the second half of the 1990s, Wolnomiejski et al. (2000) showed that the phytoplankton, in addition to occurring at a substantially lower abundance, consisted mainly of cyanobacteria and diatoms, with an admixture of chlorophytes, or was dominated by diatoms and chlorophytes, with an admixture of cyanobacteria. Aggregations of *M. aeruginosa* were short-lived only and local, occurring near the shores or in embayments as a result of wind-driven algal accumulation with water movement. In summers of the late 1990s and in the early years of the 21st century, a strong nitrate nitrogen deficiency was observed in the Lagoon (Fig. 1.3). This most likely induced a further reduction of phytoplankton abundance, including that of certain cyanobacteria. This is confirmed by results of modelling studies of Wielgat and Schernewski (2002). In addition, cyanoprokaryotes seem to show interspecific competition, e.g., an increase in the abundance of *Aphanizomenon* clearly reduces that of *Microcystis* and *vice versa* (Priymachenko and Litvinova, 1968).

In recent years, water temperature has been referred to as a key factor controlling the phytoplankton abundance, particularly that of cyanobacteria. This, however, was not evident in the multi-year phytoplankton dynamics in the Szczecin Lagoon. The heavy cyanobacterial blooms persisted over a long period of time rather than in individual years, and disappeared later on, for a long period of time as well. As shown by the comparison of mass cyanobacterial blooms in the Curonian and Vistula Lagoons at the turn of the 21st century (Aleksandrov, 2009), effects of local thermal conditions were evident in the Curonian Lagoon only. The Russian part of the Vistula Lagoon did not suffer of such mass blooms due to intensive throughput of water or effects of elevated salinity.

There are no reliable detailed quantitative data on the Lagoon phytoplankton in the period preceding that covered by this study. The available data concern only the entire seston sampled with vertical plankton net tows and measured as a mass of a material filtered by the net and sedimented (Wiktor, 1972; Mutko et al., 1994). These data had to be modified to make them amenable to comparisons with the results of the present study. According to Wolnomiejski (unpubl.), the loose sedimented seston (mainly phytoplankton) contains about 50% of external water trapped between microalgal cells and colonies. Thus, the wet weight of phytoplankton would constitute half of the volume of such sedimented material. The biomass of zooplankton (about 5 mg dm^{-3}) had to be subtracted as well. The adjusted data are shown in Table 1.49.

Table 1.49 Phytoplankton biomass [$mg_{w.w.}$ dm^{-3}] in the Great Lagoon in growing seasons of different time periods.

Years	Biomass	Source
1998-2002 average The year of highest average abundance (1999) Maximum biomass: Aug. 1999	15.5 23.2 60.1	This study
1982-1988 average The year of highest average abundance (1982) Maximum biomass: Aug. 1983	40.9 67.0 242.5	Mutko et al. (1994)
1968-1970 average The year of highest average abundance (1968) Maximum biomass: Sep. 1968	23.9 27.5 62.7	Wiktor (1972)
1957, 1959 and 1964 average The year of highest average abundance (1964) Maximum biomass: Aug. 1964	13.7 21.6 63.7	Wiktor (1972)

It seems then that the phytoplankton biomass in 1998-2002 was very similar to that in the 1950s-1960s (Fig. 1.7). On the other hand, the phytoplankton biomass was particularly high in the 1980s, at a similar level as in the early 21st century in the Curonian Lagoon (Aleksandrov, 2009). The August average for the entire period of 1982-1988 in the Szczecin Lagoon was 116.5 $mg_{w.w.}$ dm^{-3}. The period was characterised by periodic, heavy blooms of cyanobacteria (mainly *Microcystis aeruginosa*), harmful for other aquatic biota. The extremely high biomass of planktonic algae (235 $mg_{w.w.}$ dm^{-3}) in the sub-surface water in the growing season of 1971 was reported by Pełszyńska (1973), but the methods she used were not sufficiently clearly described.

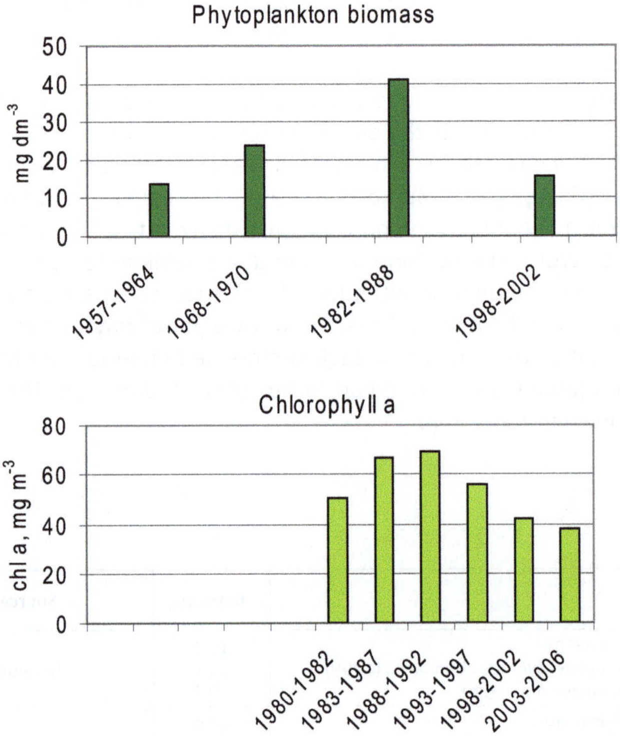

Figure 1.7 Average wet weight-based phytoplankton biomass in the Great Lagoon in growing seasons (above) and chlorophyll *a* concentrations in the whole Szczecin Lagoon (below) in different periods (phytoplankton biomass - data as in Table 1.49; chlorophyll *a* - data from RIEP, 2007).

As demonstrated by Mutko et al. (1994) and the data of the Regional Inspectorate of Environmental Protection (RIEP, 2007), chlorophyll *a* concentrations in the 1980s and in the early 1990s were higher than those in 1998-2002 (Fig. 1.7); in 1983–1992 and 1998-2002, they averaged 68 and 42 mg m^{-3}, respectively. These data most likely apply to the entire Szczecin Lagoon (together with the Kleines Haff). The maximum concentration in the present study (86.5 mg m^{-3}) was recorded in September 2000. In earlier years, the maximum concentrations exceeded the level of 200 mg m^{-3} on a number of occasions, e.g., 285 mg m^{-3} in September 1983.

The phytoplankton primary production in the Szczecin Lagoon was first measured in 1964-1966 by Wiktor (1971) using the light and dark bottle oxygen method. Her data showed the Lagoon's fertility to be already high. The gross primary production amounted to 1164 gO_2 m^{-2} yr^{-1} (about 364 gC m^{-2} yr^{-1}), which was equivalent to the net production of about 290 gC m^{-2} yr^{-1}. Wielgat (2003) followed changes in productivity of the Szczecin Lagoon in the second half of the 20th century in relation to dynamics of nutrient input by the Odra. Her dynamic biogeochemical model of the Szczecin Lagoon showed that increased nutrient concentrations in the Odra, associated with increased municipal and industrial pollution and increased mineral fertiliser use in the catchment could have resulted in increased net primary production in the Lagoon from about 100 gC m^{-2} yr^{-1} in the 1950s to more than 400 gC m^{-2} yr^{-1} in the 1990s. According to the model, the magnitude of primary production may vary by 20-30% in consecutive years. The average level of primary production remained basically unchanged from the 1970s until the late 1990s, evidencing an additional effect of nutrients deposited in the soil and groundwater of the Odra and Lagoon catchments, which contributed to the high nutrient levels in streams and rivers, despite a sharp reduction in the mineral fertiliser use in the early 1990s and the construction of numerous municipal sewage treatment plants.

1.15.2.3 Zooplankton

The mesozooplankton in the Polish part of the Szczecin Lagoon (i.e., the Great Lagoon) was first studied by Świeżawska-Wiktorowa (1957) in 1950 and 1951; she carried the study on in 1954 and 1955 (Wiktor, 1959) and in 1955-1958 (Wiktor, 1964). Until the 1990s, the mesozooplankton was tackled by a few other authors (Podolska, 1977; Wieser, 1979; Drzycimski, 1989; Piasecki, 1993; Mutko et al., 1994). Most of those studies, however, concentrated on planktonic crustaceans, which render meaningful comparisons of multi-year trends in this ecological formation of aquatic animals virtually impossible. The authors listed above recorded the presence of 46-65 zooplankton taxa (including the taxa from the Baltic Sea imported episodically with seawater incursions). The zooplankton list in the present study (1998-2002) contains 65 taxa (40 species of rotifers, 24 of crustaceans, and the zebra mussel larvae). As already mentioned, the period of study – characterised by low salinity in the Great Lagoon – did not feature brackishwater species. The taxa are similar to those reported for 1950-1951 by Świeżawska-Wiktorowa (1957), although her list contained as few as 48 taxa. At that time, however, cyclopoids of the genus *Cyclops* were not identified to species (there were most probably 7 species). Her list featured 31 rotifer and 15 crustacean taxa as well as the zebra mussel larvae. The list of Świeżawska-Wiktorowa (1957) contains no taxa from the Baltic Sea, either. In the early 1990s, the Odra zooplankton sampled near Szczecin, i.e., upstream of the river discharge into the Lagoon (Szlauer and Szlauer, 1994) showed the presence of 56 taxa: 30 rotifers, 16 cladocerans, and 10 copepods.

This study corroborated the conclusion of Wiktor (1964) on the Szczecin Lagoon zooplankton being typical of freshwater areas located close to river mouths, and consisting of a low number of species, some of which only being able to attain mass abundances under favourable conditions. The episodes of very high abundances were shown by *Leptodora kindti*. Examination of fish stomach contents occasionally (mainly in summer) revealed very high contribution of the cladoceran to the fish diet, particularly that of perch and bream. In other seasons, *L. kindti* occurred in the stomach contents of few fishes, if at all.

A meaningful quantitative analysis of zooplankton multi-year variability is rendered difficult, if not impossible, by methodological differences between various studies (different samplers or plankton nets, different mesh sizes in the latter, etc.). Quantitative data consisted usually of abundances only (e.g., in reviews referred to by Mutko et al., 1994), without any biomass estimates. In addition, the data sets were limited to crustaceans only and were fragmentary. The zooplankton biomass was treated in three previous studies only:

- in 1968-1970, the crustacean biomass was 1.7-2.7 $mg_{w.w.}$ dm^{-3} (Wiktor, 1980);
- in 1977/1978, the zooplankton biomass averaged 6.55 $mg_{w.w.}$ dm^{-3} (Wieser, 1979);
- in 1992, the zooplankton biomass averaged 1.41 $mg_{w.w.}$ dm^{-3} (Piasecki, 1993).

The mesozooplankton biomass averaged 4.8 $mg_{w.w.}$ dm^{-3} across three years of study in the period of 1998-2002 (5.54; 4.45; and 4.28 $mg_{w.w.}$ dm^{-3} in 1998, 2001, 2002, respectively). With a correction for the *Leptodora kindti* biomass, the total average biomass was 5.1 $mg_{w.w.}$ dm^{-3}. Those values are in the range reported in the previous years. The relatively low differences in average mesozooplankton biomass in the Lagoon between different years of 1998-2002 evidence a high homeostatic potential and high productivity of the area. In individual years, the abundance of new generations of fish larvae and fry were widely different, the differences being even 17-fold (Sołtysik, 1994; Wysokiński et al., 1999). As reported by Mehner and Thiel (1999) and Scheffer (2001), fish larvae and fry may drastically reduce the abundance of zooplankton. In the Great Lagoon, the abundance of perch fry alone may reach 14 thou. ind. ha^{-1} (Sołtysik, 1994). This is half the total abundance of fry of other fish species in the Lagoon.

No literature source from the pre-1998 period has provided mesozooplankton production estimate.

1.15.2.4 Macrophytes

Both during the period of the present study (1998-2002) and in the entire 40 years preceding it, there were no published studies on diversity, physiography, and abundance of higher plants in the Polish part of the Szczecin

Lagoon. Therefore, this analysis is based on an in-depth research of Garbacik-Wesołowska (1969; 1973) carried out 40 years ago. As mentioned in Section 1.6, the author's own survey of distribution of macrophytes in 2000 and 2001 did not change in any substantial way the general picture that emerged from that research. In 1983 and 1984, distribution and biomass of helophytes were assessed in a few kilometres-long stretch of the western shore of the Great Lagoon (Garbacik-Wesołowska et al., 1984). At that time, too, no significant differences were revealed relative to the earlier data collected from that area by Garbacik-Wesołowska (1973).

On the other hand, the soft plants, particularly pondweed, were notable in their variable spatial distribution in different years (author's own visual observations). Doubtless, this variability affected the biomass of those macrophytes in any given moment. The fluctuations were, however, natural and depended on water transparency, flood situation or weather conditions prevalent at the onset of the growing season.

1.15.2.5 Phytoperiphyton

This ecological category was studied over two years only (1998 and 1999). The periphytic algal biomass was similar in both years (1.5-fold difference in average biomass). In other areas, large interannual differences were quite frequent; for example, in two consecutive years, the annual average biomass of the phytoperiphyton growing on reed in the small, fertile Lake Tynwałd showed more than 7-fold differences, phytoperiphyton production differences being 11-fold (Bohr et al., 1975; Adamczak et al., 1979). However, those differences were accompanied by drastic changes in the periphytic community structure. In the Great Lagoon, domination of diatoms in the total biomass of the periphytic algae was similar in both years as well: diatoms contributed 84.6 and 91.4% to the total biomass in 1998 and 1999, respectively.

1.15.2.6 Macrozooperiphyton

Unlike that of the periphytic algae, the average biomass of the macrozooperiphyton inhabiting reed in 1998 and 1999 differed substantially between the years. The biomass in 1999 was 4.5 times that recorded in 1998. The differences were observed in both the sessile and motile organisms. The biomass recorded in 2003 was intermediate between the values revealed in 1998 and 1999. Similarly, the average mesozooperiphyton biomass in 1999 was almost 6 times that in 1998. Individual zooperiphyton samples collected at the same sites in both years showed substantial differences as well.

1.15.2.7 Macrozoobenthos

Since the mid-20th century, the macrobenthos in the Polish part of the Lagoon was studied by Wiktorowie (1954) and Wiktor (1962; 1969) in the 1950s, Kolasa (1972) and Giziński et al. (1980) in the 1970s as well as Drzycimski (1986), Masłowski (1992; 1993), and Wawrzyniak-Wydrowska (1998) in the 1980s. In the early 1990s, the macrobenthos in one of the Lagoon's embayments, Lake Nowowarpieńskie, was surveyed by Czarnecki (1993). Every year since the early 1980s, Wolnomiejski and colleagues have been studying the pelophilous fauna of the Lagoon (Wolnomiejski and Grygiel, 1989; 1992; 1994a; Wolnomiejski, 1994; Wolnomiejski, unpubl. data). The littoral macrobenthos has been studied with a much lower intensity (Wolnomiejski and Grygiel 1994a; Wolnomiejski unpubl. data from 1999-2002). In 2001, a large-scale study on distribution and quantitative assessment of *Dreissena polymorpha* population was initiated (Wolnomiejski and Woźniczka, 2003; 2008; Woźniczka and Wolnomiejski, 2005). That study was aimed at identifying changes that occurred since the previous in-depth research on the zebra mussel carried out in 1954-1967 by Wiktor (1969).

The macrobenthos of the Great Lagoon was generally found to have been abundant, but the actual abundance and biomass varied widely over time. Even in a homogenous habitat such as muddy sediment, the difference between the maximum and minimum annual average biomass of the pelophilous macrofauna was 20-fold (Wolnomiejski and Grygiel, 1989; Wolnomiejski, 1994; and unpubl. data): 181 $g_{w.w.}$ m^{-2} in 1984 *vs.* 9 $g_{w.w.}$ m^{-2} in 1987. From the mid-1990 to 2002, the differences between annual averages were much lower (about 7-fold): 74 $g_{w.w.}$ m^{-2} in 1997 *vs.* 11 $g_{w.w.}$ m^{-2} in 2002. The composition, abundance, and biomass of the pelophilous macrofauna of the Great Lagoon in different periods of the multi-year studies are summarised in Table 1.50, biomass changes being illustrated in Fig. 1.8.

Over the entire 5-year period of the present study (1998-2002), the pelophilous macrofauna showed a moderate biomass, similar to that in 1975/76 and 1987-1992. The highest biomass was recorded in 1982-1985. It seems that in the 1950s, when the macrobenthos of the Lagoon was being studied by Wiktor (1962), the pelophilous macrofauna biomass was lower than that in 1998-2002 (Fig. 1.8), although the methods used by that author could have led to some underestimation (see below). The P/B values from Table 1.27 give a production estimate in 1998-2002 at 228 $g_{w.w.}$ m^{-2} yr^{-1} (without ostracods). Like the biomass, the production is at the level similar to that in 1975/76 and in 1987-1992. On the other hand, the pelophilous macrofauna production in 1982-1985 was almost 2.5 times higher. During the period of the present study, a particularly abundant macrofauna occurred in 1998 and 2001 (the average substantially exceeded 50 $g_{w.w.}$ m^{-2}). In contrast, a clear impoverishment was recorded in 2002. There are no premises on which to base explanation of such

Table 1.50 Abundance and biomass of pelophilous macrofauna in the Great Lagoon in different periods.

Taxon \ Year	1975-1976[a]	1982-1985[b]	1987-1992[c]	1993-1997[d]	1998-2002
Abundance, ind. m^{-2}					
Chironomus f.l. *plumosus*	2099	3274	1511	2048	1379
Procladius spp.	226	498	243	112	138
other Chironomidae	14	125	44	27	30
Oligochaeta	4486	8869	3209	7373	3711
Polychaeta			13	3	
other taxa	13	558	106	33	18
Total macrobenthos	6838	13324	5126	9598	5256
[u] Biomass, $g_{w.w.}$ m^{-2}					
Chironomus f.l. *plumosus*	27.1	63.4	27.4	41.0	22.9
Procladius spp.	0.9	0.8	0.6	0.3	0.4
other Chironomidae	+	0.2	+	+	+
Oligochaeta	8.7	21.4	8.8	20.9	8.4
Polychaeta			0.2	0.2	
other taxa	+	0.7	0.2	0.2	0.1
Total macrobenthos	36.7	86.5	37.2	62.6	32.4

[a] after Giziński et al. (1980); [b] Wolnomiejski and Grygiel (1989); [c] Wolnomiejski (1994); [d] Wolnomiejski (unpublished); [u] without sporadically encountered single unionids and zebra mussels; + less than 0.1 g m^{-2}

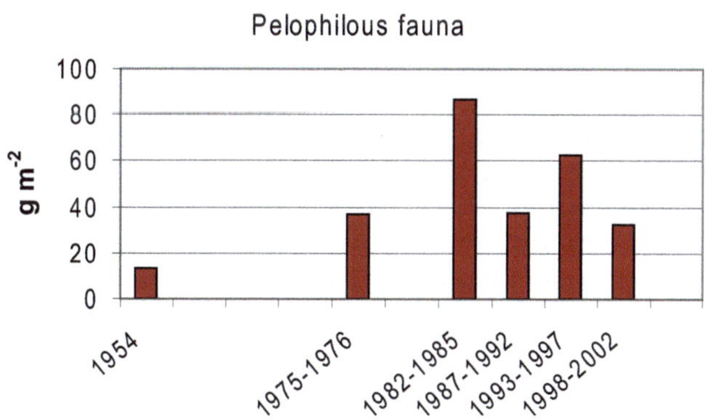

Figure 1.8 Average biomass of pelophilous macrofauna in the Great Lagoon in growing seasons of different periods, based on Table 1.50 supplemented with data of Wiktor (1962).

variability. In a Dutch estuary, Meire et al. (1994) found macrofaunal abundance to vary from year to year, although no change in environmental conditions was observed and the taxonomic structure of the macrofauna remained unchanged. Radziejewska and Chabior (2004) ascribed the year-to-year variations in the macrobenthic abundance and biomass to climatic conditions, particularly to the severity of winter, the meiobenthos being more responsive to changes in the water flow and circulation regime.

Wolnomiejski and Poleszczuk (2003) suggested that the biomass of pelophilous macrofauna is controlled by fluidisation (decomposition of micelles which consolidate the muddy sediment) produced by changes in water salinity and resulting in mobility, intensive resuspension, and transport of the sediment. The larvae of *Chironomus* sp. build tubes, which requires time and energy (e.g., Sokolova, 1983), whereas under conditions of sediment fluidisation and resuspension, the larvae are unable to feed properly. Under such conditions, the habitat of the entire pelophilous fauna is disturbed and damaged. This is indirectly shown by a close positive correlation between biomass of oligochaetes and *Chironomus* larvae. As shown by data collected over 23 years of study (1982-2005; Wolnomiejski, unpubl.), average annual biomasses of these two major pelophilous fauna components were significantly correlated ($R = 0.624$; at $p = 0.05$ and $n = 23 - 2 = 21$, the critical R value is 0.413). Biological traits, particularly reproduction and feeding, of these two groups, as well as their utilisation as food by fish differ substantially, so it would be difficult to expect an identical, correlated response to changing habitat conditions from all taxa. Most probably some random destructive mechanical effects in the sediment simultaneously eliminate both taxa from large areas of the muddy bottom. The frequent resuspension has often been regarded as responsible for paucity of benthos in shallow water bodies (e.g., Moss and Tims, 1989; Wiśniewski, 1995).

The general picture of variability of the macrofauna biomass in the entire Polish part of the Szczecin Lagoon and adjacent waters (including Lake Dąbie, the Lower Odra, the Dziwna, and the Kamieński Lagoon) was presented by Wiktor (1962), based on his research carried out in 1954. Those results are compared with data collected 1998-2002 in Table 1.51.

Despite the generally similar macrobenthos biomass in the periods compared, there were substantial differences as well. Whereas biomasses of the zebra mussel and other molluscs were very similar, the remaining forms attained clearly higher biomasses in 1998-2002. The differences are likely to stem mainly from different methodologies applied in the two periods. Wiktor (1962) used a 400 cm^2 Ekman-type grab he constructed himself and an 0.7 mm sieve. A grab of that type applied to cohesive littoral sediments did not performed sufficiently well with respect to the infauna, while the 0.7 mm sieve resulted in a loss of smaller individuals, particularly those of chironomids and oligochaetes in all the samples. In addition, Wiktor (1962) disregarded ostracods in his study. For those

Table 1.51 Comparison of composition and biomass ($g_{w.w.}$ m^{-2}) of macrobenthos in the Great Lagoon in summer 1954 and in the years 1998-2002.

Taxon	The Great Lagoon and adjacent waters, summer 1954 (Wiktor, 1962)	The Great Lagoon, growing seasons of 1998-2002 (this study)
Dreissena polymorpha	146 [a]	142
other Mollusca	16	21
Chironomidae	3	19
Oligochaeta	2	6
Remaining taxa	4	8
Total macrobenthos	171	196
(Macrobenthos without *Dreissena*)	(25)	(54)

[a] after conversion from the crude mass of Wiktor (1962) (see explanation in Chapter 1.10.1)

reasons, his conclusions on the paucity of the macrobenthos in some habitats of the Lagoon at the time of his study might be not entirely plausible.

The 1980s brought reports on a decrease in the zebra mussel resources in the Polish part of the Lagoon, compared to the earlier data of Wiktor (1962; 1969). Somewhat later, publications appeared which showed a considerable reduction of the zebra mussel abundance, and even disappearance of the bivalve, at sites it used to cover with dense beds (Drzycimski, 1989; Piesik, 1992; Masłowski, 1993). The considerable reduction of the zebra mussel abundance in the Kleines Haff was reported also by Fenske et al. (2010) who, when seeking for likely causes of the reduction, blamed it mainly on climate changes. However, studies carried out in 2000-2004 (Woźniczka and Wolnomiejski, 2005) showed the zebra mussel to have remained an extremely abundant component of the benthos in the Polish part of the Lagoon, the total resources being similar to those estimated by Wiktor half a century earlier. Nevertheless, the spatial distribution of the densest zebra mussel beds changed, most likely due to multi-year changes in water flow dynamics in the Lagoon and a change in the edaphic conditions. In some of the Lagoon's embayments, an excessive development of zebra mussel beds led occasionally to a drastic reduction in phytoplankton densities, which in turn resulted in the mass mortality of the bivalve caused by scarcity of food. A case in point was the Skoszewska Cove where an average zebra mussel biomass exceeded 1 kg m^{-2}. In autumn 2002, the excessive filtration activity of the bivalves resulted in a drastic reduction of the phytoplankton density (the chlorophyll *a* concentration of 5.5 mg m^{-3} was the record-low and the Secchi depth exceeded 4 m), whereupon the local zebra mussel population suffered mass mortality. Two and half years later, the zebra mussel biomass was still more than 9 times lower (Wolnomiejski and Woźniczka, 2008). An extensive variability of and rapid reductions in the zebra mussel abundance were reported also from other areas, e.g., from the Masurian lakes (Stańczykowska, 1961; Lewandowski, 1982) where the effects were related to deterioration of feeding conditions.

1.15.2.8 Fish Fauna

The history of fish catches in the Szczecin Lagoon over 1889-1994 as well as the magnitude and structure of catches post-World War II were described by Bartel et al. (1998). In the Great Lagoon itself, catches during the post-war period did not change drastically (Fig. 1.9). In the 1950s and 1960s, they were generally at the level of about 1500 t yr^{-1} to increase later on to more than 2000, but less than 3000 t yr^{-1}. The catches reported during the period of this study were regarded as high, although the highest were those recorded in 1973-1977.

The fish yield in the Great Lagoon was, over many years, high and usually amounted to about 50-60 kg ha^{-1} yr^{-1} (Table 1.52). An exception in this respect was observed in 1953-1967, with the yield dropping by about 40%. This marked yield reduction resulted from a drastic depletion of the roach stock in the late 1940s-early 1950s due to overexploitation caused by the use of small-mesh nets, which also caused a demise of under-sized fishes; for example, as many as 1425 t of roach were caught in 1949! That was a drastic overfishing with respect to the production potential of the Great Lagoon. Harmful were also the so-called weeding actions involving removal of those species regarded as having a low

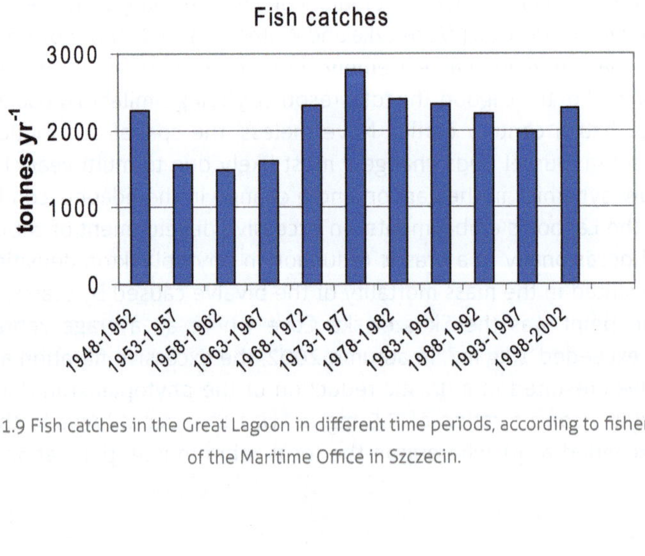

Figure 1.9 Fish catches in the Great Lagoon in different time periods, according to fishery statistics of the Maritime Office in Szczecin.

Table 1.52 Characteristics of commercial fish catches in the Great Lagoon (based on the fishery statistics of the Maritime Office in Szczecin).

Five-year period	Catch [tonnes yr^{-1}]	Yield [tonnes ha^{-1} yr^{-1}]	Major fish species[a]
1948-1952	2261	55	Rut-Abr-Sti-Ang-Eso
1953-1957	1543	38	Abr-Rut-Sti-Ang-Per
1958-1962	1487	36	Abr-Rut-Sti-Ang-Per
1963-1967	1601	39	Abr-Rut-Ang-Sti-Per
1968-1972	2324	57	Rut-Abr-Sti-Ang-Per
1973-1977	2793	68	Rut-Abr-Per-Sti-Ang
1978-1982	2410	59	Rut-Per-Abr-Sti-Ang
1983-1987	2339	57	Rut-Per-Abr-Sti-Ang
1988-1992	2216	54	Rut-Per-Abr-Sti-Ang
1993-1997	1981	48	Rut-Per-Abr-Sti-Ang
1998-2002	2298	56	Rut-Abr- Per-Sti-Ang
Average	**2114**	**52**	**Major species accounted for about 90% of total catch**

[a] Rut - roach; Abr - common bream; Sti - pike-perch; Per - perch; Ang - eel; Eso - pike

commercial value, e.g., roach (Pęczalska and Kraczkiewicz, 1972). According to Olsza (1973), in the 1950s and in the early 1960s, it was necessary to minimise catches of all the fish in summer, because then their meat exuded a phenolic off-flavour rendering their market value unacceptably low.

Since 1973, the structure of catches has been similar in each year (Table 1.52). They were dominated by roach, followed by bream and perch (in varying order). On the other hand, the contribution of large predators, pike and pike-perch, to the total catch was observed to decrease systematically. This trend resulted from increased eutrophication, from the fisheries specifically targeting those species, and from a poor fishing gear selectivity resulting in mass mortality of fry and juveniles (cf. Section 1.12.3.2). Pike was the first to suffer mass mortality: in 1948-1952, annual pike catches averaged 100 t, a four-fold decrease being observed 20 years later. With the demise of pike, the fishing pressure on pike-perch increased markedly, which, with time, resulted in a significant change in the proportion between pike-perch and perch, a small predator. The pike-perch to perch weight ratio in 1968-1972 catches was 3:1 (with 376 t pike-perch being caught a year), while the ratio of 1:5 was recorded in 1998-2002 (with 101 t pike-perch being caught a year). Those changes were accompanied by a distinct increase in the perch stock biomass. According to Wysokiński (1998), the increase resulted mainly from the excessive reduction of the pike-perch stock, pike-perch predation being the most important control of perch abundance.

Quantitative and structural changes in the fish fauna have been reported from other southern Baltic lagoons and coastal lakes, the changes in some of them being still more extensive. The causes of the changes were similar, too. According to Skurzak (2009), in the mid-1950s, predatory fish contributed 40% to the total catch in Lake Gardno, their contribution at present amounting to 8%. Borowski et al. (1998) observed a systematic decrease in fish yield in the Polish part of the Vistula Lagoon, the decrease from 1948 to 1995 being three-fold. The underlying causes included overfishing (excessive fishing effort) and destruction of fry and juveniles by non-selective fishing gear. Repechka et al. (1998) reported the fish yield in the Curonian Lagoon in the first half of the 20th century (60-80 kg ha^{-1} yr^{-1}) to be similar to that in the Great Lagoon. However, it decreased four times due to the disappearance of migratory fishes (due to the damming of rivers in the catchment area), high level of pollution, and inappropriate fisheries management. Regrettably, there are no reliable long-term data from large coastal water bodies on losses of and modifications in the fish resources due to the feeding pressure of cormorants, the population of which has been observed to increase extremely dynamically.

The review of multi-annual variability of different components of the Great Lagoon ecosystem, presented in this section, allows to conclude that the individual ecosystem components differed in their respective responses to changes in nutrient supply to the Lagoon in the second half of the 20th century. Concentrations or biomasses of certain components changed in parallel to the changing nutrient supply, thus indicating close trophic couplings. This was the case with phosphate concentrations and phytoplankton densities, and most likely also the biomass of the pelophilous fauna and zooplankton. Quantitative parameters of other ecosystem components did not respond in this manner. For example, the biomasses of macrophytes or of the zebra mussel recorded during this study did not differ noticeably from the levels reported in the 1950s. Most probably, the trophic coupling played a minor role, some other factor(s) being limiting, e.g., a lack of a suitable substrate. In certain groups, in addition to the trophic coupling, some other environmental variables or effects could be of importance, e.g., the changing pressure of fisheries in the case of fishes or, hypothetically, the periodically changing sediment consistency in the case of the pelophilous fauna. It seems, however, that the period of 1998-2002 covered by this study and coinciding with a gradual alleviation of anthropogenic pressure on the Lagoon was devoid of more serious ecological stressors, which enabled the ecosystem to function effectively and efficiently during that time.

1.16 The Great Lagoon Biotope and its Biota: a Synthesis

In 1960-1990, the prevalent opinion held that the Szczecin Lagoon, particularly its largest part – the Great Lagoon (410 km^2) – was in a very bad (even catastrophic) state, although evidence provided by local hydrobiological research was far too insufficient to support such opinions. Those opinions were influenced by the constant supply of pollutants transported with the riverine discharge and by fears that those pollutants would accumulate in the area. Phenolic contamination of Odra river water and fish meat, maintained from 1950s through 1970s (Mutko et al., 1994), exaggerated those apprehensions. However, throughout that time, commercial fish catches were invariably high. The high fish yield (averaging 52 kg ha^{-1} yr^{-1} in 1948-2002 and 56 kg ha^{-1} yr^{-1} during the period of this study) provided indirect evidence that the ecosystem was functioning properly, that the environmental conditions were satisfactory (despite pollution and excessive eutrophication), and that different food web components were abundant. Of a key importance was the high homeostatic potential of the Lagoon, resulting primarily from the hydrographic situation, particularly the connection with the Pomeranian Bay and other extensive areas of the Odra mouth as well as the constant riverine flow, periodic influxes of the relatively clean Baltic water, and the high water dynamics preventing stagnation in the near-bottom water layer. The high biological diversity (50 species) of the fish fauna results from conditions favouring fish migrations all over the Odra river mouth area (including the Pomeranian Bay, Kleines Haff, Świna straits, and branches and canals of the downstream Odra reaches); the migrations also allow the fish to select habitat and edaphic conditions most suitable for different age classes. The good oxygen conditions of the entire water column enhance abundances of the zooplankton and zoobenthos constituting the fish food resources, and make it possible for fish (particularly the benthivores) to exploit the entire surface of the bottom. In 1998-2002, commercial catches were dominated by benthivores, particularly roach (36%) and bream (27%); taken together, benthivorous fishes contributed 64% to the total catch. Fishes utilised the primary production with 1.1% efficiency, the secondary production of zooplankton and benthos combined being utilised with 8.7% efficiency. These efficiency indices are very high. The ratio of total fish production to the predatory fish production (4.2) was within the range reported from other water bodies. The predators were markedly dominated by perch (23% of the total catch). The low contribution of pike-perch (as little as 4% of the total catch) resulted from the excessive fishing pressure on the species. It was the fishing mortality, along with additional mortality caused by non-selective fishing gear and poaching (particularly with respect to pike-perch), rather than pollution and polytrophy that was mainly responsible for excessive elimination of the most commercially valuable species.

The periodic incursions of the brackish Baltic water into the Lagoon are decisive for the oligo-mixohaline nature of its water. During the period covered by the present study, the salinity was low (averaging 1 PSU and peaking at 2.1 PSU). Earlier, the maximum salinity reached 5 PSU. The highest salinity tolerance was exhibited by fishes. The majority of local fish species are capable of occurring both in the Lagoon and in the Pomeranian Bay; the species from the Baltic Sea are, to some extent, capable of occurring periodically in the Lagoon. On the other hand, macrophytes and invertebrates consist of the freshwater taxa. The Szczecin Lagoon salinity is not limiting for freshwater invertebrates, but is much below the lower tolerance limit of most species from the Baltic Sea. The latter, both in the plankton and in the benthos, were only sporadically encountered in the Lagoon. The planktonic organisms brought in with the Pomeranian Bay water incursions die off very rapidly.

The general levels of various chemicals in the Lagoon as well as the dynamics of concentrations of individual compounds during a year depend primarily on the variable chemical load of the Odra, a major factor functionally responsible for the trophic status and pollution of the area. The Odra catchment (120 thou. km^2) encompasses the most industrialised areas of Poland and also those with intensive agriculture. The river mouth area features Szczecin, a large city the wastewater management of which had for years been far from appropriate. Therefore, the potential pollutant load is substantial. The relatively high riverine runoff is responsible for extremely high nutrient loads (per unit surface area) supplied to the Lagoon. However, effects of the river-borne chemical load do not accumulate as much as they do in areas subjected to constant supply of pollutants, but lacking a possibility of water exchange. The Szczecin Lagoon functions as a natural water treatment plant by virtue of temporary retention, conditioning, and metabolising of the supplied pollutants before they are transported to the Baltic Sea. In addition, the sediment-forming processes and bottom erosion are in a specific equilibrium. Erosion is caused by the very intensive water dynamics bringing about resuspension of soft sediment, hampering resedimentation, and enhancing export of suspended particulates to the Pomeranian Bay. Resuspension is enhanced by sediment fluidisation associated with salinity changes. In addition to the high water dynamics intensifying chemical inactivation of pollutants, the Lagoon's water supports an abundance of organisms which constitute the so-called biofilter. Particularly important in this respect are beds of the zebra mussel (*Dreissena polymorpha*). The total population of the mussel (about 60 thou. t in the Great Lagoon alone) is theoretically capable of filtering the entire water volume of the Lagoon within about 40 days. The mussels filter out not only phytoplankton, but also the inorganic suspended particulates with their adsorbed chemical load. The zebra mussel's faeces and pseudofaeces are deposited in the sediment, increasing its organic enrichment. All those processes enhance the ecosystem's homeostatic performance and, to a large extent, eliminate or alleviate potential effects of the high chemical load and results of excessive eutrophication by limiting phytoplankton blooms or hydrogen sulphide and ammonium formation near the bottom. The Szczecin Lagoon is a water body of a high, but variable, trophic status. In 2000, the mean Trophic State Index (TSI) (Carlson, 1977) as determined with three parameters (water transparency, chlorophyll *a* concentration, and total phosphorus concentration) was 69 (on a scale of 0 to 100).

The high homeostatic potential of the Lagoon is a prerequisite of high biological production of all the food web components investigated, although abundances and biomass of some of them show extensive year-to-year fluctuations. Compared to that in other inland and coastal Baltic water bodies, the biomass of the ecosystem components studied was high or very high. The relevant data are summarised in Table 1.53.

Table 1.53 Biomass and production of the Great Lagoon food web components in 1998-2002.

Component \ Parameter	Biomass in entire Great Lagoon, $g_{w.w.}\ m^{-2}$	Production in entire Great Lagoon, $gC\ m^{-2}\ yr^{-1}$
Phytoplankton	58.9	400
Hydromacrophytes (above-ground tissues)	461.9	66.1
Phytoperiphyton	28.0	22
Mesozooplankton	19.5	25.7
Mesozooperiphyton	1.4	2.7
Macrozooperiphyton	5.7	3.0
Macrobenthos (without *Dreissena polymorpha* and Unionidae)	42.5	18.6
Dreissena polymorpha and Unionidae	154.5	3.8
Meiobenthos	42.0	16.3
Ichthyofauna	45.5	4.5

The avifauna of the Lagoon was regarded as consumers of lower trophic levels and the bird production was not estimated.

Contributions of the three basic categories of primary producers (phytoplankton, macrophytes, and phytoperiphyton) to the total annual primary production (488 gC m^{-2}), 82, 13.5, and 4.5%, respectively, are typical of eutrophic water bodies. The efficiency with which the phytoplankton primary production is utilised for the macrobenthos production was very high (5.6%). The efficiency with which the zooplankton (mesozooplankton only) utilised the primary production (6.4%) and the combined efficiency of the zooplankton and benthos (12.0%) were much lower and remained in mid-range of average values reported in the literature.

The results obtained, compared with corresponding data from other water bodies and with those characterising the multi-year period in the Lagoon, demonstrated the majority of the parameters studied in the Great Lagoon in 1998-2002 to be within ranges reported from other highly productive water bodies and did not represent extreme situations. In 1998-2002, the Great Lagoon did not succumb to mass invasions or excessive development of populations of alien invertebrates, particularly *Cercopagis pengoi* and *Marenzelleria* sp. In other Baltic coastal areas, such invasions drastically altered the structure of zooplanktonic and benthic communities. The abundance of the fish fauna (particularly the commercially exploited species) and its composition evidenced a successional stage within a series of slow changes brought about mainly by fishing operations. No excessive pressure of cormorants upon the fish resources was evident yet. The abundance of those birds in the period discussed was several times lower than at present, 10 years later. During the period of this study (1998-2002), the anthropogenic pressure in the Lagoon was observed to ebb; there seem to have been no serious environmental stress, whereby the ecosystem could function efficiently. Results of the hydrobiological research pursued in 1998-2002 present thus a reliable basis on which to assess the ecosystem functioning and construct a food web model.

1.17 Annex. Empirical-virtual Model of the Great Lagoon Pike-perch Population

The model presented is a reconstruction of body weight growth and reduction of fish abundance (mortality) during a life span of one generation, starting from hatching. Among the fish species living in the Szczecin Lagoon, only pike-perch was amenable to such reconstruction, as sufficient empirical data covering the two earliest life stages (larvae and fry) were available for this species only. The empirical data were collected in 1994 by Szkudlarek-Pawełczyk (2001; 2003) with respect to larvae and Sołtysik (1995) with respect to fry.

The empirical data used for model development included:

- larval abundance (Szkudlarek-Pawełczyk, 2001; 2003), averaging about 780 thou. ind. km^{-2} in mid-May (about 2 weeks post-hatch);
- abundance of fry (age group 0) from July through December (Sołtysik, 1995): 51 thou. ind. km^{-2} in July and a monthly average of 23 thou. ind. km^{-2} within July-December (Table 1.54).

- fry body length (Sołtysik, 1995) (Table 1.54).
- body length (L)-weight (W) relationship for larvae and fry of the Lagoon pike-perch (Wolnomiejski, unpubl.):

$$W_{w.w.}\,[mg] = 2.9343 \times (L\,[cm])^{3.3685}$$

- multi-year weight-length relationship for local juvenile and adult pike-perch (Garbacik-Wesołowska and Boberski, unpubl.):

$$W_{w.w.}\,[g] = 0.0061 \times (L\,[cm])^{3.1251}$$

- average body length of juvenile and adult pike-perch in successive age classes, as derived from length measurements and age readings of local pike-perch in 1991-1999 (Garbacik-Wesołowska and Boberski, unpubl. data; Draganik et al., 2000).

All those data sets were used to:

- determine body weight increments; a linear length growth of larvae and fry from 0.5 (after hatch in early May) to 9.2 cm *l.t.* (a July average), i.e., the length empirically determined by Sołtysik (1995). Length increments in age group 0 were determined using data of Sołtysik (1995) and assuming growth to be linear. The growth rate of fry aged 0+ depended on the empirically determined standard length of the smallest individuals of age group 1 (20 cm *l.t.*), determined from multiannual measurements of Garbacik-Wesołowska and Boberski (unpubl. data);
- determine mortality rate; the abundance decreased from 780 thou. ind. km^{-2} in mid-May to 51 thou. ind. km^{-2} in July, i.e., the value determined empirically by Sołtysik (1995). This mortality rate (0.045 d^{-1}) was extrapolated over the period from early- to mid-May. As the dynamics of fry abundance presented by Sołtysik (1995) was apparently distorted (after a period of abundance decrease, from July through October, the abundance unnaturally increased in November and December; Table 1.54), as a likely result of mass, short-term migrations of fry within the river mouth system (Wysokiński et al., 1999), the data in the model were corrected. The correction involved applying an identical, decreasing trend in abundance (Z = 0.014 d^{-1}) from July, when the maximum abundance was recorded, through December when the measurements were terminated. However, the period's average abundance was retained, in agreement with the average abundance as in the data set of Sołtysik (1995). That the correction is justified is confirmed by the general agreement between the fry total abundance from July through December in the data set reported by Sołtysik and the results of the model (Table 1.54). From December through April of the subsequent year, a marked drop in mortality of the fry aged 0+ had to be assumed (a 2-fold decrease was adopted), as no stationary gear, causing mass mortality of grown-out fry, is set in winter (cf. Section 1.12.3).

Table 1.54 Abundance (N) and mean body length (*longitudo totalis, l.t.*) of pike-perch fry in the Great Lagoon in consecutive months of 1994, after Sołtysik (1995) and produced by the model.

Month	Data of Sołtysik (1995)		Data produced by the model	
	N [ind. km^{-2}]	Mean *l.t.* [cm]	N [ind. km^{-2}]	Mean *l.t.* [cm]
July	51 318	9.2	51 000	9.2
August	36 053	12.1	33 056	10.8
September	14 264	13.6	21 426	12.4
October	9 922	15.2	14 083	14.0
November	12 171	16.2	9 128	15.6
December	14 419	17.2	6 000	17.3
Average	**23 024**	**13.9**	**22 449**	**13.2**

The survival rates of juvenile and adult pike-perch in the Great Lagoon in 1994-1998 were determined based on data reported by Draganik (2000). According to him, age classes 3-10 showed survival rates of 42-65% yr^{-1}; the mid-range value (53% yr^{-1}) was used. An identical value was used for asp by Backiel (1963 in Nagięć, 1964). For pike-perch aged 2+ and older, Nagięć (1964) used the survival rate of 43% yr^{-1}, but regarded it as an underestimate. For pike-perch age groups 3-9, Wilkońska (1996) assumed an annual survival rate of 49% yr^{-1}, a value similar to those reported for bream by Mooij et al. (1996) and Kakareko (2000).

Results produced by the model are shown in Tables 1.55 and 1.56. The virtual population model allows to determine abundance, biomass, and production as well as the annual P/B of three major developmental stages: larvae and fry (age classes 0 and 0+), juveniles (age classes 1-2) and adults (age classes 3-13). The exploited stock consisted of adult pike-perch, i.e., fish aged 3 years and older, measuring, on the average, about 39 cm (in 1994, the protected size in the Great Lagoon was set for pike-perch at 40 cm, and only in 2000 it was raised to 45 cm). The annual P/B values of fry and larvae, juveniles, and adults were 7.11, 0.99, and 0.352, respectively. For the fish aged 3-6 years, which formed the core (89%) of the exploited pike-perch stock, the annual P/B was 0.4, 1.2 being the annual P/B of the entire pike-perch population in the Lagoon.

Results produced by the model may be verified by comparison with true commercial catches and yields. According to the official records, the total pike-perch yield in 1994 was 356 kg km^{-2}, i.e., 26.7 and 76% of the biomass and production calculated by the model. Those values are fully acceptable, although production efficiency would be very high. The results corroborate opinions, expressed by, i.a., Garbacik-Wesołowska et al. (1995) and Draganik (2000), that

Table 1.55 Reconstruction of pike-perch growth, mortality, and production in the first year of life (age groups 0 and 0+) in the Great Lagoon, based on assumptions and references given in the text. N - abundance, $N_{t-1} \times \exp(-Z \times \Delta t)$; Z - mortality, $(\ln N_{t-1} - \ln N_t) / \Delta t$; S - survival, exp (-Z), in %%); L - body length (*longitudo totalis, l.t.*); W - body weight; B_t - biomass on a given day, N × W; G - growth rate, $(\ln W_t - \ln W_{t-1}) / \Delta t$; B_{avg} - average biomass between dates; P - production, $B_{avg} \times G \times \Delta t$. Values of N, L, W, and B_t refer to a given date, remaining values referring to periods between dates.

Date [yr-mo-d]	N [ind. km^{-2}]	Z [d^{-1}]	S [% d^{-1}]	L [cm]	W [g]	Bt [kg km^{-2}]	G [d^{-1}]	Bavg [kg km^{-2}]	P [kg km^{-2} mo^{-1}]	P/B [yr^{-1}]
1994-05-01	1458 655	0.045	95.6	0.5	0.00028	0.4	0.361	18	89	
1994-05-15	780 000	0.045	95.6	2.2	0.0448	35	0.101	118	371	
1994-06-15	195 040	0.045	95.6	5.7	1.03	201	0.054	233	375	
1994-07-15	51000	0.014	98.6	9.2	5.2	264	0.017	279	151	
1994-08-15	33056	0.014	98.6	10.8	8.9	294	0.015	298	139	
1994-09-15	21426	0.014	98.6	12.4	14.1	303	0.014	302	123	
1994-10-15	14083	0.014	98.6	14.0	21.3	300	0.012	290	106	
1994-11-15	9128	0.014	98.6	15.6	30.7	280	0.012	270	94	
1994-12-15	6000	0.007	99.3	17.3	43.4	261	0.004	248	28	
1995-01-15	4830	0.007	99.3	17.9	48.7	235	0.004	223	25	
1995-02-15	3887	0.007	99.3	18.5	54.4	212	0.004	203	22	
1995-03-15	3196	0.007	99.3	19.1	60.6	194	0.003	183	19	
1995-04-15	2572	0.007	99.3	19.7	67.3	173	0.003	168	9	
1995-05-01	2300			20.0	70.8	163				
Annual:								**218**	**1550**	**7.11**

Table 1.56 Reconstruction of pike-perch growth, mortality, and production in consecutive years of life, based on data from Table 1.55 and assumptions given in text. For explanation see Table 1.55. Values of N, L, W, and B_t refer to the first day of life of a new age group, remaining values referring to whole age groups.

Age group	N [ind. km^{-2}]	Z [yr^{-1}]	S [% yr^{-1}]	L [cm]	W [g]	Bt [kg km^{-2}]	G [yr^{-1}]	Bavg [kg km^{-2}]	P [kg km^{-2} yr^{-1}]	P/B [yr^{-1}]
Juveniles:										
1	2300	0.63	53	20	71	163	1.37	253	346	
2	1225	0.63	53	31	279	342	0.72	358	257	
3	652			39	572	373				
age groups 1 and 2 total:								**610**	**603**	**0.99**
Adults (=exploited stock):										
3	652	0.63	53	39	572	373	0.52	353	182	
4	347	0.63	53	46	959	333	0.38	297	114	
5	185	0.63	53	52	1406	260	0.34	228	78	
6	99	0.63	53	58	1978	195	0.26	165	43	
7	52	0.63	53	63	2561	134	0.24	113	27	
8	28	0.63	53	68	3252	91	0.18	74	13	
9	15	0.63	53	72	3888	58	0.17	47	8	
10	8	0.63	53	76	4603	37	0.08	29	2	
11	4	0.63	53	78	4992	21	0.08	17	1	
12	2	0.63	53	80	5404	12	0.08	10	1	
13	1	0.63	53	82	5837	7				
Adults total:								**1331**	**469**	**0.352**

the fishing pressure on pike-perch was excessive and from year to year led to a gradual reduction of the stock and reduced yields and catches. Thus, by referring the pike-perch biomass and production produced by the model to true catches and yields it can be concluded that the model correctly reproduces biological production of the entire pike-perch population in the Lagoon during a (standard) year.

In an earlier study (Draganik, 2000), the abundance of the exploited pike-perch stock in the Polish part of the Szczecin Lagoon in 1994-1998 was estimated with the CAGEAN model. Although the growth rate, mortality, average length, and weight in different age classes were very similar to those assumed in the model described above, the resultant stock abundance data are not acceptable. The pike-perch stock annual production in 1994, as calculated with the CAGEAN, would amount to as little as 60% of the true catch.

Part 2. A MODEL OF THE GREAT LAGOON FOOD WEB

Zbigniew Witek and Norbert Wolnomiejski

2.1 Introduction

Mathematical models are becoming an indispensable tool in studies on complex ecological systems. Models make it possible to integrate a variety of information on the system and allow to control the consistency of the information. More importantly, such models provide better and deeper insights into the functioning of an ecosystem. Knowledge on mechanisms of functioning of ecosystems enhances their understanding and allows to predict their changes as well as to improve management of their resources. However, a mathematical ecosystem model has to be based on a sufficiently vast set of empirical data describing different components of habitats and communities in the ecosystem to be modelled.

The first attempt at a synthesis of knowledge on the functioning of the Szczecin Lagoon ecosystem based on research on different ecosystem components was made by Wiktor (1967). She described, in general terms, the food web structure of and the energy flow in the ecosystem, and assigned various groups of organisms to individual trophic levels. The rather low level of knowledge on the ecosystem and the early stage of ecosystem research methodology at that time prevented application of advanced analyses or quantitative assessments. Despite the subsequent continuous efforts to broaden the knowledge on habitats and communities of the Lagoon, the functioning of its entire ecosystem has not been tackled by researchers. Although descriptions of mathematical modelling of primary production and element cycling in the Szczecin Lagoon were published in the 2000s (Humborg et al., 2000; Wielgat, 2003; Wielgat and Witek, 2004), only the phytoplankton was considered, no other community component being taken into account.

It is becoming increasingly urgent to develop mathematical models of aquatic ecosystems not only for the sake of knowledge itself, but also for practical reasons. Overfishing observed in numerous fishing grounds worldwide and the frequently resultant elimination of the species exploited, advanced environmental degradation and irreversible biological damage (Marten, 1979; Sissenwine and Rosenberg, 1993; Fogarty and Murawski, 1998; Pauly et al., 1998; Clark, 2006) have led to realisation that the fisheries policy and governance have to be rooted in the ecosystem context. During the last two decades, the concept of the 'Ecosystem-Based Fishery Management' has been recognised worldwide (NMFS, 1999; Pikitch et al., 2004; FAO, 2005). The concept involves development of models as one of major activities indispensable for effective policy for sustainable management of aquatic ecosystems. The European Commission, too, has planned to reform the Common Fishery Policy in 2012 so that the knowledge base of legal regulations pertaining to fisheries is augmented by including the knowledge on the functioning of marine ecosystems (POSTNOTE, 2010).

The aim of this part of the study was to develop a static, mass-balance food web model of the central part of the Szczecin Lagoon (known as the Great Lagoon) based on a relatively large set of empirical data which could be collected in the area in 1998-2002, and which were described in the Part 1. Certain less or poorly known biotic components of the ecosystem required that the empirical data be supplemented by compilation of information found in the literature. The model may be treated as a starting point from which to develop more complex ecosystem-portraying tools which could aid decision making process involved in management of the area. The authors' intention was also to use the model to highlight those parts of the ecosystem the knowledge of which is still insufficient, and which require suitably planned, intensive studies.

2.2 Description of the Model

The food web of the Great Lagoon is described by a bottom-up static model, i.e., a model constructed starting from the bottom of the trophic pyramid up. A static mass-balance model is a system in which the internal structure of the community remains stable, the supply of matter/energy from the outside being balanced by the matter/energy losses and accumulation in the sediment. The organisms making up the Lagoon communities as well as the non-living organic matter were divided, for the purpose of the model, into trophic-functional categories generally termed the system components. These include also dead organisms (carcasses) and faeces. The matter/energy flow in the ecosystem was described by a matrix of trophic links between individual system components, energy budget equations, and a few additional simple formulae to describe sedimentation and export. The matter flow in the ecosystem was expressed in gC m^{-2} yr^{-1}. The model, written in an MS Excel spreadsheet, can be easily modified or adjusted to specific conditions of other ecosystems.

System components. The model consists of three basic groups of components: autotrophs, heterotrophs, and different categories of non-living organic matter. A part of non-living organic matter is allochthonous. Autotrophs together with the allochthonous non-living organic matter constitute the first trophic level, that of producers, which lies at the base of the trophic pyramid. The primary production of the autotrophs and the magnitude of allochthonous organic matter supply are based in the model on empirical data and belong to the so-called model drivers. All the system components may be consumed by heterotrophs. The model estimates the secondary production of all heterotrophic components as well as the "production" of autochthonous non-living organic matter. It is the task of the model operator to adjust model coefficients so that the heterotrophic production the model estimates matches the values determined in direct empirical studies.

Matrix of trophic links. The matrix of trophic links determines how the production (or the supply from the outside) of each component is utilized within the ecosystem. Coefficients F_{ij} define the part of the production (supply) of component i utilised by component j (by ingestion). To some extent, a component is allowed to utilise its own production (feeding on members of that component trophic-functional category). Selection of appropriate values of F for individual system components is a basic task of the operator during model parameterization.

Live components, in addition to being susceptible to mortality caused by grazing or predation, may suffer additional 'natural' mortality not associated with feeding of other components, and produced by, e.g., autolysis, disease, senility etc. Dead organisms resulting from natural mortality (carcasses) are a separate component in the model. In the case of fishes, in addition to natural mortality, the model makes allowance for that caused by fishing.

Sedimentation, export and accumulation. In the pelagic compartment of the ecosystem, the part of phytoplankton production, allochthonous organic matter supply, dead organisms and faeces which have not been utilised by the remaining ecosystem components may be sedimented or exported (export being understood as removal and transfer from the Lagoon to the Pomeranian Bay). The phytoplankton sedimenting to the benthic part of the ecosystem, the deposited allochthonous organic matter, and the sinking dead organisms and faeces are treated by the model as separate system components available as food to the benthos, fishes and birds. The unused part of the production of all the remaining system components (bacterio- and zooplankton) the sedimentation of which is not allowed in the model is – in addition to natural mortality – subjected to export only. In the benthic compartment of the ecosystem, the unused dead organisms, faeces, and the unused sedimented phytoplankton and allochthonous material are accumulated in the bottom sediments. It is only insect larvae that can be exported from the benthic compartment (via emergence of imagines). Fish migrations to/from other parts of the Odra river mouth system are not taken into account in the model. It is assumed instead that the emigration of fish from the Great Lagoon is counterbalanced by immigration of fish from outside of the Lagoon. Aquatic birds, the terminal links in the food web, spend a part of their lives outside the Lagoon's aquatic ecosystem and are treated by the model primarily as consumers; their production is regarded as exported from the ecosystem.

Energy budget equations. As already mentioned, the (net) primary production and the supply of allochthonous organic matter, making up the base of the trophic pyramid, have to be entered in the model as data from independent studies. The production (*P*) of the remaining, heterotrophic system components is calculated from:

$$P = C \times AE \times K_2,$$

where *C* is the amount of food consumed; *AE* is the food assimilation efficiency coefficient; and K_2 is the coefficient associated with the amount of assimilated food used for growth; it is also termed the net production efficiency. Consumption of component *j* is a sum of products $F_{ij} \times P_i$ of all the food items (*i*) of the component in question:

$$C_j = \sum_{i=1}^{n} (F_{ij} \times P_i),$$

where *n* is the number of system components in the food of component *j*. Coefficients *AE* and K_2 are parameters which have to be entered as data from independent studies. The knowledge of *C*, *AE*, and K_2 allows easy estimation of the remaining components of the energy budget, i.e., respiration (*R*):

$$R = C \times AE \times (1 - K_2)$$

and the non-assimilated portion of food (faeces, *FU*):

$$FU = C \times (1 - AE).$$

The system of energy budget equations is constructed by proceeding from the base of the trophic pyramid up towards its top. The food sources for the first-order consumers are the primary production and the non-living organic matter.

System components feeding on representatives of their own group (within-group consumption). Development of energy budget equations becomes somewhat complicated when system components feeding on representatives of their own trophic-functional category have to be dealt with. These include protists collectively treated as a single system component in which, e.g., numerous ciliates feed on heterotrophic flagellates and predatory ciliates feed on non-predatory ones. The within-group consumption is the case also in predatory fish. In such cases, to reflect the consumption of such system component in the model, only the food originating from other system components, C_{nk}, is taken into account so that:

$$C_j = C_{nk}.$$

The consumption calculated this way is lower than the sum of consumption by all the elements (individuals) a system component contains. The production of such a component is calculated in the model as a sum of production based on food from both other system components (P_{nk}) and from the system component in question (P_k) minus consumption within the group (C_k):

$$P_j = P_{nk} + P_k - C_k.$$

The production estimated this way is lower than the summary production of all the elements making up a given system component; it is the net production available as food to other system components. On the other hand, respiration and excretion of the unassimilated part of the food of a system component whose elements feed within their own category is equal to, respectively, summary respiration and summary excretion by all the elements of the component.

The coefficients F_{jj} pertaining to within-group feeding ($i = j$), are located along the diagonal of the trophic linkages matrix. Their meaning differs somewhat from that of all the remaining coefficients F_{ij}, as they denote a part of the production of component i (= j) based on the food derived only from other system components, which is consumed by component j (= i):

$$C_{jj} = C_k = F_{jj} \times P_{nk}.$$

Export of pelagic components. For objects passively floating in the water, export to the Pomeranian Bay (*Exp*) may be expressed as a ratio between water outflow (*Q*) and concentration of a given system component. As the component's absolute biomass (*B*) in the Lagoon is equal to the product of the Lagoon volume (*V*) and the component's concentration, the export to biomass ratio is equal to the ratio between the outflow and the Lagoon volume:

$$Exp/B = Q/V.$$

By standardizing export by production of a given component, we get:

$$Exp/P = (Q/V) / (P/B).$$

It is worth noting that, even at a total lack of mortality, the production rate (*P/B*) of the components floating passively in the water has to be higher than the water exchange ratio (*Q/V*). Should the opposite be the case, export would exceed production, whereby the component could not persist in the Lagoon.

Removal of individual objects from the Lagoon can be mediated by numerous agents. A higher density during the period of lower water outflow (e.g., in summer) and lower density during strong outflow (e.g., in winter) may, on the scale of one year, reduce losses produced by transport from the Lagoon compared to situations with stable flow and constant concentration of objects. In addition, export may be substantially modified by vertical migrations, whereby the objects (e.g., the zooplankton) during the day stay on or near the bottom, where the flow is reduced or is even directed opposite to that in the upper water layer.

Sedimentation. Equations describing sedimentation pertain to those system components which passively float in the water and do not resist gravitational settling (some phytoplankton, allochthonous particulate matter, faeces, remains of organisms). The bacterioplankton, zooplankton, and dissolved organic matter were assumed not to sediment. The equations define the part of production of a component not consumed in the water column (*NC*) which settles out (*S*) to the benthic part of the ecosystem:

$$S = NC \times k,$$

where k denotes the fraction subjected to sedimentation. The remaining part $(1 - k)$ is exported. The coefficients k should be entered into the model independently. Unfortunately, they are not easily amenable to measurements. In view of the lack of appropriate data, hypothetical values were assumed and fitted during model parameterization.

Iteration. As calculations of the amount of faeces and dead heterotrophs involve numerous feedbacks (because both faeces and dead organisms may be consumed by certain system components), it was necessary to carry out those calculations by iteration.

Results of calculations carried out by the model were checked by comparing the total matter supply and loss in the entire ecosystem. The supply consists of primary production of all autotrophic system components and the input of allochthonous organic matter. The loss consists of total respiration of all the living system components, export, catches, and accumulation in the sediment. The difference between the supply and the loss of matter should be zero.

Results of modelling. The basic results of model calculations are the magnitude of consumption (*C*), production (*P*), respiration (*R*), amount of unassimilated food (*FU*), additional mortality (*M*), sedimentation (*S*), accumulation in sediment (*Acum*), and export (*Exp*) of individual system components. In addition, the results of modelling include food composition of individual heterotrophic system components. Such data form a basis with which to estimate various ecological indicators describing the ecosystem under study.

Food composition. A table containing data on food composition of individual system components may be used as an auxiliary tool in model parameterization. The percent contribution of component *i* to the diet of component *j* ($C\%_{ij}$) is calculated from

$$C\%_{ij} = (F_{ij} \times P_i) / C_j \times 100.$$

The food composition calculated by the model can be compared with results of independent studies, but the values of $C\%_{ij}$ cannot be directly changed. The food composition in the model can be modified only by appropriate selection of values of F_{ij}.

Parameterization. The model is parameterized by adjusting the values of terms in energy budgets of individual system components, calculated by the model (X_m) to the values produced by empirical studies (X_e). This usually involves selection of F_{ij} (or other parameters present in the model) such that the differences between the values compared do not exceed a preset level. The difference is measured by the relative error (*RE*) expressed as a percentage:

$$RE = (X_m - X_e) / X_e \times 100.$$

In this work, the differences were adjusted to the level below 1% of the empirical value ($RE < 1\%$).

Production of invertebrates is an energy budget term which is easiest to determine independently of the model. Although production of aquatic ecosystem components is seldom measured directly, it is possible to determine production of numerous species indirectly, based on the empirically determined biomass (*B*) and published *P/B* ratios. The biomass of numerous components of the Great Lagoon biota, particularly that of the benthos (cf. Part 1), is relatively well known. In addition, there is ample literature (cited profusely in Part 1) on the P/B ratios of various invertebrates inhabiting temperate fresh- and brackishwater areas. It should be mentioned that production of a component calculated from biomass (*B*) and the *P/B* ratio is the gross production (a sum of production of all the individuals in a given component of the food web); therefore, during parameterization, rather than comparing it with the net production calculated by the model, it is compared with gross production (P_g) calculated as

$$P_g = R \times K_2 / (1 - K_2).$$

In the case of birds, rather than calculating the relative error of production, it was calculated for consumption determined by the model and by independent analyses.

The knowledge on certain components of the Great Lagoon ecosystem proved too scant for empirical determination of their biomass or any element of energy budget. Such components include primarily pelagic, benthic, and periphytic bacteria, fungi, and protists. Their production could be estimated only tentatively from proportions between various components and from patterns known from other ecosystems. In those cases, the values of energy budget terms carry the highest load of uncertainty.

2.3 The Great Lagoon Food Web

2.3.1 Defining the Food Web Components

The structure of the Great Lagoon food web defined for the purpose of developing the model is based primarily on taxonomic groups and ecological formations identified in Part 1. However, in some cases the division was modified.

The entire phytoplankton was treated as a single component. The mesozooplankton was divided into rotifers, non-predatory cladocerans, and copepods as well as the predatory cladoceran *Leptodora kindti*, a species differing from the remaining mesozooplankton by its large body size, feeding mode, and importance as food for other organisms. The macrozooplankton was represented by the mysid *Neomysis vulgaris*. In addition, two other components, important for the functioning of the entire ecosystem, were defined in the pelagic compartment of the system: the bacterioplankton and the protozooplankton. Those microorganisms were not discussed in Part 1 due to the absence of relevant data from the Great Lagoon. However, they are extremely important in any aquatic ecosystem and cannot be ignored in a food web model. The macrophytes were divided into two components: 1) emergent plants, with stems protruding above the water surface, and 2) submerged plants, including those with floating leaves. The periphytic community on macrophytes was divided into phyto-, mesozoo-, and two categories of macrozooperiphyton: sessile and motile macrozooperiphyton. Despite the absence of relevant empirical data from the Great Lagoon, two additional components were defined within the periphyton: 1) periphytic bacteria and fungi, and 2) periphytic protists (protoperiphyton).

Among benthic invertebrates, the meiobenthos was treated as a single component, whereas the macrobenthos was divided into six groups: 1) *Chironomus* sp. larvae, 2) macrobenthic ostracods, 3) oligochaetes, 4) the zebra mussel (*Dreissena polymorpha*), 5) other molluscs, and 6) remaining zoobenthos including chironomid larvae other than those of *Chironomus*, larvae of other insects, crustaceans, hirudineans, hydracarids, etc. The majority of organisms grouped under the 'remaining zoobenthos' are littoral forms. Additionally, microbiological components were defined in the benthos as well: 1) microphytobenthos, 2) benthic bacteria and fungi, and 3) benthic protists, although no relevant quantitative data were available from the Great Lagoon. The microphytobenthos consists of unicellular and colonial algae inhabiting the littoral bottom; they are taxonomically and functionally very similar to the phytoperiphyton.

The fish fauna was divided into three categories: planktivores, benthivores, and predators, each being subdivided into two components, one consisting of larvae and fry, and the other composed of juveniles and adults. The final group included in the model are birds, treated as four components: herbivores, molluscivores, omnivores, and piscivores. Humans are a 'hidden' component of the food web, their contribution to the energy flow in the system being expressed only as a fish yield.

The food web includes also, as components, different forms of non-living organic matter. The model identifies three forms of allochthonous organic matter: 1) allochthonous dissolved organic matter (DOM), 2) allochthonous particulate organic matter (POM) in the water column, and 3) allochthonous organic matter settled out from the water column to the bottom sediment. In addition, some categories of autochthonous organic matter were defined. The pelagic compartment in the system supports two pools of autochthonous organic matter: faeces and remains of pelagic organisms. The benthic part contains four pools: 1) sedimented phytoplankton, 2) faeces of the benthos and faecal pellets sedimented from the water column, 3) animal remains, including those sedimented from the water column, and 4) remains of benthic and periphytic plants.

A total of 45 components were thus identified in the Great Lagoon food web (Fig. 2.1). Five components are autotrophic (phytoplankton, emergent macrophytes, submerged and floating leaved macrophytes, phytoperiphyton, and microphytobenthos), 31 components representing various functional-trophic groups of heterotrophs (7 belonging to the plankton, 5 to the periphyton, 9 to the benthos, 6 to fish, and 4 to the waterfowl); the remaining 9 components cover non-living organic matter (3 and 6 involve allochthonous and autochthonous organic matter, respectively).

It has to be borne in mind that the division adopted cannot fully reflect the complexity of the Great Lagoon biota. Particularly simplified is the part of the food web situated close to the base of the trophic pyramid, as only the elementary trophic-functional categories were identified, without setting off some narrower, more specialised groups or individual fractions of non-living organic matter. The benthic macrofauna and top consumers, i.e. fish and birds, were treated in more detail, as the three animal groups are represented by half of the live system components.

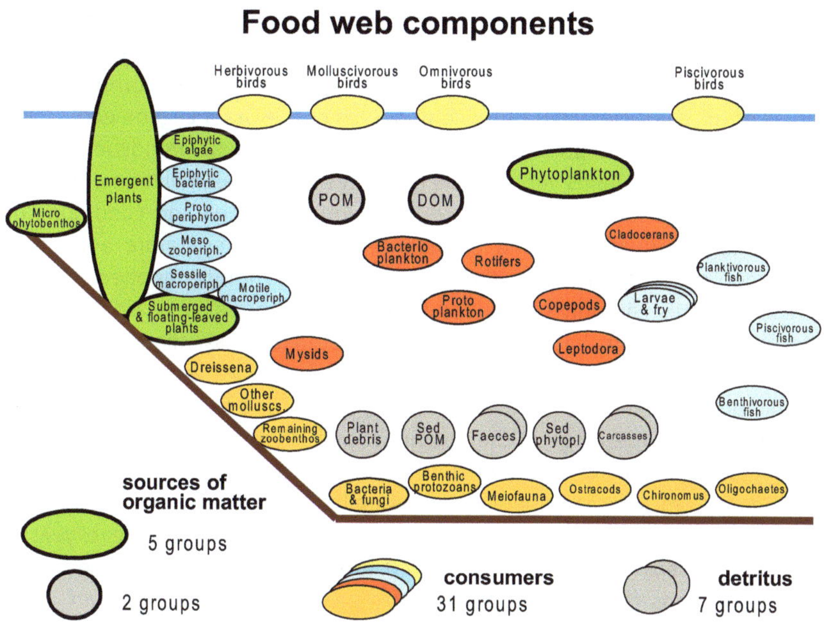

Figure 2.1 Components of the Great Lagoon (Szczecin Lagoon) food web. Components forming the source of organic matter to the ecosystem are marked with thick contours; grey-filled circles denote components composed of non-living organic matter.

2.3.2 Bioenergetic Coefficients

The energy budget of all the heterotrophic components of the food web required that bioenergetic coefficients *AE* and K_2 be entered into the model for each group. As there are no published data on the values of those coefficients for organisms living in the Great Lagoon, the coefficients were derived from studies in other ecosystems. In the first place, the data collected in the Vistula Lagoon, an area ecologically similar to the Szczecin Lagoon, were used. In the Vistula Lagoon, bioenergetic analyses of the youngest developmental stages of certain fish species and the mysid *Neomysis integer* were carried out by Maciejewska and Margoński (2001), Maciejewska and Opaliński (2002; 2004; 2010) and Opaliński et al. (2004). Maciejewska and Opaliński (2010) reported values of *AE* and K_2 for the youngest developmental stages of herring, smelt, three-spined stickleback, and perch. The food assimilation efficiency coefficient (*AE*) was found to range within 0.40-0.69 (averaging 0.50), the net production efficiency coefficient (K_2) ranging within 0.17-0.31 (averaging 0.26). Maciejewska and Opaliński (2002) supplied data necessary for determining *AE* for *N. integer*. As shown by those data, food assimilation efficiency was similar between different size classes of the mysid and ranged within 0.66-0.81 (averaging 0.75). The corresponding values of the coefficients for other components of the Great Lagoon ecosystem had to be looked for in publications pertaining to other areas worldwide.

The literature contains very diverse estimates of food assimilation efficiency (AE) and net production efficiency (K_2), depending on age and conditions of the animals studied, type and abundance of food, temperature and other experimental conditions, and the measurement technique applied. Certain general patterns can only be derived from review papers. Based on data from dozens of publications, Valiela (1995) demonstrated food assimilation to depend largely on the type of food consumed (detritus, plants, animals). Although assimilation efficiency estimates of detritus, plants, and bacteria as food have covered a very wide range in the individual published studied (from a few to 100%), assimilation efficiency of detritus in most publications was reported as ranging from a few to about 60%; assimilation efficiency of macrophytes ranged from 20 to 80%, whereas unicellular organisms (bacteria, algae, yeasts) were assimilated with an efficiency of 40 to more than 80%. In none of the publications was the assimilation efficiency of animal food lower than 40%, most observations indicating more than 80%. Based on those general patterns, for most of the system components whose diet relies on phytoplankton and bacteria, and much less so on animals and detritus, the model calculations assumed *AE* = 0.6 (60%). *AE* was assumed to be 0.7 for mysids (Maciejewska and Opaliński, 2002) and for benthivorous fish whose diet, in addition to animals, includes a substantial contribution of detritus. A lower *AE* (0.4) was assumed for those components relying more on detritus or macrophytes. Based on Maciejewska and Opaliński (2010), *AE* for fish larvae and fry was assumed to be 0.5. On the other hand, elevated *AE* values were assumed for planktivores and piscivores (0.8-0.9). For bacteria and fungi, the model assumed *AE* = 1, as the food assimilated is the food consumed by those organisms with external digestion (Table 2.1).

Utilisation of the assimilated food for growth (K_2) and respiration in the animal world was reviewed by Humphreys (1979) who examined 235 energy budgets published by other authors for natural animal populations, both aquatic and terrestrial. He demonstrated that poikilotherms can be divided, in terms of their utilisation of food for growth, into 3 groups: fish and social insects, invertebrates (except insects), and non-social insects. Within those groups, terrestrial and aquatic animals did not differ in the degree to which they converted their food for growth. The highest average K_2 (40.7%) was typical of non-social insects (including those living in aquatic habitats); a lower average K_2 (25%) was shown by non-insect invertebrates, the lowest (9.77%) being typical of fish and social insects. Non-insect invertebrates could be additionally subdivided into

Table 2.1 Food assimilation efficiency (*AE*) and net growth efficiency (K_2) of heterotrophic components of the food web (based on energy units).

Functional group:	***AE***	K_2
Plankton:		
Bacterioplankton	1	0.33
Protozooplankton	0.6	0.5
Rotifers	0.6	0.3
Non-predatory cladocerans	0.6	0.3
Copepods	0.6	0.3
Leptodora kindti	0.8	0.3
Mysids	0.7	0.3
Periphyton:		
Epiphytic bacteria and fungi	1	0.33
Protozooperiphyton	0.6	0.5
Mesozooperiphyton	0.6	0.3
Sessile macrozooperiphyton	0.8	0.3
Motile macrozooperiphyton	0.6	0.3
Benthos:		
Benthic bacteria and fungi	1	0.33
Benthic protists	0.6	0.5
Meiobenthos	0.4	0.3
Chironomus sp. larvae	0.6	0.4
Ostracods > 0.5 mm	0.4	0.3
Oligochaetes	0.4	0.3
Dreissena polymorpha	0.6	0.3
Other molluscs	0.6	0.3
Remaining zoobenthos	0.6	0.4
Ichthyofauna:		
Larvae and fry of planktivorous fish	0.5	0.25
Juvenile and adult planktivorous fish	0.8	0.2
Larvae and fry of benthivorous fish	0.5	0.25
Juvenile and adult benthivorous fish	0.7	0.2
Larvae and fry of piscivorous fish	0.5	0.25
Juvenile and adult piscivorous fish	0.9	0.2
Avifauna:		
Herbivorous birds	0.4	0.015
Molluscivorous birds	0.4	0.015
Omnivorous birds	0.8	0.015
Piscivorous birds	0.9	0.015

3 categories, depending on the type of food consumed: detritivores, showing the highest K_2 (averaging 36.2%); carnivores, of intermediate net production efficiency (27.6%), and herbivores showing the lowest efficiency (20.8%). Like other homoiotherms, birds were characterised by a very low K_2 (averaging 1.29%).

Fish in Humphreys's (1979) study were represented by a relatively low amount of data (9 sets). Later on, in connection with rapid development of studies on trophic relationships among fishes, new compilations of bioenergetic information, based on larger data sets, were made available. One of the most popular sources is the FishBase web site (www.fishbase.org). The FishBase data show the food conversion for growth in fishes to be higher than that shown by Humphreys (1979). Values of K_1 which describe gross production efficiency (production to food consumption ratio, *P/C*) for fish species occurring in inland waters and in the Baltic were in most cases within 0.1-0.2. Values of K_2 that can be derived from those data, assuming 80% food assimilation efficiency, were ranging, in more than 75% of the data, within 0.1-0.3 (10-30%; median of 21%).

Neither the FishBase, nor the paper by Humphreys (1979) contain data on K_2 of unicellular organisms (protists, bacteria). With regard to protists, there is a fairly convergent opinion that their net production efficiency is higher than that of invertebrates and fishes. Estimates published by Callow (1977), Khlebovich (1979), and Fenchel and Finlay (1983) ranged within 0.44-0.60. On the other hand, production efficiency of bacteria is lower. Schwaerter et al. (1988) estimated the average production efficiency of bacteria in lakes at 29%, whereas Middelboe et al. (1992) arrived at an estimated 21-45% as the bacterial production efficiency in marine and coastal waters. Numerous authors pointed to a dependence of bacterial production efficiency on the nutrition value of the organic matter utilised by bacteria. At high C:N and C:P ratios in the organic matter, evidencing nitrogen- and phosphorus-deficient food, bacterial production may be minimal, the food used being almost entirely respired; on the other hand, at low C:N and C:P ratios in the organic matter, bacterial production efficiency may even exceed 40% (Hopkinson et al., 1989; Coffin et al., 1993; Kuparinen and Heinanen, 1993).

Based on all those data, the model assumed a fairly simplified pattern of K_2 variation among heterotrophic components of the Great Lagoon food web (Table 2.1). K_2 = 0.3 was assigned to all the invertebrates, except insects. A higher efficiency was assigned to insects (0.4) and protists (0.5). The earliest developmental stages of fish as well as fish juveniles and adults received somewhat lowered values (0.25 and 0.2, respectively). The lowest K_2 value was assigned to birds (0.015). The bacterial K_2 = 0.33 was entered to the model after initial parameterization, taking into account the assumed total community respiration in which bacterial respiration is the dominant component.

2.3.3 Compilation of Microbial Production Estimates in the Great Lagoon

The microbiological literature concerning the Great Lagoon lacks references dealing with quantitative estimates of bacteria, protists or microalgae for 1998-2002, as well as for any other period. Therefore, microbial production had to be estimated based on studies conducted in other areas under similar climatic conditions.

Bacterioplankton. Bacterial production can be approximated using the review of Cole et al. (1988) who demonstrated that in the photic zone of both fresh and marine areas it amounts to about 20% of the average net phytoplankton production. In the Gulf of Gdańsk, Ameryk et al. (2005) estimated bacterial production at 14.3% of the gross phytoplankton production. Therefore, bacterial production in the Great Lagoon was assumed to amount to 80 gC m^{-2} yr^{-1}, i.e., 20% of the net primary production or 16% of the gross primary production. In addition to allochthonous and autochthonous non-living organic matter, phytoplankton production was the major food source for the bacterioplankton.

Protozooplankton. In the Baltic Sea and temperate lakes, the protozooplankton is made up by three major groups of unicellular organisms: ciliates, heterotrophic dinoflagellates, and nanoplanktonic heterotrophic flagellates (Azam et al., 1983; Fenchel, 1987; Witek et al., 1993; Wetzel, 2001). In the Gulf of Gdańsk, the annual production of planktonic ciliates was estimated at 4-5% of the gross phytoplankton primary production (Witek, 1998); the production of heterotrophic dinoflagellates and heterotrophic nanoflagellates was estimated at 5-6% of the gross primary production and at 10-33% of the bacterial production, respectively (Witek, 1995). Studies on the protozooplankton conducted in brackish coastal lakes, the Gardno and the Łebsko, showed planktonic ciliates to be abundant and heterotrophic dinoflagellates to occur at much lower densities (Rychert et al., 2012). The potential ciliate production in the two coastal lakes was estimated by the authors referred to at 12-13% of the gross primary production. In Lake Michigan, the production of heterotrophic flagellates and ciliates corresponded to about 40% of the bacterial production (Carrick et al., 1992). Taking those data into account, the total protozooplankton production in the Great Lagoon was estimated at 50 gC m^{-2} yr^{-1}, which is equivalent to 10% of the gross phytoplankton primary production or about 60% of the bacterial production.

The protozooplankton in the Lagoon was assumed to feed primary on phyto- and bacterioplankton; in addition, feeding within the protozooplankton was regarded as relatively important, as heterotrophic flagellates of the protozooplankton are an important food sources for other protists, the ciliates.

Microphytobenthos. Masses of live microalgae are found on large areas of the Lagoon's bottom, particularly on muds (Wolnomiejski et al., 2000). These are, however, mainly phytoplankton (particularly diatoms), sedimented upon the bottom or "glued" to the sediment surface. Some of them may return to the water column, resuspended by waves and currents. Except for most cyanobacteria, more than half of those sedimented algae contain live chloroplasts. Particularly well-retained chloroplasts were seen in coccal chlorophytes, indicating their survival, for a certain period of time, on the muddy bottom beyond the littoral zone. However, the only microalgae found exclusively on the muddy bottom of the main Lagoon basin were the cyanobacteria *Phormidium tenue* (Wolnomiejski et al., 2000).

The littoral supports a specific community of microalgae forming the so-called biofilm or a microbiotic sediment layer which constitute an additional, substantial pool of organic matter on the bottom and produce various substances (mainly polysaccharides and proteins) consolidating the sediment particles (Johnson et al., 1989; Madsen et al., 1993; McLusky and Elliott, 2004). Unfortunately, those microalgae in the Great Lagoon have not been studied. Owing to the potential importance of benthic microalgae for primary production, particularly on the shallow bottom (e.g. Wasmund and Kowalczewski, 1982; Graneli and Sundbäck, 1986), even theoretical estimates of microalgal production merit inclusion into food web characterisation.

To arrive at a hypothetical estimate of microphytobenthos production, a compilation can be made from published information and from the authors' own observations and assumptions as to the role of sediment type, the presence of macrophytes and zebra mussel beds, exposure of the bottom to destructive effects of wave action etc., which will theoretically determine what part of the bottom is amenable for colonisation by and capable of supporting the microphytobenthos. The following major premises were taken into account:

- photic zone depth – Secchi disk visibility relationships (Kajak, 2001);
- bottom area amenable to colonisation by and development of strictly benthic microalgae in the Great Lagoon; this area is about 120 km^2 (40 km^2 of the unvegetated littoral bottom, 60 km^2 of the phytolittoral bottom, and 20 km^2 of bottom in embayments, within 2-4 m depth range);
- an opinion expressed by McLusky and Elliott (2004) that the production of estuarine microphytobenthos may be similar to that of the phytoplankton in the water column overlying the bottom supporting the microphytobenthos in question;
- opinions of other authors quoted by Wetzel (2001) that phytoplankton production among the littoral vegetation drops by an average of 42% (by more than 50% among pondweed) due to shading, CO_2 and nutrient deficiency, and the release of plant inhibitors (mainly polyphenolic compounds) by macrophytes.

A compilation based on the premises listed above yields a microphytobenthos production estimate of ca. 35 gC m^{-2} yr^{-1}.

Another way of estimating that production may involve a relationship developed by Boulion (2004a) in his review paper on primary production estimation in littoral and open parts of various lakes in the temperate zone of the northern hemisphere. He found the microphytobenthos production (PP_{mphb}) to be a function of the proportion of littoral in the surface area of the entire lake (A_{litt} / A_{lake}) and the phytoplankton production (PP_{phpl}), which can be described by the equation:

$$PP_{mphb} = 0.32 \times A_{litt} / A_{lake} \times PP_{phpl}.$$

Assuming the littoral surface area to account for 25% of the Great Lagoon bottom, and the net phytoplankton production to be 400 gC m^{-2} yr^{-1}, the microphytobenthos production was estimated at 32 gC m^{-2} yr^{-1}. This value was used in the Great Lagoon food web model.

Benthic bacteria and fungi. The production of benthic bacteria and fungi in the Great Lagoon has not been studied so far. However, Cole et al. (1988) assessed the relationship between bacterial production and sediment organic carbon content. Their analysis was based on data collected in riverine and coastal marine ecosystems. The relationship was described with the equation:

$$\log(SBP) = 0.69 \times \log(SOC) - 0.15 \quad (r^2 = 0.66),$$

where SBP is the sedimentary bacteria production (µgC $g_{d.w.}^{-1}$ d^{-1}) and SOC is the sediment organic carbon content (mgC $g_{d.w.}^{-1}$).

According to Osadczuk (2004), organic matter accounted for an average of 9.1% of dry weight of the silty fraction (< 63 µm) in the Szczecin Lagoon's muddy sediments, and for 5.9% of that fraction in sandy sediments. The silty fraction in the muddy and sandy sediments averaged 84.9 and 22% of the total sediment weight, respectively. The organic carbon content calculated from these data amounted to 7.7% (77 mgC $g_{d.w.}^{-1}$) and 1.3% (13 mgC $g_{d.w.}^{-1}$) of the total sediment weight in the muddy and sandy sediments, respectively. Using the equation developed by Cole et al. (1988) shown above, the bacterial production rate in the Great Lagoon can thus be estimated at about 14 and 4 µgC $g_{d.w.}^{-1}$ d^{-1} in the muddy and sandy sediments, respectively.

Information on bacterial production normalised to 1 g sediment dry weight cannot be, in a straightforward manner, utilised in estimates of production per unit bottom surface of a lagoon. It is necessary to know the amount of dry sediment per unit surface in a layer of a certain thickness. The amount of dry sediment per 1 m^2 bottom can be expressed as:

$$DM = 10\,000 \times h \times BD \times (100 - W) / 100,$$

where: DM, sediment dry weight (g m^{-2}),
h, sediment layer thickness (cm)
BD, sediment bulk density (g cm^{-3}),
W, sediment moisture content (% wet weight).

Sediment bulk density depends on the moisture content as well as on contents of organic and mineral matter. Assuming an approximately identical density of water and organic matter (about 1 g cm^{-3}), and setting the mineral density at about 2.6 g cm^{-3}, the sediment bulk density can be calculated as:

$$BD = 260 / (100 + 1.6 \times (W + OM - W \times OM/100)),$$

where OM is the organic matter content (% sediment dry weight).

The mean organic matter content in the silty fraction of the muddy and sandy sediments in the Szczecin Lagoon was estimated at 23.3 and 14.4% dry weight, respectively (Osadczuk, 2004). This yields 19.8 and 3.2% of the total sediment weight, respectively. No data on the Great Lagoon moisture content were found in the available literature. According to Graca et al. (2006), the moisture content of silty-clayey and clayey-silty sediments of the Gulf of Gdańsk ranged within 55-90%, the sandy sediment most frequently showing the moisture content of 15-30%. Assuming an average moisture content of muddy sediments of 70% and organic matter content of 20%, the sediment bulk density can be assumed as about 1.2 g cm^{-3}. Following such assumptions, a 1 cm thick layer of muddy sediment on 1 m^2 bottom surface would contain about 3.5 kg sediment dry weight. On the sandy bottom of an average moisture content of 25% and organic matter content of 3%, the sediment bulk density can be estimated at about 1.8 g cm^{-3}, 1 cm thick sediment layer of 1 m^2 bottom surface containing 13.5 kg sediment dry weight. Before these data can be used to estimate the annual bacterial production, it is necessary to make further assumptions, this time with regard to the thickness of the biologically active sediment layer and depth distribution of bacterial production in the sediment. For example, at a 1-5 cm thick active layer and average annual bacterial production rate in this layer of 14 and 4 μgC $g_{d.m.}^{-1}$ d^{-1} in the muddy and sandy sediments, respectively, a similar annual bacterial production can be expected in both sediment types (18-100 gC m^{-2} yr^{-1}). As the above estimates are very tentative and the expected annual production range is very wide, the sediment bacterial and fungal production have not been adjusted in the model to any concrete level.

It was assumed that the sediment bacteria and fungi feed on all the forms of non-living organic matter and sedimented phytoplankton, present in the bottom deposits.

Benthic protists. The lack of local studies and the paucity of literature information on benthic protists made it impossible to estimate their production without using the model. The protozoobenthic organisms were assumed to feed on the sedimented phytoplankton, sediment-dwelling bacteria and fungi as well as on organic remains and faeces.

Periphytic bacteria, fungi, and protists. The absence of empirical and literature data on periphytic microorganisms, too, made it impossible to assess their production in the Great Lagoon without the recourse to the model. It was assumed that periphytic bacteria and fungi feed mainly on phytoperiphyton and dead tissues of macrophytes serving as substrate, whereas periphytic protists feed mainly on phytoperiphyton and periphytic bacteria and fungi; fine particles in suspension, containing phyto- and bacterioplankton as well as non-living organic matter, contribute to the diet as well.

2.3.4 Primary Production and the Supply of Allochthonous Organic Matter

Organic matter in an ecosystem originates from the autotrophic primary production and is also supplied from the outside of the system. Part 1 discussed production of different autotrophic system components in the Great Lagoon in 1998-2002. Those data were used as drivers in the model described in this part.

The gross primary production (GPP) of the phytoplankton was estimated at 500 gC m^{-2} yr^{-1}. Assuming the phytoplankton respiration to use up 20% GPP, the net primary production was estimated at 400 gC m^{-2} yr^{-1}. A similar GPP loss on respiration was assigned to the remaining groups of aquatic plants, except helophytes (Table 2.2). In the latter, with their leaves and upper part of the stem protruding from the water, most of the gas exchange proceeds in the atmosphere. It was therefore assumed that helophyte respiration in water accounts for as little as ¼ of the total respiration, therefore the helophyte stipulated GPP in the water was assumed to be about 75.5 gC m^{-2} yr^{-1}.

The major supplier of allochthonous organic matter to the Great Lagoon is the River Odra. The annual supply with the Odra discharge can be estimated from monitoring data of the Regional Inspectorate of Environmental Protection (RIEP) in Szczecin. The average annual supply in 1998-2002, as determined for the Krajnik Dolny monitoring site, was slightly above 230 thou. tC yr^{-1} (Anon., 1991-2003). The RIEP data do not distinguish between the soluble and particulate fractions of organic matter. Analyses performed in River Vistula, where the two fractions were distinguished between, showed the dissolved fraction to account for 60% of the total organic matter load discharged by the river into the Baltic, the remaining part consisting of the particulate fraction (Pempkowiak and Kupryszewski, 1980). Assuming similar proportions for the Odra, the dissolved organic carbon (DOC) load discharged by the Odra in the Lagoon in 1998-2002 may be estimated at about 138 thou. tC yr^{-1}.

However, it cannot be assumed that the entire particulate organic carbon (POC) load transported by the Odra and recorded at Krajnik Dolny reaches the Lagoon. The Krajnik Dolny site is located about 70 km away upstream from the Odra mouth. Along this section, the river flow slows down and the bed forks out to two branches: the Eastern and the Western Odra, the branches being connected with a system of transverse canals and wetlands. The Eastern Odra passes through Lake Dąbie before it reconnects with the Western Odra to form a common mouth area (the Roztoka Odrzańska) in the southern part of the Szczecin Lagoon. Nowak (1980) calculated that as little as 30% of the suspended particulates transported by the Odra upstream of the forking point reaches the Szczecin Lagoon, the remaining 70% being sedimented out en route. Assuming those proportions to hold, the POC load discharged by the Odra into the Lagoon in 1998-2002 can be estimated at somewhat less than 28 thou. tC yr^{-1}.

The city of Szczecin is an additional source of organic matter for the Lagoon. In the period covered by this study, the city's wastewater management was in a very bad shape. In 2001, of about 90 sewage outlets, only 2 were equipped with mechanical and biological sewage treatment devices (Jurkowski et al., 2003). In his report on the Szczecin Lagoon water quality, Mutko (1994) referred to the 1985 data showing the contribution of municipal and industrial sewage from the city of Szczecin to the total organic pollution in the Szczecin Lagoon to be estimated at about 24%. Assuming this contribution and rounding it up to 25% to include sewage discharged by other settlements situated along the Lagoon shores, it can be demonstrated that, with the Odra's organic load identical to that in 1998-2002, the municipal and industrial wastes could have contributed additional 55 thou. tC yr^{-1}. Assuming an even 50% reduction of that load as a result of diminution of total wastewater production and progress in sewage treatment in 1985-2000, the total organic load from municipal and industrial sources can be estimated at about 27 thou. tC yr^{-1}.

The total organic matter supply to the Great Lagoon in 1998-2002 may be thus estimated at about 193 thou. tC yr^{-1}, about 152 thou. tC yr^{-1} of which were supplied in the dissolved form, 41 thou. tC yr^{-1} being contained in suspended particulates (assuming the DOC:POC ratio of 1:1 in the organic matter of municipal and industrial origin). When related to the Great Lagoon surface area (410 km^2), the organic matter supply amounts to about 370 g DOC m^{-2} yr^{-1} and about 100 g POC m^{-2} yr^{-1}, the combined value being comparable with the Lagoon's primary production (Table 2.2).

Table 2.2 Sources of organic matter in the Great Lagoon in 1998-2002.

Organic matter source	**gross [gC m^{-2} yr^{-1}]**	**net [gC m^{-2} yr^{-1}]**
Primary production of phytoplankton	500	400
Primary production of emergent macrophytes	75.5	71.7
Primary production of submerged and floating-leaved macrophytes	14.5	11.6
Primary production of microphytobenthos	40	32
Primary production of epiphytic algae	27.5	22
Supply of allochthonous particulate organic matter (POM)		100
Supply of allochthonous dissolved organic matter (DOM)		370

The sum total of gross primary production of all the autotrophic components of the Great Lagoon ecosystem in 1998-2002 was estimated at 658 gC m^{-2} yr^{-1}, the supply of allochthonous organic matter being estimated at 470 gC m^{-2} yr^{-1}; thus the total input to the ecosystem was 1128 gC m^{-2} yr^{-1}. However, consumers are capable of utilising the net primary production only (537 gC m^{-2} yr^{-1}), for which reason the true organic matter supply to the heterotrophic part of the system was about 1000 gC m^{-2} yr^{-1}. Almost half of that supply (47%) was in the form of allochthonous organic matter, 40% were contributed by the phytoplankton production, 8% by macrophytes, the remaining 5% being contributed by production of the microphytobenthos and phytoperiphyton.

2.3.5 Production of Heterotrophic Components, Used in Model Parameterization

As already mentioned in the section describing the model, it processes all organic matter fluxes in the ecosystem on an annual basis. Production of most heterotrophic ecosystem components was estimated in Part 1, based primarily on the biomass averaged for the growing season (210 days, from April through October) and seasonal P/B values. When estimated by multiplying the average seasonal biomass by the seasonal P/B, the production value in fact represented that of the growing season. A possible discrepancy between the annual and the growing season production may introduce certain inaccuracy to the model. In autumn-winter (November-March), when water temperature is usually lower that 5 °C, life processes of poikilotherms slow considerably down, and some species (e.g. numerous copepods: Bosselmann, 1975; Alekseev, 1987) enter the diapause state. Under such conditions, the assumption of the annual production being equal to that of the growing season does not err too much. However, certain species continue growing also after the growing season has terminated, as evidenced by, e.g., the development of a winter generation of *Chironomus plumosus* in Lake Jeziorak (Giziński and Wiśniewski, 1971). In those cases, the annual production is higher than that during the growing season. Another sources of error in production estimates from seasonal values of biomass and P/Bs may be sought in differences between the length of the growing season used in the in biomass estimates and in the P/B estimates. In the temperate zone, depending on climatic conditions (and also on resources in a researcher's disposal), the field season can take from 6 to 12 months during a year. If, for example, the production estimate is based on the biomass averaged over a period shorter than that used for the P/B estimate, the biomass may easily be relatively higher than that estimated over a longer period of time, covering more autumn-winter months, because biomass of numerous aquatic organisms (e.g., rotifers, cladocerans) decreases in autumn-winter. Then the production will be ultimately overestimated. Such inaccuracies are difficult to avoid in a compilation-based work, but the interpretation of modelling results has to take into account a possibility of such errors.

Table 2.3 summarises production estimates for those heterotrophic system components for which it was calculated in Part 1 or estimated in Section 2.3.3 above. All the production estimates were treated as annual values, production of each system component being understood here as a sum total of production of all the individuals making up the component. These production estimates were used to parameterize the model.

The highest production in the plankton was typical of bacteria and protists (80 and 50 gC m^{-2} yr^{-1}, respectively). Their respective production was an order of magnitude higher than the production of mesozooplanktonic components (rotifers, cladocerans, and copepods; 1-13 gC m^{-2} yr^{-1}). Owing to the lack of reliable biomass data, production of mysids was not included in the table.

Production values of the mesozooperiphyton, sessile and motile macrozooperiphyton were similar (0.8-2.8 gC m^{-2} yr^{-1}) and indicative of a several-fold lower production of zooperiphyton, compared with production of the zooplankton or that of the zoobenthos. The lack of appropriate data made it impossible to estimate the production of periphytic bacteria and protists without the assistance of the model.

In the benthic community, production of the meiofauna (16 gC m^{-2} yr^{-1}) was only slightly lower than that of the total macrozoobenthos (22 gC m^{-2} yr^{-1}). A distinct component of the latter was *Chironomus* sp. the production of which (11.7 gC m^{-2} yr^{-1}) was several times higher than production of any of the remaining components (0.6-3.7 gC m^{-2} yr^{-1}). Production of benthic bacteria and protists could not be estimated without the recourse to the model.

Table 2.3 Secondary production in the Great Lagoon in 1998-2002, estimated from field observations and theoretical assumptions. Production of each component is considered here to be the sum total of production of all individuals contributing to the component.

Food web component	**Production [gC m^{-2} yr^{-1}]**
Plankton:	
Bacterioplankton	80
Protozooplankton	50
Rotifers	5.95
Non-predatory cladocerans	13.01
Copepods	5.48
Leptodora kindti	1.28
Periphyton:	
Mesozooperiphyton	2.75
Sessile macrozooperiphyton	0.80
Motile macrozooperiphyton	1.30
Benthos:	
Meiofauna	16.30
Chironomus sp. larvae	11.71
Oligochaetes	2.43
Macrobenthic ostracods	0.60
Dreissena polymorpha	3.70
Other molluscs	0.60
Remaining zoobenthos	3.39
Ichthyofauna:	
Larvae and fry of planktivorous fish	0.17
Juvenile and adult planktivorous fish	0.32
Larvae and fry of benthivorous fish	1.01
Juvenile and adult benthivorous fish	1.91
Larvae and fry of piscivorous fish	0.41
Juvenile and adult piscivorous fish	0.63

Among the system components representing adult and juvenile fish, the highest production (1.9 gC m^{-2} yr^{-1}) was typical of benthivorous fish. In contrast, production of the planktivorous fish was low (about 0.3 gC m^{-2} yr^{-1}), whereas production of predatory fish corresponded to ⅓ of the benthivore production (in excess of 0.6 gC m^{-2} yr^{-1}). Noteworthy is the substantial production of larvae and fry, estimated jointly for all the fish categories at about 1.6 gC m^{-2} yr^{-1}.

With regard to the waterfowl, the model was parameterized with their annual consumption (Table 2.4) rather than annual production. The consumption values estimated for herbivorous, molluscivorous, and piscivorous birds were similar, from 0.15 to 0.20 gC m^{-2} yr^{-1}, consumption of omnivores being lower by 2 orders of magnitude (0.002 gC m^{-2} yr^{-1}), but only with respect to the food originating directly in the Lagoon. As mentioned in Part 1, the omnivorous birds (gulls, terns) obtain a large part of their food outside of the Lagoon ecosystem (at garbage dumps, in town, etc.).

Table 2.4 Food consumption of avifauna in the Great Lagoon, estimated from field observations and theoretical assumptions.

Birds' trophic group:	**Consumption gC m^{-2} yr^{-1}**
Herbivorous birds	0.163
Molluscivorous birds	0.149
Omnivorous birds	0.002
Piscivorous birds	0.203

2.3.6 Export of Pelagic Ecosystem Components to the Pomeranian Bay

Transport of planktonic ecosystem components to the Pomeranian Bay was estimated based on the assumptions discussed in Section 2.2. Assuming the water outflow to the Pomeranian Bay (*Q*) to be balanced out by the sum total of riverine inflow and to average 18.4 km^3 yr^{-1}, and approximating the Great Lagoon volume (*V*) at 1.6 km^3, the water exchange rate (*Q/V*) in the Great Lagoon can be estimated at about 11.5 yr^{-1}. As the concentration of planktonic organisms is strongly season-dependent, and during the growing season (April-October) is usually several-fold higher than at other times, export of plankton varies as well and peaks during the growing season. The outflow in

April-October amounts to about 50% of the annual outflow (Majewski, 1980). Some planktonic organisms (cladocerans, copepods, *Leptodora*) spend about half of the diurnal cycle near the bottom, away from the main current. It may be thus assumed, for simplicity, that the effective volume of water exporting the plankton (Q_e) equals only half, or even ¼, of the total volume of water outflowing from the Lagoon. The part of the planktonic production that is exported (*Exp/P*) can be calculated as:

$$Exp/P = (Q_e/Q) \times (Q/V) / (P/B)$$

Table 2.5 lists the parts of planktonic production exported, as estimated based on the premises described above.

Table 2.5 Estimation of the proportion of pelagic production exported from the Great Lagoon. P/B, annual production/biomass ratio; Q_e/Q, part of the outflow effectively transporting plankton; *Exp/P*, part of production exported. The Lagoon volume V = 1.6 km^3; the total water outflow Q = 18.4 km^3 yr^{-1}.

Functional group:	P/B	Q_e/Q	Exp/P
Phytoplankton	75	0.5	0.077
Bacterioplankton	50[a]	0.5	0.115
Protozooplankton	150[b]	0.5	0.038
Rotifers	60	0.5	0.096
Non-predatory cladocerans	27	0.25	0.106
Copepods	10	0.25	0.288
Leptodora kindti	15	0.25	0.192

[a] guess estimate (compare with annual average P/B of bacterioplankton in Lake Constance, Germany = 0.13 day^{-1} (Straile, 1998 in Wetzel, 2001))
[b] guess estimate (compare with P/B of combined heterotrophic nanoflagellates and ciliates in the Gulf of Gdańsk (the Baltic Sea) = 149 yr^{-1} (Witek, 1995))

Mysids were assumed to be capable of resisting the outflowing current, therefore their passive export was not included. Mysids are even known of arriving to the Lagoon from the Pomeranian Bay (cf. Section. 1.14.4).

In addition to live ecosystem components (the plankton), non-living organic matter is exported as well. It was assumed that the entire DOM load not used in the water column is exported to the Pomeranian Bay. With respect to POM (allochthonous POM, faeces, detritus), 90% of its remains not utilised by consumers were arbitrarily regarded as sedimenting to the bottom (k = 0.9), the remaining 10% being exported to the Pomeranian Bay.

2.3.7 Directions of Matter and Energy Flows in the Ecosystem of the Lagoon

The model operator's basic task when calibrating the model was to select the F_{ij} values which would form the matrix of trophic connections and define proportions at which the production of various food web components would be utilised by consumers (Tables 2.5 - 2.11). It was also necessary to determine the fate (export, natural mortality, fishing, etc.) of that part of the production which was not utilised by consumers. The parameterization procedure was started with components forming the basis of the trophic pyramid, and proceeded along the matter flux direction towards the top of the pyramid. The F_{ij} indices are rarely directly determined, and were not dealt with in any study focusing on the Szczecin Lagoon or on other Baltic coastal water bodies. Estimation of F_{ij} values is, on the other hand, aided by results of research on food composition and food rations of aquatic organisms. In the Great Lagoon, such studies were conducted with respect to some fish species (Wolnomiejski and Grygiel, 1994b; 1994c; 2002) and *Chironomus* sp. larvae (Wolnomiejski et al., 2000). For the remaining aquatic organisms, relevant were studies carried out in other, similar ecosystems, and the theoretical knowledge. Thus, the extent to which production of individual ecosystem components was utilised was estimated by successive approximations in which the values of F_{ij} were selected so that production, food composition, catches (landings), export etc. of possibly largest number of components corresponded to production, food composition etc. determined from empirical data or from theoretical knowledge. Effort was made not to violate the matter conservation principle. It should be realised that, at a high number of all the theoretically possible trophic links between food web components and their scalar nature, there is an infinite number of possible trophic link matrices which would reflect the reality more or less accurately. The picture proposed in this work represents the current knowledge and the authors' intuitions. As new data are collected in the future, and as the knowledge on ecosystems of the type represented by the Great Lagoon increases, the picture is likely to be modified and refined.

2.3.7.1 Utilisation of Plankton Production and Non-living Organic Matter Supply in the Water Column

The trophic link matrix developed from the selection of F_{ij} values shows the phytoplankton production to be utilised to the highest degree (20 and 25%) by the bacterio- and protozooplankton, respectively, the metazoan zooplankton utilising a somewhat lower fraction (about 17%) (Table 2.6). Almost ⅕ of the

Table 2.6 Utilisation of phytoplankton primary production and non-living organic matter supply in the pelagic compartment. Values in each column add up to 1.

PRODUCER OR FOOD SOURCE: / CONSUMER:	Phytoplankton	Allochthonous POM	Allochthonous DOM	Faeces, pelagic	Carcasses and remains of pelagic organisms
Plankton:					
Bacterioplankton	0.2	0.2	0.22	0.4	0.45
Protozooplankton	0.25	0.05		0.05	0.07
Rotifers	0.04	0.0009		0.005	0.007
Non-predatory cladocerans	0.09	0.005		0.03	0.04
Copepods	0.042	0.0005		0.01	0.04
Mysids	0.0003	0.0001		0.0001	0.001
Periphyton:					
Epiphytic bacteria and fungi		0.002	0.001	0.005	0.005
Protozooperiphyton	0.001	0.001		0.006	0.008
Mesozooperiphyton	0.0085	0.002		0.005	0.005
Sessile macrozooperiphyton	0.0005	0.0005		0.0005	0.003
Motile macrozooperiphyton	0.0035	0.0005			
Benthos:					
Chironomus sp. larvae	0.017	0.001			
Dreissena polymorpha	0.024	0.01		0.01	0.06
Other molluscs	0.0016	0.0007		0.0016	0.005
Ichthyofauna:					
Larvae and fry of planktivorous fish	0.00001				
Larvae and fry of benthivorous fish	0.0001				
Larvae and fry of piscivorous fish	0.00003				
Natural mortality	0.05				
Export	0.077	0.073	0.779	0.048	0.031
Sedimentation	0.194	0.653		0.429	0.275

phytoplankton production was sedimented on the bottom, about 8% being exported to the Pomeranian Bay.

The efficiency of allochthonous organic matter utilisation in the Great Lagoon pelagic compartment was low. The major consumer utilising about 20% of the organic matter supply was the bacterioplankton. Zooplankton and -benthos utilised as little as few per cent of the allochthonous POM supplied. About 65% of the total POM supply were sedimented, whereas almost 80% of the allochthonous DOM supplied were exported to the Pomeranian Bay.

Consumers in the ecosystem relied more on faeces and remains of organisms produced in the water column than they did on the allochthonous organic matter, as the bacterio- and zooplankton as well as bivalves together used up more than half of the amount produced. The non-utilised detritus was sedimented, some small amounts being exported from the system.

The bacterioplankton was the most important pelagic consumer of non-living organic matter and was, in turn, consumed primarily by protists responsible for using up half of the bacterial production (Table 2.7). A high proportion (about 30%) of the bacterioplankton production was also used up by the metazoan zooplankton, mainly cladocerans.

In the zooplankton, the trophic link matrix was somewhat complicated by within-group utilisation (feeding on representatives of the same trophic group). This phenomenon is particularly pronounced in protists when they are treated as a single food web component. It was assumed in the model, that protists feeding on other protists utilised 30% of the protist production based on consumption of other food web components. Representatives of the metazoan zooplankton were the most important consumers of the net protozooplankton production (i.e., the sum total of production of all protists minus consumption of protists feeding on other protists); taken together, they utilised about 85% of that production. Only a slight proportion of the protist production (less than 4%) was exported, the fast growth rate of protists compensating for most losses due to export.

The production of rotifers, the finest fraction of the metazoan zooplankton, was primarily (more than 40%) utilised by the larger zooplankters: copepods and the predatory cladoceran *Leptodora kindti*. Fish, primarily larvae and fry, utilised about 25% of the rotifer production, a few per cent each being available to the sessile zooperiphyton (*Cordylophora*) and the zebra mussel. Productions of non-predatory cladocerans and copepods were utilised to a similar degree. The predatory cladoceran *L. kindti* removed from several to 20%, the highest proportion (about 50%) being used up by fish. They utilised also almost the entire available production of *L. kindti* and mysids. Export of the metazoan zooplankton production was estimated from 10 to almost 30%, whereas export of mysids was considered non-existent because, as already mentioned, they are capable of resisting the outflowing current and may even enter the Great

Table 2.7 Utilisation of bacterio- and zooplankton production. Values in *italics* indicate the within-group utilisation. Values in each column (without within-group utilisation) add up to 1.

PRODUCER: / **CONSUMER:**	**Bacterioplankton**	**Protozooplankton**	**Rotifers**	**Non-predatory cladocerans**	**Copepods**	***Leptodora kindti***	**Mysids (production + import)**
Plankton:							
Protozooplankton	0.5	*0.3*					
Rotifers	0.06	0.3	*0.05*				
Non-predatory cladocerans	0.22	0.357					
Copepods	0.02	0.2	0.2	0.06	*0.04*		
Leptodora kindti			0.21	0.203	0.14		
Mysids			0.015	0.005	0.005	0.015	
Periphyton:							
Protozooperiphyton	0.0025						
Mesozooperiphyton	0.002						
Sessile macrozooperiphyton	0.001	0.01	0.04	0.015	0.015		
Benthos:							
Dreissena polymorpha	0.025	0.035	0.05	0.01	0.01		
Other molluscs	0.0016	0.0016	0.0016				
Ichthyofauna:							
Larvae and fry of planktivorous fish			0.065	0.05	0.025	0.03	0.01
Juvenile and adult planktivorous fish			0.03	0.03	0.021	0.07	0.53
Larvae and fry of benthivorous fish			0.107	0.23	0.3	0.11	0.01
Juvenile and adult benthivorous fish				0.05	0.05	0.46	0.08
Larvae and fry of piscivorous fish			0.047	0.1	0.076	0.05	0.23
Juvenile and adult piscivorous fish				0.018	0.018	0.02	0.08
Birds:							
Herbivorous birds							0.011
Molluscivorous birds							0.0045
Omnivorous birds							0.00077
Natural mortality	0.05	0.06	0.14	0.123	0.052	0.05	0.044
Export	0.115	0.038	0.098	0.106	0.288	0.192	0

Lagoon from the Pomeranian Bay. This active migration of mysids to the Lagoon was factored in into the model as their additional supply, in addition to the autochthonous production.

To summarise utilisation of the production and organic matter supply in the pelagic compartment of the ecosystem, it has to be emphasised that the largest part (65-95%) of the production of live components is utilised as food by other ecosystem components (Fig. 2.2). The autochthonous non-living organic matter was a less important food source, but more than 50% of its supply was consumed in the water column. The allochthonous organic matter was utilised the least. Most of the allochthonous DOM was exported, unused, to the Pomeranian Bay, whereas most of the allochthonous POM sedimented onto the bottom. In addition, an estimated 30-40% of autochthonous organic matter and about 20% of the phytoplankton production sedimented out. From a few to almost 30% of the planktonic production was exported to the Pomeranian Bay.

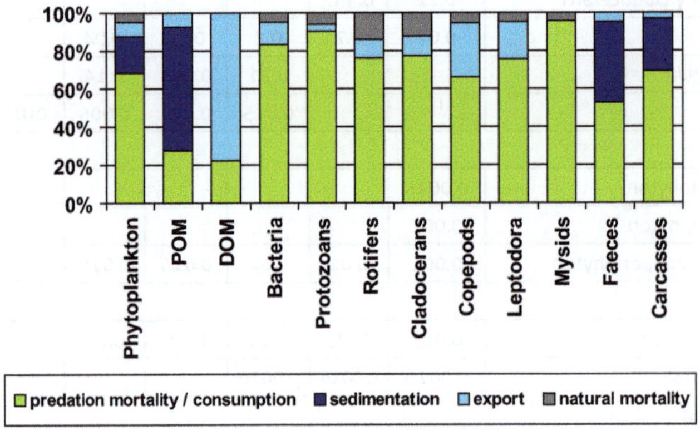

Figure 2.2 Utilisation of production of planktonic organisms and non-living organic matter in the pelagic compartment of the ecosystem.

2.3.7.2 Utilisation of Periphyton Production

Trophic relationships in the Great Lagoon periphyton belong to those known the least, therefore the periphyton components' production utilisation coefficients (Table 2.8) are mostly based on the authors' intuitions or are guess estimates. It was assumed that the phytoperiphyton production was utilised primarily by the periphytic bacteria and fungi (about 40%), protists (about 20%), and meso- and macrozooperiphyton (about 35% taken together). In addition to the periphyton, also littoral invertebrates could utilize some (about 5%) of the

Table 2.8 Utilisation of periphytic production. Values in *italics* indicate the within-group utilisation. Values in each column (without within-group utilisation) add up to 1.

CONSUMER: \ PRODUCER:	Epiphytic algae	Epiphytic bacteria and fungi	Protozooperiphyton	Mesozooperiphyton	Sessile macrozooperiphyton	Motile macrozooperiphyton
Plankton:						
Mysids	0.001	0.001	0.001	0.001		
Periphyton:						
Epiphytic bacteria and fungi	0.37					
Protozooperiphyton	0.18	0.65	*0.15*			
Mesozooperiphyton	0.26	0.17	0.88	*0.1*		
Sessile macrozooperiphyton	0.005	0.02	0.05	0.4		0.25
Motile macrozooperiphyton	0.09	0.1	0.02	0.55	0.4	*0.25*
Benthos:						
Other molluscs	0.002	0.001	0.001	0.001	0.001	
Remaining zoobenthos	0.048	0.02	0.01	0.01	0.002	0.001
Ichthyofauna:						
Larvae and fry of planktivorous fish	0.0001	0.0002	0.0005	0.001	0.005	0.006
Juvenile and adult planktivorous fish	0.0001	0.0001	0.0001	0.0001		0.0001
Larvae and fry of benthivorous fish	0.005	0.001	0.001	0.001	0.025	0.08
Juvenile and adult benthivorous fish	0.0005	0.0005	0.0005	0.00015	0.03	0.48
Larvae and fry of piscivorous fish					0.005	0.025
Juvenile and adult piscivorous fish						0.06
Birds:						
Herbivorous birds	0.0002	0.0002	0.0002	0.0002	0.001	0.0015
Molluscivorous birds	0.00005	0.00005	0.00005	0.00005	0.0003	0.0002
Emergence						0.04
Natural mortality	0.04	0.04	0.04	0.04	0.53	0.06

epiphytic algal production. The periphytic bacteria and fungi were assumed to be consumed primarily (65% of production) by periphytic protists, almost the entire remainder being used by the zooperiphyton. The mesozooperiphyton was assigned, in the model, a role of the major periphytic protist consumer (consumption of about 90% of the net production) assisted, to some extent, by the macrozooperiphyton. The within-group protist consumption was included in the model as well. The internal cycling played a role in the mesozooperiphyton, too, the macrozooperiphyton being considered the major consumer of its net production. The sessile macrozooperiphyton production was assumed to be consumed primarily by the motile macrozooperiphyton (40%, mainly gammarid predation on *Cordylophora*), but about 5% of the production could be also used by fish (mainly those feeding on the juvenile zebra mussel). The fish were also considered the most important consumer of the motile macrozooperiphyton (about 2/3 of the net production). In the motile macrozooperiphyton, of importance was probably the within-group predation. The motile macrozooperiphyton consisted of, i.a., insect larvae. Emergence of the adult insects and the associated energy losses could amount to about few per cent of production of this component.

With the assumptions as above, more than 90% of the periphyton production was used as food, the natural mortality amounting to a few percent of production (Fig. 2.3). An exception was the sessile macrozooperiphyton the production of which was much less (about 60%) utilised; chitinous covers, exoskeletons, shells etc. of those animals protect them from predation and increase the importance of natural mortality.

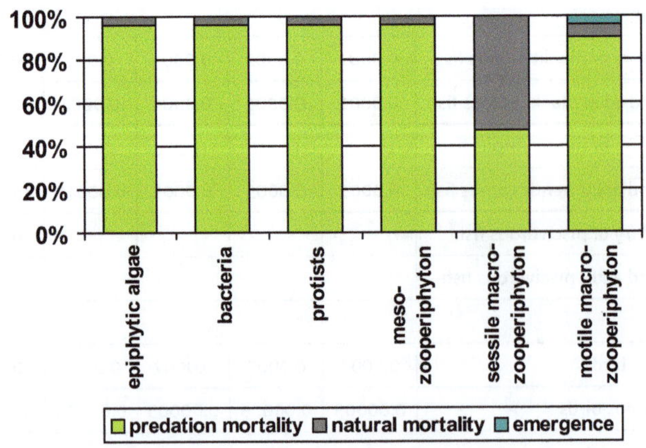

Figure 2.3 Utilisation of periphytic production.

2.3.7.3 Utilisation of Benthic Production and Non-living Organic Matter Supply in the Sedimentary Compartment of the Ecosystem

Based on the general hydrobiological knowledge (Wetzel, 2001), the benthic vegetation of the Great Lagoon was assumed to be, when fresh, poorly utilised by consumers (Table 2.9; Fig. 2.4). No animal was a consumer of the rush (emergent) plants. Production of soft (submerged and floating-leaved) macrophytes was used mostly by littoral and periphytic invertebrates, but their consumption engaged as little as a few per cent of the production. A relatively high utilisation rate was assumed for the microphytobenthos (about ¾ of the production), the meio- and macrobenthos being deemed the major microalgal grazers.

Almost the entire helophyte production, more than 90% of the production by soft macrophytes, and about one-quarter of the microphytobenthos production contributed to the phytodetritus pool which served as food mainly for benthic bacteria and fungi and for the meiobenthos. The sedimented phytoplankton, treated in the model as a separate food web component, was most probably consumed by benthic animals more efficiently than the phytobenthos. The major consumers of sedimented phytoplankton were, in addition to bacteria, *Chironomus* sp. larvae and the meiobenthos. The importance of sedimented phytoplankton in the diet of *Chironomus* sp. larvae was pointed out by Wolnomiejski et al. (2000). Like in the pelagic ecosystem compartment, the major consumers of non-living organic matter in the sediment were bacteria and fungi. The model assumed that, in addition to plant remains, bacteria and fungi used up 50% of the sedimented allochthonous matter and 60-70% of the sedimented faeces and animal remains. Among invertebrates, the meiobenthos was attributed the highest utilisation of the non-living organic matter (Table 2.9). It is likely that animal remains were a form of non-living organic matter consumed most readily by benthic animals.

The non-living organic matter not utilised by consumers is accumulated in bottom sediments. For the purposes of the model, it was assumed that the highest accumulation rate (more than 30% of the supply to the sediment) was typical of the allochthonous matter, followed by faeces and plant debris (several percent of the supply each). Only a small part (a few per cent) of the sedimented phytoplankton and animal remains eventually were accumulated in the sediment (Fig. 2.4).

Benthic bacteria, fungi and protists were most likely consumed primarily by the meiobenthos. In the model, the meiobenthos was attributed about 70% utilisation of production of those components (Table 2.10). Of the remaining invertebrates, a larger part in the overall consumption of benthic bacteria, fungi,

Table 2.9 Utilisation of phytobenthos primary production and non-living organic matter supply in the benthic ecosystem compartment. Values in individual columns add up to 1.

PRODUCER OR FOOD SOURCE: / CONSUMER:	Emergent plants	Submerged and floating-leaved plants	Microphytobenthos	Sedimented phytoplankton	Sedimented allochthonous POM	Plant debris	Faeces, sedimented and benthic	Carcasses and remains, sedimented and benthic
Plankton:								
Mysids			0.01	0.005	0.0001	0.001	0.0003	0.001
Periphyton:								
Epiphytic bacteria and fungi	0.005	0.005				0.05	0.001	0.0001
Mesozooperiphyton						0.02		
Motile macrozooperiphyton		0.02	0.0025			0.0066		
Benthos:								
Benthic bacteria and fungi			0.01	0.3	0.5	0.6	0.7	0.68
Benthic protists			0.025	0.05	0.03	0.01	0.01	0.05
Meiofauna			0.36	0.2	0.1	0.1	0.09	0.1
Chironomus sp. larvae			0.1	0.34	0.02	0.015	0.0082	0.005
Macrobenthic ostracods			0.042	0.005	0.001	0.002	0.0035	0.01
Oligochaetes			0.075	0.02	0.03	0.01	0.019	0.02
Dreissena polymorpha				0.04	0.002			
Other molluscs			0.022	0.0033	0.0007	0.0015	0.0012	0.0015
Remaining zoobenthos		0.02	0.1	0.008	0.005	0.02	0.001	0.05
Ichthyofauna:								
Larvae and fry of planktivorous fish		0.0002	0.00005	0.00005				
Juvenile and adult planktivorous fish		0.0005	0.00005	0.00001				
Larvae and fry of benthivorous fish		0.01	0.004	0.00001				0.001
Juvenile and adult benthivorous fish		0.02	0.006	0.0002		0.001		0.005
Larvae and fry of piscivorous fish				0.0001				
Juvenile and adult piscivorous fish								0.0001
Birds:								
Herbivorous birds		0.0105				0.0002		
Molluscivorous birds		0.0011				0.00005		
Natural mortality	0.995	0.91	0.24					
Deposition in bottom sediments				0.028	0.311	0.163	0.166	0.076

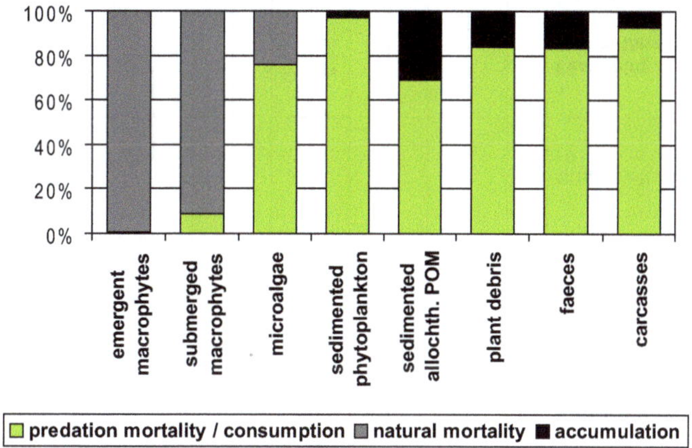

Figure. 2.4 Utilisation of benthic primary production and non-living organic matter supply in the benthic compartment of the ecosystem.

and protists could be played by the pelophilous taxa: oligochaetes, ostracods, and chironomid larvae. Pelophilous macroinvertebrates were most likely also the major consumers of the meiobenthos. The model assumed *Chironomus* sp., macrobenthic ostracods, and oligochaetes to jointly utilise about 50% of the meiofaunal production. About 10% of the meiobenthic production could be used up by littoral invertebrates, a few per cent each – by the zooplankton (mainly *Leptodora*) and benthivorous fish.

As opposed to the ecological categories discussed so far, for which estimates of production utilisation stemmed mainly from literature and theoretical knowledge, utilisation of production by macrobenthic components could be estimated, to a large extent, on results of fish feeding studies, carried out in the Great Lagoon (Wolnomiejski and Grygiel, 1994b; 1994c; 2002). Production of the macrobenthic invertebrates was utilised primarily by benthivorous fish. They were assumed to consume an estimated 85-90% of the production of *Chironomus* sp. larvae and macrobenthic ostracods. Fish were most probably the main predators on the zebra mussel, other molluscs, and the remaining littoral zoobenthos, and consumed 15-50% of their production. Studies on food composition of fish in the Great Lagoon failed to show any substantial contribution of oligochaetes to the fish diet. It was therefore assumed that oligochaetes fell prey primarily to other invertebrates; assuming the oligochaete contribution to the fish food not exceeding 2%, fish would consume less than 20% of the oligochaete production. In addition to benthivorous fish, an important role in consumption of certain macrobenthic components could be played by predatory fish. They (perch, eel) were assumed to consume several percent each

Table 2.10 Utilisation of production of benthic bacteria, protozoans and fauna. Values in *italics* indicate the within-group utilisation. Values in each column (without within-group utilisation) add up to 1.

CONSUMER: \ PRODUCER:	Benthic bacteria and fungi	Benthic protists	Meiofauna	*Chironomus* sp. larvae	Macrobenthic ostracods	Oligochaetes	*Dreissena polymorpha*	Other molluscs	Remaining zoobenthos
Plankton:									
Copepods			0.01						
Leptodora kindti			0.052						
Mysids			0.015						0.005
Periphyton:									
Sessile macrozooperiphyton									0.18
Motile macrozooperiphyton									0.135
Benthos:									
Benthic protists	0.1	*0.05*	0.05						
Meiobenthos	0.72	0.72	*0.05*						
Chironomus sp. larvae	0.03	0.05	0.3			0.01			
Macrobenthic ostracods	0.01	0.05	0.045			0.05			
Oligochaetes	0.067	0.05	0.16						
Other molluscs	0.0035		0.018						
Remaining zoobenthos	0.011	0.05	0.12	0.01	0.04	0.3	0.1	0.1	*0.3*

cont. Table 2.10 Utilisation of production of benthic bacteria, protozoans and fauna. Values in *italics* indicate the within-group utilisation. Values in each column (without within-group utilisation) add up to 1.

CONSUMER: \ PRODUCER:	Benthic bacteria and fungi	Benthic protists	Meiofauna	*Chironomus* sp. larvae	Macrobenthic ostracods	Oligochaetes	*Dreissena polymorpha*	Other molluscs	Remaining zoobenthos
Ichthyofauna:									
Larvae and fry of planktivorous fish			0.002	0.004	0.005	0.001	0.0002	0.001	0.02
Juvenile and adult planktivorous fish			0.0005	0.023	0.02	0.003	0.002	0.005	0.015
Larvae and fry of benthivorous fish			0.046	0.074	0.44	0.06	0.005	0.11	0.08
Juvenile and adult benthivorous fish	0.0005	0.005	0.05	0.675	0.43	0.12	0.15	0.42	0.3
Larvae and fry of piscivorous fish				0.052	0.01	0.005		0.005	0.025
Juvenile and adult piscivorous fish				0.032	0.001	0.001	0.001		0.03
Birds:									
Molluscivorous birds							0.02	0.022	0.015
Emergence				0.08					0.06
Natural mortality	0.06	0.07	0.13	0.05	0.05	0.45	0.72	0.34	0.14

of production of *Chironomus* sp. larvae and littoral invertebrates other than the zebra mussel. In the case of non-mollusc littoral zoobenthos ('remaining zoobenthos': insects, crustaceans, and leeches), within-group predation could be of importance (estimated at 30% of production based on consumption of other components). A few per cent of production of molluscs and other littoral invertebrates were consumed by molluscivorous birds.

A part of the aquatic insect larvae production is utilised for metamorphosis and emergence of the imagines. Energy losses associated with emergence of adult insects, metamorphosis, and export of exuvia to the Pomeranian Lagoon were estimated at about 8% of production of *Chironomus* sp. and a somewhat lower percent of production of the remaining (mainly littoral) zoobenthos and motile macrozooperiphyton (Tables 2.8 and 2.10), a substantial contribution to which is provided by insect larvae.

Natural mortality of some benthic components (bacteria, protists, *Chironomus* sp. larvae, ostracods) was estimated at a low level of a few percent of production. However, natural mortality of molluscs, particularly *D. polymorpha,* and oligochaetes, was estimated to be high, at 35-70% of their production. This high natural mortality is most likely related to poor availability of those animals to consumers. Natural mortality of the meiobenthos and the remaining, primarily littoral, zoobenthos, assumed intermediate values at a level of several percent of production (Fig. 2.5).

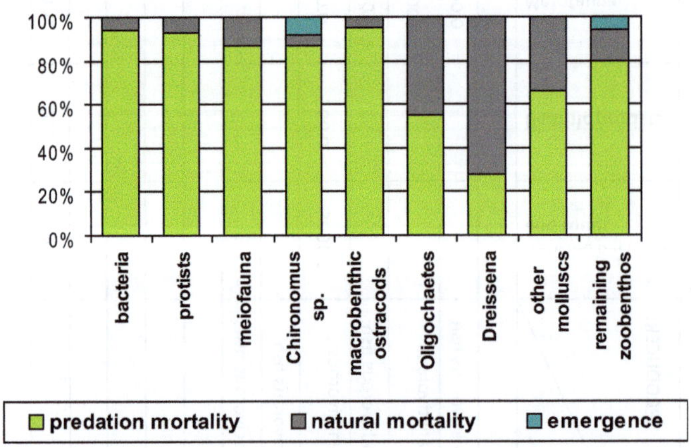

Figure 2.5 Utilisation of production of benthic bacteria, protists, and zoobenthos.

2.3.7.4 Utilisation of Fish Production

Production of larvae and fry of planktivorous, benthivorous, and predatory fish was utilised in a similar manner. The largest part of larval and fry production (about 60-70%) was consumed by juvenile and adult predatory fish. Predation pressure of planktivorous fish was estimated at a few per cent of larvae and fry production; the main prey of planktivorous fish (primarily smelt) was the ichthyoplankton. Predation on larvae and fry by piscivorous birds was estimated at a few per cent of production as well, but the major prey was the fry. About 10% of production of fish larvae and fry constituted recruitment to that part of the population consisting of juveniles and adult fish. Natural mortality of fish larvae and fry was at a level of as little as a few per cent (Table 2.11).

Production of older plankti- and benthivorous fish was utilised in a similar manner. Those fish groups were mainly preyed upon by predatory species

Table 2.11 Utilisation of fish production. Values in *italics* indicate the within-group utilisation. Values in each column (without within-group utilisation) add up to 1.

PRODUCER: / CONSUMER:	Larvae and fry of planktivorous fish	Juvenile and adult planktivorous fish	Larvae and fry of benthivorous fish	Juvenile and adult benthivorous fish	Larvae and fry of piscivorous fish	Juvenile and adult piscivorous fish
Ichthyofauna:						
Juvenile and adult planktivorous fish	0.05		0.05		0.05	
Juvenile and adult benthivorous fish						
Larvae and fry of piscivorous fish	0.13		0.15		*0.08*	
Juvenile and adult piscivorous fish	0.594	0.682	0.585	0.622	0.7	*0.2*
Birds:						
Omnivorous birds	0.0005		0.00078		0.0003	
Piscivorous birds	0.09	0.07	0.061	0.04	0.06	
Recruitment	0.1		0.1		0.1	
Natural mortality	0.04	0.17	0.05	0.16	0.09	0.68
Fishing mortality		0.0785		0.18		0.32

(60-70% of production) and by piscivorous birds (a few per cent of production). Fishing mortality in planktivorous and benthivorous fish was less than 10% and less than 20% of their production, respectively. Natural mortality of plankti- and benthivorous fish was estimated at dozen or so per cent. In the case of predatory fish, within-group predation could be important; it was estimated at 20% of production of those fish feeding on other food web components. Fishing mortality concerned 32% of the net production of juvenile and adult predatory species, the remaining types of mortality affecting as much as 68% of the production.

With respect to juvenile and adult fish, natural mortality shown in Table 2.11 includes also the reproductive material deposited during spawning in the environment. Although fish spawning is usually seen as production of new biomass, actually the newly hatched larvae account only for a small fraction of the reproductive material deposited; a much larger part of that material dies off during embryonic development or becomes food for other animals (Szczerbowski, 1993). In the juvenile and adult fish taken together, production of the reproductive materials accounts for about 10% of the total production of those fish (Section 1.12.5.6). In the present model, natural mortality of plankti- and benthivorous fish due to diseases, damages, senility, etc. concerns a few percent of production of juveniles and adults. In the case of predatory fish, natural mortality due to such causes would concern as much as over 50% of their production (Fig. 2.6). Such a high natural mortality of predatory fish does not seem plausible. The discrepancy would point out to real catches much exceeding the reported landings.

With regard to the waterfowl, their entire production was assumed to be exported from the system.

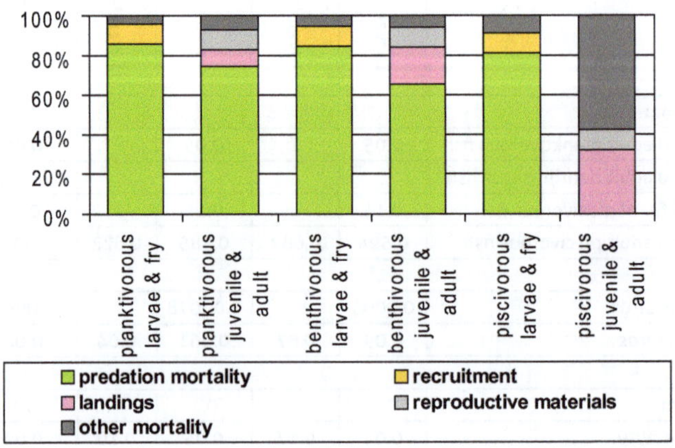

Figure 2.6 Utilisation of fish production. Within-group predation not shown.

2.3.8 Energy Flow Budget in the Great Lagoon Ecosystem

Basic elements of the energy budget of all the food web components in the Great Lagoon, produced by the model, are shown in Table 2.12. Additionally, the table contains the empirical biomass data and the P/B values adopted in this study. It can be seen that in some cases, the production estimated by the model deviates strongly from that produced by the biomass multiplied by P/B (cf. Table 2.3). The discrepancies are due to within-group consumption discussed earlier, which concerns certain food web components. Due to this effect, net production of a component available as food to other components is lower than the algebraic sum of production of all the individuals making up the component. Similarly, consumption of those components affected by within-group consumption, given in Table 2.12, is lower than the algebraic sum of consumption of all the individuals making up the component, as it concerns only the uptake of food derived from other food web components.

The model allowed to determine production of those components, for which biomass and production rate data were missing. The most important food web components, characterized only with the aid of the model, include benthic bacteria and fungi; according to the model, their production was by several per cent higher than that of the bacterioplankton. On the other hand, production of the benthic protists turned out to be almost 6 times lower than production of the pelagic protists.

When comparing biomasses of individual food web components, it is evident (Fig. 2.7) that the highest biomass in the Great Lagoon was that of vascular plants, primarily the emergent vegetation (mainly reed). Macrophyte biomass was several times higher than that of the phytoplankton, and accounted for almost 80% of the total biomass of the Great Lagoon flora and fauna combined (but without the biomass of bacteria and protists). The highest contribution to the animal biomass was provided by the zebra mussel (about ¼ of the animal biomass). The total zoobenthos accounted for almost 60% of the faunal biomass; fish, zooplankton, zooperiphyton, and waterfowl contributed 27, 8, 5, and 0.1%, respectively. Among the fish, the highest biomass was attributed to the benthivorous species. The biomass of larvae and fry accounted for as little as 7% of all the fish biomass in the ecosystem. Compared to the zoobenthos and ichthyofauna, biomass of the zooplankton was low, 7 times lower than the biomass of the zoobenthos and 3 times lower than that of fish. Thus, a larger part of the plant and animal biomass in the Great Lagoon was confined to the bottom, primarily to the littoral zone.

The energy flow in the Great Lagoon was controlled by unicellular organisms. Among autotrophs, the dominant role was played by the phytoplankton (74% of net primary production), bacteria and fungi being of major importance in the heterotrophic part of the biota. Bacterial and fungal consumption, secondary

Table 2.12 Biomass and energy budgets of the Great Lagoon food web components. B, average biomass (empirical data); P/B, production to biomass ratio; C, consumption; P, production (net); R, respiration; FU, faeces and exuvia. C, P, R, and FU are estimates produced by the model.

	B	P/B	C	P	R	FU
	gC m^{-2}	season^{-1}	gC m^{-2} yr^{-1}			
Plankton:						
Allochthonous organic matter, POM				100.0		
Allochthonous organic matter, DOM				370.0		
Phytoplankton				400.0	100.0	
Bacterioplankton			243.5	80.4	163.2	
Protozooplankton			153.3	36.3	50.1	66.9
Rotifers	0.10	60	32.6	5.6	13.8	13.2
Non-predatory cladocerans	0.48	26.9	72.0	13.0	30.2	28.8
Copepods	0.55	10	30.2	5.3	12.8	12.2
Leptodora kindti	0.085	15	5.4	1.3	3.0	1.1
Mysids[a]	0.034	3.3	1.5	0.3	0.8	0.5
Faeces, pelagic				122.5		
Carcasses and remains, pelagic				29.1		
Periphyton:						
Epiphytic algae	1.570	14		22.0	5.5	
Epiphytic bacteria and fungi			14.6	4.6	9.8	
Protozooperiphyton			8.8	2.4	2.7	3.7
Mesozooperiphyton	0.089	31	14.9	2.5	6.4	6.1
Sessile macrozooperiphyton	0.400	2	3.3	0.8	1.9	0.7
Motile macrozooperiphyton	0.324	5.3	6.9	1.0	3.0	2.9
Benthos:						
Sedimenting allochthonous POM				65.3		
Sedimenting phytoplankton				77.8		
Sedimenting pelagic faeces				52.6		
Sedimenting pelagic carcasses and remains				8.0		
Emergent plants	90.6	0.79		71.7	3.8[b]	
Submerged and floating-leaved plants	13.9	0.84		11.6	2.9	
Microphytobenthos	2.1	15		32.0	8.0	
Benthic bacteria and fungi			277.8	91.7	186.1	
Benthic protists			20.8	6.0	6.3	8.4

cont. Table 2.12 Biomass and energy budgets of the Great Lagoon food web components. B, average biomass (empirical data); P/B, production to biomass ratio; C, consumption; P, production (net); R, respiration; FU, faeces and exuvia. C, P, R, and FU are estimates produced by the model.

	B	P/B	C	P	R	FU
	gC m^{-2}	season^{-1}	gC m^{-2} yr^{-1}			
Meiobenthos	2.04	8	134.8	15.5	38.0	81.3
Chironomus sp. larvae	1.46	8	48.8	11.7	17.6	19.4
Macrobenthic ostracods	0.22	2.7	5.0	0.6	1.4	3.0
Oligochaetes	0.49	5	20.3	2.4	5.7	12.2
Dreissena polymorpha	3.70	1	20.6	3.7	8.6	8.2
Other molluscs	0.54	1.11	3.3	0.6	1.4	1.3
Remaining zoobenthos	0.56	6.05	13.2	2.4	5.1	5.7
Plant debris				90.6		
Faeces [c]				217.0		
Carcasses and remains [d]				22.4		
Ichthyofauna:						
Larvae and fry of planktivorous fish	0.034	5	1.36	0.17	0.51	0.68
Juvenile and adult planktivorous fish	0.440	0.733	1.89	0.32	1.21	0.38
Larvae and fry of benthivorous fish	0.202	5	8.11	1.01	3.04	4.05
Juvenile and adult benthivorous fish	2.600	0.733	13.61	2.01	7.62	4.08
Larvae and fry of piscivorous fish	0.082	5	3.24	0.38	1.23	1.64
Juvenile and adult piscivorous fish	1.055	0.6	3.40	0.55	2.54	0.35
Birds:						
Herbivorous birds	0.0022		0.163	0.00098	0.064 [e]	0.098
Molluscivorous birds	0.002		0.149	0.00090	0.059 [e]	0.090
Omnivorous birds	0.000043		0.002	0.000024	0.0016 [e]	0.0004
Piscivorous birds	0.0017		0.202	0.00273	0.179 [e]	0.020

[a] in addition to the autochthonous population shown in the table, the model included also the input of mysids from the Pomeranian Bay, estimated at 0.65 gC m^{-2} yr^{-1}

[b] respiration attributed to the submerged part of plants

[c] total: sedimented from the water column as well as derived from periphyton, benthos, fish and birds

[d] total: sedimented from the water column as well as derived from periphyton, benthos and fish

[e] outside the aquatic environment

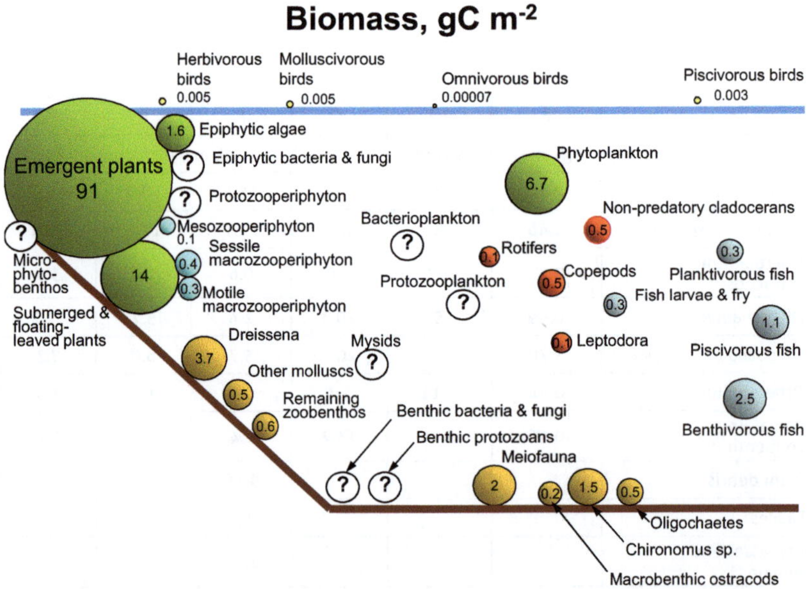

Figure 2.7 Biomass of various components of the Great Lagoon food web.

production, and respiration were, both in the water column and in the sediment, higher than those of all the other food web components. Out of about 1000 gC m^{-2} yr^{-1} organic matter supplied to the system and available to consumers, heterotrophs respired about 580 gC m^{-2} yr^{-1}. Bacteria and fungi accounted for more than 60% of that, further 10% being used by protists. Thus, metazoan animals utilised less than 30% of the organic matter supply respired by heterotrophs.

The role of benthic and planktonic communities in the overall matter and energy flow in the ecosystem was similar. Benthos and plankton were found to be responsible for utilisation of 46 and 47% of the organic matter used up the heterotrophic part of the biota, respectively, the periphyton and fish utilising 4 and 3%, respectively. The higher biomass of the benthic organisms was counterbalanced by a higher metabolic rate of the plankters.

From the standpoint of fisheries management in the Lagoon, it may be interesting to compare production of benthic invertebrates and plankters (Fig. 2.8), as both groups constitute the basic food of non-predatory fish. Production of the metazoan zooplankton in the Great Lagoon was estimated at about 25 gC m^{-2} yr^{-1}, production of the zoobenthos amounting to about 37 gC m^{-2} yr^{-1}. After deducting the production of rotifers and meiobenthos,

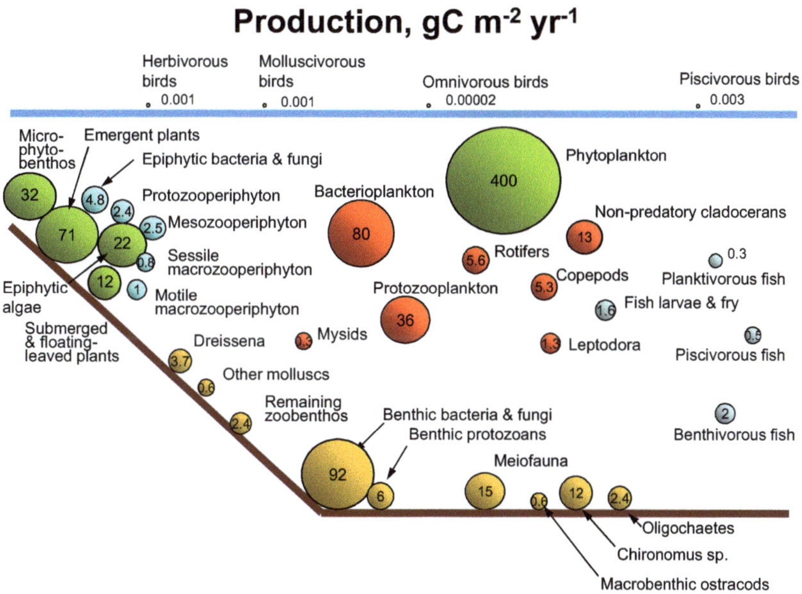

Figure 2.8 Production of various components of the Great Lagoon food web.

which owing to their small size, do not contribute much to the fish food, production of the crustacean zooplankton and the macrozoobenthos may be estimated at about 20 and about 22 gC m^{-2} yr^{-1}, respectively. As the two values are close, the substantial (several-fold) domination of landings by benthivorous fish species over planktivorous ones may be puzzling. The apparent contradiction is explained by the fact that a considerable part of production of the crustacean zooplankton is consumed by larvae and fry of all the fish species, including benthivores and predators. The amount of food consumed by larvae and fry of all the fish species was almost 10 times that consumed by adult and juvenile planktivores. An additional effect is the zooplankton export to the Pomeranian Bay. As mentioned above, the export accounted for 10 to 30% of the crustacean zooplankton production.

The Great Lagoon ecosystem produced about 420 gC m^{-2} yr^{-1} of non-living organic matter in the form of faeces and remains of plants and animals. These, together with allochthonous organic matter, constituted a substantial food resource for decomposers (Fig. 2.9). Detritus consumption in the Lagoon was estimated at about 590 gC m^{-2} yr^{-1} and was similar to the biomass consumption estimated at about 570 gC m^{-2} yr^{-1}. The detritus consumption was dominated (about 75%) by bacteria.

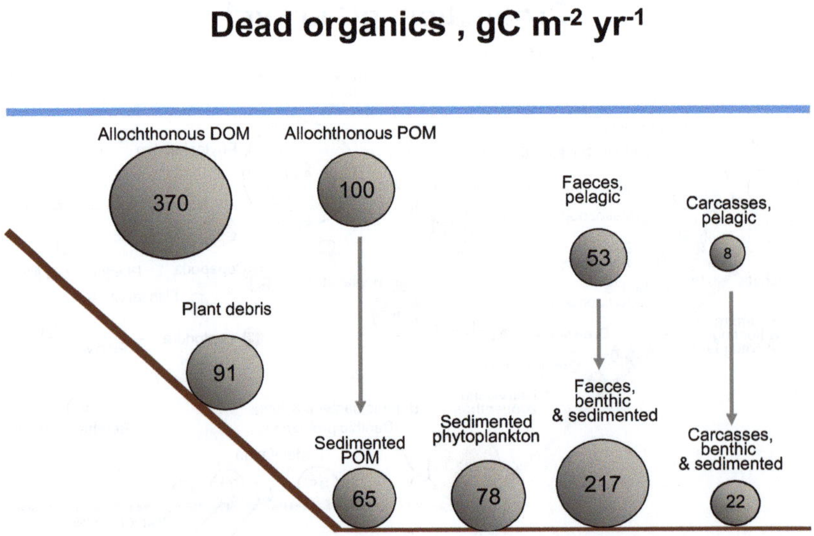

Figure 2.9 Input and production of non-living organic matter in the Great Lagoon. Note internal detritus fluxes between certain components due to sedimentation.

2.3.9 Composition of Food Consumed by Individual Components of the Lagoon Biotic Community

The food composition as shown by the model, is a result of selecting the values of F_{ij} and production of individual food web components, adopted for the purpose of modelling. It was in a few cases only (*Chironomus* sp., fish) that the food composition as defined by the model could be compared with the empirical data from the Great Lagoon. However, even in those cases the verification could not be deemed decisive due to the presence of numerous factors affecting results of food composition studies (year and season, area, age of animals examined, identifiability of diet items and their amenity to weighing, etc.). Thus, the food composition of ecosystem components determined by the model should be treated with utmost caution and regarded as hypothetical only, and in need of further verification. On the other hand, it does not seem possible to obtain exhaustive information on animal feeding solely by analysing alimentary tract contents. It cannot be expected that diet items such as protists, bacteria, naked rotifers, and other small invertebrates, organic remains, etc. would be correctly identified and quantified by microscopic examination only. Moreover, it does not seem possible to analyse the food of all the trophic web components in synchrony. Thus, determination of food composition of all the trophic web components by means of a model in which energy budgets of all the components are balanced may be treated as a valuable method with which to complement empirical studies. Application of the stable isotope technique would be extremely helpful for the analysis of trophic relationships (Lajtha and Michener, 1994; Fry, 2006), but the stable isotope-based studies are only at their infancy in Poland (Sokołowski, 2009) and have not been so far carried out in the Great Lagoon.

When estimating food composition, it had to be remembered that numerous components of gut contents, invisible or rapidly digested (particularly the microorganisms) had to be consumed in passing, e.g., when an animal was grazing on the phytoperiphyton or microphytobenthos. The food of suspension feeders consists not only of an appropriate seston fraction, e.g., the sedimenting phytoplankton, but contains also a rich bacterial flora decomposing the phytodetritus. When feeding on plants, herbivorous birds take up also the periphyton, including invertebrates. Thus the data on food composition of the aquatic organisms (based mainly on Monakov, 1974; 1998) had to be appropriately adjusted.

In the model, the bacterioplankton fed on allochthonous DOM, allo- and autochthonous POM as well as phytoplankton-derived food, the three items being taken up in more or less similar proportions (Table 2.13). The food of protists, rotifers, non-predatory cladocerans, and copepods was dominated (about 50% or more) by the phytoplankton. The bacterioplankton was an important diet item for protists, rotifers, and cladocerans (15-25% contribution to the diet). The third (in addition to the phyto- and bacterioplankton) major food item for rotifers, cladocerans, and copepods were protists. Zooplankton, primarily non-predatory cladocerans, contributed 85% to the food of the predatory cladoceran *Leptodora kindti*. Meiobenthic animals could be an additional food item for *L. kindti*, which the cladoceran could pick up from the sediment surface. Mysids showed the most diverse food. The major components were microalgae which contributed more than half of the food and were collected from the near-bottom water layer (sedimenting phytoplankton), from the sediment surface (microphytobenthos), from the water column (phytoplankton), and from aquatic vegetation (phytoperiphyton). The animal food (primarily the meiobenthos) accounted for about a quarter of the food, the remainder consisting of particulate organic matter of diverse origin.

In the periphyton, all the components except the sessile macrozooperiphyton derived most of their food from the periphyton itself (Table 2.14). The major food sources for periphytic bacteria and fungi included phytoperiphyton and dead macrophyte tissues. Food of pelagic origin contributed less than 10% of the entire food. Periphytic protists fed primarily on phytoperiphyton and periphytic bacteria and fungi, a few per cent of their food only being obtained from the water column. According to the model, the mesozooperiphyton fed primarily on the phytoperiphyton, protists, and dead macrophyte tissues. About 30%

Table 2.13 Food composition (%) of the planktonic components of the food web.

FOOD ITEM: \ CONSUMER:	Bacterioplankton	Protozooplankton	Rotifers	Non-predatory cladocerans	Copepods	*Leptodora*	Mysids
Planktonic:							
Allochthonous POM	8.2	3.0	0.3	0.7	0.2		0.7
Allochthonous DOM	33.4						
Phytoplankton	32.9	60.9	48.6	50.0	55.3		7.8
Bacterioplankton		24.5	14.6	24.6	5.3		
Protozooplankton		6.6	33.1	18.0	23.9		
Rotifers			0.9		3.7	22.1	5.5
Non-predatory cladocerans					2.6	49.1	4.2
Copepods					0.7	13.7	1.7
Leptodora							1.3
Faeces, pelagic	20.1	3.7	1.9	5.1	4.0		0.8
Carcasses, pelagic	5.4	1.2	0.6	1.6	3.8		1.9
Epiphytic:							
Epiphytic algae							1.4
Epiphytic bacteria & fungi							0.4
Protozooperiphyton							0.2
Mesozooperiphyton							0.2
Benthic:							
Sedimented allochthonous POM							0.4
Sedimented phytoplankton							25.3
Microphytobenthos							20.5
Meiofauna					0.5	15.0	15.1
Remaining zoobenthos							0.8
Plant debris							5.9
Faeces							4.2
Carcasses and remains							1.5

Table 2.14 Food composition (%) of the epiphytic components of the food web.

FOOD ITEM: \ CONSUMER:	Epiphytic bacteria & fungi	Protozooperiphyton	Mesozooperiphyton	Sessile macrozooperiphyton	Motile macrozooperiphyton
Planktonic:					
Allochthonous POM	1.4	1.1	1.3	1.5	0.7
Allochthonous DOM	2.6				
Phytoplankton		4.4	22.4	6.4	19.7
Bacterioplankton		2.2	1.1	2.4	
Protozooplankton				10.9	
Rotifers				6.7	
Non-predatory cladocerans				5.8	
Copepods				2.4	
Faeces, pelagic	4.3	8.1	4.0	1.8	
Carcasses, pelagic	1.0	2.6	1.0	2.6	
Epiphytic:					
Epiphytic algae	56.5	43.4	37.7	3.3	27.8
Epiphytic bacteria & fungi		34.4	5.4	2.9	6.8
Protozooperiphyton		3.9	13.6	3.5	0.7
Mesozooperiphyton			1.6	29.6	19.1
Sessile macrozooperiphyton					4.5
Motile macrozooperiphyton				7.4	3.5
Benthic:					
Emergent plants	2.5				
Submerged & floating-leaved plants	0.4				3.3
Microphytobenthos					1.1
Remaining zoobenthos				13.2	4.6
Plant debris	31.4		11.9		8.4

of the mesozooperiphyton food were obtained from the water column (mainly the phytoplankton). The motile macrozooperiphyton fed on phytoperiphyton, phytoplankton, mesozooperiphyton, live and dead macrophyte tissues, and to some extent also on periphytic bacteria and fungi as well as macrozooperiphytic animals. Pelagic components (phytoplankton) accounted for about 20% of the motile macrozooperiphyton food. On the other hand, the food of the sessile macrozooperiphyton represented primarily by the predatory hydrozoan *Cordylophora caspia*, consisted largely of zooplankters. Cladocerans, copepods, rotifers and planktonic protists contributed jointly more than 25% of the food. Further several per cent were contributed by the phytoplankton, bacterioplankton, and non-living seston. Of the periphytic components, the most important was the mesozooperiphyton. Motile littoral and macrozooperiphytic invertebrates could be important (more than 20%) in the diet as well. Taken together, the animal food (protists and invertebrates) accounted for almost 80% of the sessile macrozooperiphyton food.

In the benthos (Table 2.15), the sediment-dwelling bacteria and fungi fed mainly on the autochthonous organic matter in the sediment, primarily faeces and phytodetritus.

In their turn, benthic bacteria and fungi were fed on primarily by the benthic protists and meiofauna. Sedimented phytoplankton and detritus was an additional, important food item for protists and meiobenthos. Those items were consumed also by the pelophilous macrobenthos. Two-thirds of the food of larval *Chironomus* sp. consisted of the phytoplankton, mainly the algae that sedimented on the bottom from the water column (Wolnomiejski et al., 2000). According to the model, the food of the macrobenthic ostracods and oligochaetes was more diverse, but the detritus, benthic bacteria and fungi, and microphytobenthos were the major constituents as well, and the meiofauna being an additional food source. The littoral macrozoobenthos relied on the pelagic and periphytic food to a much larger extent than the pelophilous macrofauna did. The zebra mussel consumed organic seston half of which was the phytoplankton. About one-third of the food of other molluscs was derived from the water column (mainly the phytoplankton), most of the food being collected from the bottom (mainly the microphytobenthos). The food of the remaining littoral zoobenthos was most diverse and consisted primarily of microphytobenthos, phytodetritus, meiobenthos, and phytoperiphyton.

According to the model, the zooplankton was the dominant item in the food of planktivorous fish and all the fish larvae and fry (Table 2.16). Almost all the fish groups consumed substantial amounts of *Chironomus* sp. larvae and pupae. Larval *Chironomus* sp. accounted for more than half of the food of juvenile and adult benthivorous fish. The larvae were the major food item for bream and ruffe in the Great Lagoon (Wolnomiejski and Grygiel, 1994c; 2002). Those fish not adapted to collecting their food from the bottom can catch the pupae of *Chironomus* sp. and other insects in the water column when the

Table 2.15 Food composition (%) of the benthic components of the food web.

CONSUMER: / FOOD ITEM:	Benthic bacteria and fungi	Benthic protozoans	Meiofauna	*Chironomus* sp. larvae	Macrobenthic ostracods	Oligochaetes	*Dreissena polymorpha*	Other molluscs	Remaining zoobenthos
Planktonic:									
Allochthonous POM				0.2			4.9	2.1	
Phytoplankton				13.9			46.7	19.2	
Bacterioplankton							9.8	3.9	
Protozooplankton							6.2	1.7	
Rotifers							1.4	0.3	
Non-predatory cladocerans							0.6		
Copepods							0.3		
Faeces, pelagic							6.0	5.9	
Carcasses, pelagic							8.5	4.4	
Epiphytic:									
Epiphytic algae								1.3	7.6
Epiphytic bacteria and fungi								0.1	0.7
Protozooperiphyton								0.1	0.2
Mesozooperiphyton								0.1	0.2
Sessile macrozooperiphyton									0.01
Motile macrozooperiphyton									0.01
Benthic:									
Sedimented allochthonous POM	11.8	9.3	4.8	2.7	1.3	9.6	0.6	1.4	2.3
Sedimented phytoplankton	8.4	18.4	11.5	54.2	7.8	7.7	15.1	7.7	4.5
Submerged and floating-leaved plants									1.7
Microphytobenthos	0.1	3.8	8.5	6.3	26.9	11.8		21.1	23.0
Benthic bacteria and fungi		43.5	48.7	5.6	18.3	30.2		9.6	7.2
Benthic protists		1.4	3.2	0.6	6.0	1.5			2.2
Meiobenthos		3.7	0.6	9.5	13.9	12.2		8.3	13.3
Chironomus sp.									0.8
Macrobenthic ostracods									0.2
Oligochaetes				0.05	2.4				5.2
Dreissena polymorpha									2.7
Other molluscs									0.4
Remaining zoobenthos									5.3
Plant debris	19.6	4.3	6.7	2.8	3.6	4.5		4.1	13.0
Faeces	54.7	10.3	14.4	3.6	15.2	20.3		7.8	1.6
Carcasses and remains	5.5	5.3	1.7	0.2	4.5	2.2		1.0	8.0

Table 2.16 Food composition (%) of fish.

FOOD ITEM: \ CONSUMER:	Larvae and fry of planktivorous fish	Juvenile and adult planktivorous fish	Larvae and fry of benthivorous fish	Juvenile and adult benthivorous fish	Larvae and fry of piscivorous fish	Juvenile and adult piscivorous fish
Planktonic:						
Phytoplankton	0.3		0.5		0.4	
Rotifers	26.9	8.9	7.4		8.1	
Non-predatory cladocerans	47.7	20.5	36.8	4.8	39.6	6.6
Copepods	9.7	5.8	19.4	1.9	12.2	2.7
Leptodora	2.8	4.8	1.7	4.3	2.0	0.7
Mysids	1.0	37.0	0.2	0.8	9.3	3.0
Epiphytic:						
Epiphytic algae	0.2	0.1	1.4	0.1		
Epiphytic bacteria and fungi	0.1		0.1			
Protozooperiphyton	0.1					
Mesozooperiphyton	0.2					
Sessile macrozooperiphyton	0.3		0.2	0.2	0.1	
Motile macrozooperiphyton	0.4		1.0	3.5	0.8	1.7
Benthic:						
Sedimented phytoplankton	0.3			0.1	0.2	
Submerged and floating-leaved plants	0.2	0.3	1.4	1.7		
Microphytobenthos	0.1	0.1	1.6	1.4		
Benthic bacteria and fungi				0.3		
Benthic protists				0.2		
Meiofauna	2.3	0.4	8.8	5.7		
Chironomus sp. larvae	3.4	14.2	10.7	58.1	18.6	10.7
Macrobenthic ostracods	0.2	0.6	3.3	1.9	0.2	
Oligochaetes	0.2	0.4	1.8	2.1	0.4	0.1
Dreissena polymorpha	0.1	0.4	0.2	4.1		0.1
Other molluscs		0.2	0.8	1.9	0.1	
Remaining zoobenthos	3.6	1.9	2.4	5.4	1.9	2.1
Plant debris				0.7		
Carcasses and remains			0.3	0.8		0.1
Fish:						
Larvae and fry of planktivorous fish		0.4			0.7	2.9
Juvenile and adult planktivorous fish						6.2
Larvae and fry of benthivorous fish		2.7			4.6	16.9
Juvenile and adult benthivorous fish						35.9
Larvae and fry of piscivorous fish		1.0			0.9	7.5
Juvenile and adult piscivorous fish						3.1

pupae are migrating upwards toward the water surface. Of other food items for benthivorous fish, those most important were molluscs and other littoral invertebrates as well as the meiobenthos, consisting mainly of ostracods. Fish accounted for about three quarters of the diet of juvenile and adult predatory fish, the zooplankton and benthos (mainly the larval *Chironomus* sp.) contributing several per cent each. In the Great Lagoon, those components are important in the food of juvenile perch (Wolnomiejski and Grygiel, 1994b). According to the model, periphytic organisms did not play any major role in fish food; they contributed as little as about 3% to the diet of benthivorous fish (Table 2.16). This category of organisms was not accounted for at all in the empirical studies on food composition of the Great Lagoon fish. The macrozooperiphyton consists mainly of invertebrate taxa occurring also in the macrobenthos, particularly insect larvae and molluscs. Practically, the provenance (macrobenthos or macrozooperiphyton) of a fish food item cannot be identified with any certainty.

The food of herbivorous birds consisted primarily of soft macrophytes. As already mentioned, periphyton were most probably taken up together with them (Table. 2.17). The molluscivorous birds feed primarily on animal

Table 2.17 Food composition (%) of birds.

CONSUMER: / FOOD ITEM:	Herbivorous birds	Molluscivorous birds	Omnivorous birds	Piscivorous birds
Planktonic:				
Mysids	8.9	4.0	50.7	
Epiphytic:				
Epiphytic algae	2.7	0.7		
Epiphytic bacteria and fungi	0.6	0.2		
Protozooperiphyton	0.3	0.1		
Mesozooperiphyton	0.3	0.1		
Sessile macrozooperiphyton	0.5	0.2		
Motile macrozooperiphyton	0.9	0.1		
Benthic:				
Submerged and floating-leaved plants	74.7	8.6		
Dreissena polymorpha		49.6		
Other molluscs		8.9		
Remaining zoobenthos		24.6		
Plant debris	11.1	3.0		
Fish:				
Larvae and fry of planktivorous fish			4.2	7.6
Juvenile and adult planktivorous fish				11.1
Larvae and fry of benthivorous fish			39.4	30.5
Juvenile and adult benthivorous fish				39.7
Larvae and fry of piscivorous fish			5.6	11.2

food, more than 80% of their diet being accounted for by invertebrates, particularly the zebra mussel. The omnivorous birds fed, in addition to the food obtained on land, on mysids and fish larvae and fry. Their diet relied primarily on organisms or waste found on land (not included in the table) as well as on terrestrial insects found in water. The piscivorous birds fed exclusively on fish; half of their food consisted of fry, while juvenile and adult fish made up the other half.

2.3.10 Trophic Levels

The trophic position (trophic level, TL) of each consumer category in the Great Lagoon food web was calculated as a weighted average of trophic levels of all constituents of its food increased by 1. Non-living organic matter, similarly to all the primary producers, were assigned to trophic level 1.

The majority of the food web components were placed between trophic levels 2 and 3 (Fig. 2.10). This TL range was occupied by herbivorous birds, protists, and almost all the invertebrate components. Molluscivorous birds, sessile macrozooperiphyton, *Leptodora* as well as plankti- and benthivorous fish were placed between trophic levels 3 and 4. The top consumers (TL between 3.9 and 4.5) consisted of omni- and piscivorous birds as well as predatory fish.

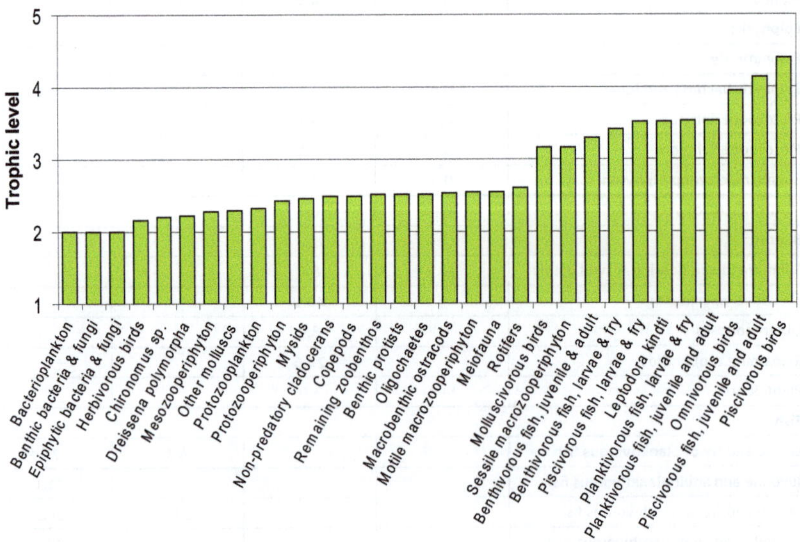

Figure 2.10 Trophic level of consumers in the food web.

2.3.11 Biogeochemical Macrofluxes

The model is a good tool with which to calculate various biogeochemical fluxes (Table 2.18). Some of the calculated fluxes may be useful for comparisons with empirical measurements carried out at the ecosystem level, and for this reason they may be helpful also for the quality assessment of the model and for its improvement. Other fluxes or interactions between them are used to characterise the ecosystem and may be applied in the between-ecosystems comparisons.

In addition to the primary production, it is relatively easy to determine empirically fluxes such as pelagic and benthic community respiration, sedimentation, deposition in sediment, and export from the system, but only few of those fluxes have been quantified in the Great Lagoon.

The pelagic community respiration in the Great Lagoon was measured in the 1960s by Wiktor (1971) when she determined primary production using the light and dark bottle-based oxygen method. The respiration that may be estimated from her measurement was about 750 gO_2 m^{-2} yr^{-1}, which is approximately equal to 280 gC m^{-2} yr^{-1}. Considering the increase in eutrophication level of the

Table 2.18 Biogeochemical fluxes in the Great Lagoon. Model estimates.

	gC m^{-2} yr^{-1}
Total gross primary production	657.5
Total net primary production	537.5
Supply of allochthonous organic matter	470.0
Total system throughput[a] (TST)	2638.3
Total respiration	704.5
Pelagic community respiration[b]	373.9
Benthic community respiration[c]	278.2
Sedimentation	203.7
Deposition in sediment	74.9
Catches/Landings	0.56
Export	348.5

[a] sum total of consumption, export (outflow, emergence, catches), respiration, and transformation to detritus of all components
[b] respiration of phyto-, bacterio-, protozoo- and metazoan zooplankton (without mysids)
[c] respiration of microphytobenthos, benthic bacteria, fungi, protists, meio-, and macrozoobenthos (without respiration of macrophytes and periphyton)

Great Lagoon taking place from the 1960s until the end of the 20th century, the pelagic community respiration estimated by the model (about 370 gC m^{-2} yr^{-1}) can be regarded as very plausible.

Regarding the deposition in sediment, Leipe et al. (1998) as well as Meyer and Lampe (1999) demonstrated the lack of recent increase in the sediment thickness in the Szczecin Lagoon and explained this lack by sediment resuspension and removal, the processes counterbalancing sedimentation and matter input. However, maintenance of the navigability of the fairway intersecting the Lagoon calls for continuous dredging, which results in an annual dredge spoils of about 1 million m^3 deposited on land (Maritime Office in Szczecin, unpubl. data). Assuming that 1 m^3 of the sediment dredged out contains about 200 kg dry matter, and that 1 kg dry sediment is equivalent to 107 $gC_{org.}$ (Minning, 2004), the amount of organic carbon removed by dredging can be estimated at 52 gC m^{-2} yr^{-1}. Therefore, the deposition estimated by the model (about 75 gC m^{-2} yr^{-1}) seems realistic as well.

The lack of any other empirical studies on biogeochemical fluxes in the Great Lagoon prevents any more comprehensive verification of model estimates. Therefore, the need for such studies is very urgent, both for a better understanding of the processes themselves and for the assessment, and improvement, of the performance of models describing the ecosystem.

2.3.12 Ecological Efficiency

There are numerous ways with which to characterise the trophic structure of ecosystems or their individual components, and which utilise various terms of energy budgets. The simplest indicators of ecosystem maturity were proposed by Odum (1969; 1971). More complex tools, rooted in information theory and utilising computer techniques, constitute the so-called ecosystem network analysis (Ulanowicz, 1986; Wulff et al., 1989). Their application is progressively increasing in ecological modelling (Christensen, 1995; Christensen et al., 2008). This section highlights only the simplest indicators, easy to calculate with the model results discussed above.

The ecotrophic efficiency (*EE*) is an indicator characterising the degree to which production of various ecosystem components is utilised in this ecosystem:

$$EE = (M_2 + Exp + Acum + Y) / P = 1 - M_0,$$

where M_2 is mortality due to consumption; *Exp* is export (removal, emergence); *Acum* is accumulation in the ecosystem; *Y* denotes fish catches; and M_0 is natural mortality.

Among autotrophs, the highest ecotrophic efficiency was – according to the model –typical of microalgae, primarily the phytoperiphyton (Fig. 2.11). The phytoperiphyton ecotrophic efficiency of more than 0.95 may raise some reservations, as field observation show that, at the end of the growing season, much of the periphytic biomass is deposited on the bottom where it most likely dies off and enters the non-living organic matter pool. This process, a form of natural mortality, has not been quantified so far, but considering the relatively high periphyton biomass at the end of the growing season, it is likely that it exceeds 0.04 (4%) of the phytoperiphyton production suggested by the model. The relatively low *EE* (about 0.75) of the phytoplankton seems plausible, as it is related to sedimentation some of the phytoplankton biomass to the bottom and its transfer to the deposited phytoplankton pool. Although observations show some of the deposited phytoplankton cells to be viable for some time (Wolnomiejski et al., 2000), the deposited phytoplankton was in the model classified with the non-living organic matter, as most of the phytoplankton reached the bottom of the relatively deep central part of the Lagoon where there are no conditions for photosynthesis. The very low ecotrophic efficiency of macrophytes, particularly the emergent ones, is also consistent with expectations.

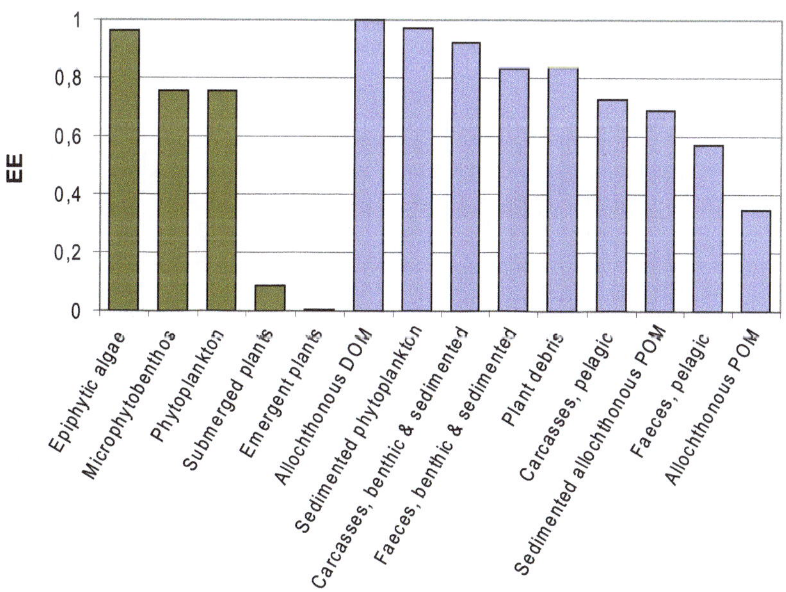

Figure 2.11 Ecotrophic efficiency (*EE*) of autotrophs and non-living organic matter.

Of the components representing non-living organic matter, the highest utilisation in the ecosystem was typical of the autochthonous components, particularly the deposited phytoplankton and animal remains (Fig. 2.11). The high (1.0) *EE* of the allochthonous DOM resulted from the export of the total pool of unused DOM. In reality, as little as 22% of the DOM input to the Great Lagoon was consumed. The allochthonous POM, too, was poorly utilised (*EE* of 0.35) while in the water column. Only when it sedimented to the bottom, its utilisation increased.

Most of heterotrophic components showed a high ecotrophic efficiency, exceeding or close to 0.9 (Fig. 2.12). Particularly high *EEs* were shown for periphytic components (the mesozoo- and protozooperiphyton as well as periphytic bacteria and fungi), which –in the light of the comments above pertaining to the phytoperiphyton – allows to conclude that either the selection of F_{ij} values for the periphyton or the production of some periphytic components entered into the model had not been appropriate. The drawback here was the lack of empirical data on zooperiphyton feeding, including the contribution of pelagic components to the food of the periphytic ones, as well as a paucity of research on periphyton production rate. The ecotrophic efficiency of the bacterio- and protozooplankton as well as benthic bacteria, fungi, and protists (about 0.95) stems also from assumptions adopted by the authors without any empirical backing.

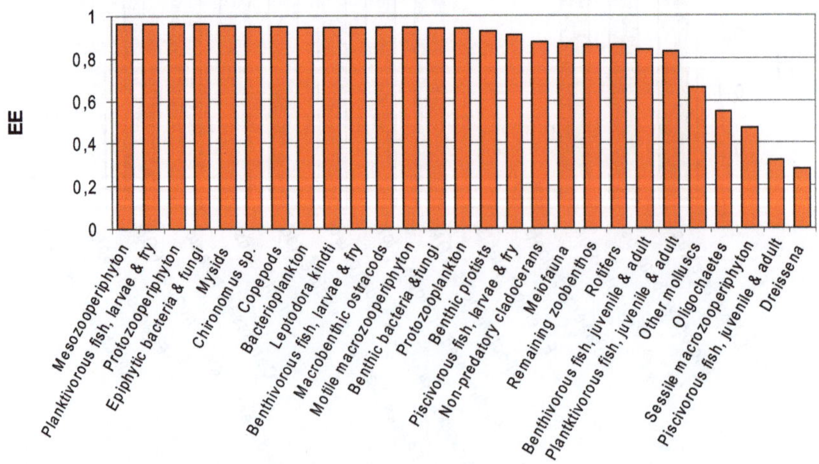

Figure 2.12 Ecotrophic efficiency (*EE*) of heterotrophs.

Six components showed somewhat lower ecotrophic efficiencies, between 0.9 and 0.8. Those components included cladocerans, rotifers, the meiobenthos, littoral benthos (other than molluscs) as well as plankti- and benthivorous fish. In the case of invertebrates (the first four components), reservations identical to those put forth for the periphyton can be presented, namely insufficient knowledge on the feeding of potential consumers of those categories as well as inadequate estimates of their production rates. On the other hand, somewhat lower *EEs* of adult and juvenile plankti- and benthivorous fish stemmed from assigning their reproductive materials to production destined to the detritus.

Low ecotrophic efficiency (0.3-0.7) was shown by molluscs (including the zebra mussel) and the sessile macrozooperiphyton, which can be explained by their effective protection from predation in the form of shells and external skeletons. The relatively low *EE* of oligochaetes (0.55) is difficult to explain. Perhaps the oligochaetes, living among the sediment particles, are protected from predation better than other organisms and succumb to natural mortality much more than other animals. Most of the oligochaete biomass is confined to the muddy bottom where they are available only to the obligatory benthophages, mainly bream. On the other hand, bream show a distinct preference towards feeding on large, more caloric pelophilous larvae of *Chironomus* which are much more abundant in the sediment. However, adoption of a too high oligochaete production in the model cannot be ruled out.

Particularly low ecotrophic efficiency was shown for the predatory fish. As already mentioned, this is most likely an artefact produced by underestimation of the true catches of those fish.

The most straightforward indicators characterising the ecosystem maturity is the net system production (GPP-R) and the primary production to respiration ratio (GPP/R) (Odum, 1971). The primary production to respiration ratio of the Great Lagoon close to unity (Table 2.19) evidences the ecosystem to have reached a high level of maturity. The negative net system production results from a substantial supply of organic matter from the outside. The absolute value of net production is low and corresponds to 7% of GPP.

Certain ecological indicators used to characterise ecosystems pertain to total biomass of the entire biota. However, not all the components of the Great Lagoon biota were amenable to biomass determination. To provide at least crude estimates of those indicators, the missing biomasses of the phytoplankton, bacteria, and protists were approximated in a simplified manner, assuming their P/B values at 75, 50, and 150 yr^{-1} (as in Table 2.5). Thus, the total biomass of the Great Lagoon biota was estimated at about 143 gC m^{-2}. An example of an indicator related to the total biomass is B/TST, i.e., biota biomass referred to unit ecosystem energy flow. It is believed that the B/TST

Table 2.19 Ecological indicators characterising the Great Lagoon ecosystem.

Net system production, GPP-R, gC m^{-2} yr^{-1}	- 47
Total GPP/R, unitless	0.933
Total biomass, B gC m^{-2}	142.7
Total NPP/B, yr^{-1}	3.77
Total biomass/total system throughput, yr	0.054
Detritus consumption/biomass consumption, unitless	1.03
Efficiency of fisheries, Y/NPP, unitless	0.001
Fisheries yield, tonnes ha^{-1}	56

increases with the increasing ecosystem maturity (Christensen et al. 2005). In the Great Lagoon, the annual estimate of B/TST was about 0.05. According to Christensen et al. (2005), the increasing maturity of the ecosystem may be expected to be also accompanied by an increasing primary production to total biomass ratio (NPP/B). In the Great Lagoon, the ratio was about 3.8 yr^{-1}. In the Lagoon, the value of the two indicators is strongly affected by the presence of macrophytes the biomass of which accounted for about 80% of the total biota biomass. Under such conditions, the utility of ecosystem maturity indicators based on the total biomass seems doubtful, as they could lead to contradictory conclusions. Particularly the later index (NPP/B) appears questionable. For example, the absence of macrophytes does not have to be a sign of ecosystem maturity, and their abundance does not have to signify the immaturity of an ecosystem.

As already mentioned in Section 2.3.8, consumption of non-living organic matter in the Great Lagoon was very similar to the biomass consumption, the ratio between the two being 1.03.

The fishing efficiency of ecosystem can be expressed in a variety of ways. Catches amounting to 0.56 gC m^{-2} yr^{-1}, equivalent to 56 kg of fish caught from a hectare during the year, evidence a high production available to fisheries. With respect to the net primary production, the fishing efficiency was 0.1%. Considering that the fish catches in the Great Lagoon were most probably underestimated, the true value of the indicators pertaining to the fishing efficiency could be higher than those reported in Table 2.19.

2.4 Discussion

2.4.1 The Model Concept Problem

When setting off to develop a food web model, a researcher has to decide on the number and identification of functional groups, and on the way the equations in the model will be linked. The functional group identification for the purposes of a food web model will be understandably based on feeding mode of an organism (organisms) and the role of the latter as a food for other consumer groups. Important are also considerations such as the habitat occupied by an organism (water column, sediment, littoral zone, other organisms, etc.) as well as taxonomic affinity and size, translating into biological traits and metabolic rates. An additional subjective factor affecting the model structure is the level of knowledge of biology of individual organism categories and personal interests of a researcher. The number of model components is usually a trade-off between the required level of detail of the model and its operating capacity. A low number of components oversimplifies the trophic web structure, whereby individual characteristics of the ecosystem become obliterated and the individual components, combining very different species, become very abstract units the bioenergetic parameters of which require summing up or averaging, and may change depending on proportions in the species composition of a component. On the other hand, a high number of components identified, in addition to increasing the number of terms in the model, increases – in a geometric progression – the number of possible interactions between the components, thereby complicating the control of the system being modelled and extending the time necessary for preparing successive scenarios.

The 45 components of the Great Lagoon food web model, including 36 live components, is a very high number for mass-balance models of aquatic ecosystems, as can be seen from the review of a number of relevant publications which appeared during the recent two decades (Fig. 2.13). Particularly the number of components involving non-living organic matter (9) is higher than that in other models (usually 1-3). In the Great Lagoon model presented, the number of species included in different live components varies greatly. Some components consist of a high number of species, or even higher taxa (phytoplankton, meiobenthos, bacteria, protists) which may be even linked by significant within-group trophic interactions (e.g., protists). Other components are made up by a few taxonomically related species (e.g., copepods, ostracods), still others consisting of a single species each (*Leptodora kindti*, *Chironomus* sp., *Dreissena polymorpha*). The reason underlying such was, on the one hand, inadequate knowledge on some groups of organisms and the necessity of reducing the number of components, and – on the other – the intention of having a closer look at selected taxa, particularly benthic invertebrates, fish, and birds. In the case of fish, a group of organisms very important for the ecosystem functioning and undergoing a particularly extensive ontogenetic size changes accompanied by changes in food composition and feeding mode, different life stages were assigned to different food web components (larvae and fry *versus* juveniles and adults).

When constructing a food web model, energy budget components may be linked together starting from the bottom of the trophic pyramid bottom (a 'bottom-up' approach), whereby the production is transferred, as food, to consumers at higher levels, according to the energy flow direction. This approach was adopted in, e.g., food web models of a number of estuaries (Baird and Ulanowicz, 1993) and Atlantic US coastal ecosystems (Link et al., 2006; the EcoNetwrk model). This is the approach adopted in this work as well. Another possibility is to arrange the equations starting from the top of the trophic pyramid (a 'top-down' approach)

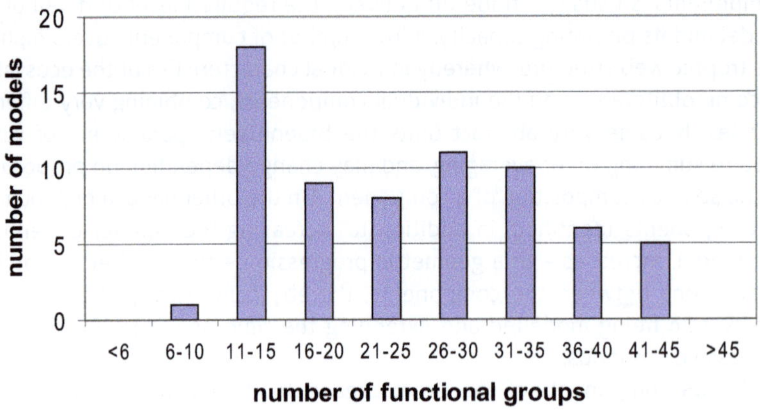

Figure 2.13 Number of food web functional groups in 68 aquatic ecosystem models presented in 48 publications.

Data sources: Angelini and Agostinho, 2005; Arreguin-Sanchez et al., 2003; Baird and Ulanowicz, 1993; Barausse et al., 2009; Bradford-Grieve et al., 2003; Cerco et al., 2010; Chen et al., 2008; Coll et al., 2006; Cornejo-Donoso and Antezana, 2008; Cruz-Escalona et al., 2007; Darwall et al., 2010; Duan et al., 2009; Fetahi and Mengistou, 2007; Guenette et al., 2008; Harvey et al., 2003; Hauputhantri et al., 2008; Heymans and Baird, 2000; Heymans et al., 2007; Kissler, 2009; Kitchell et al., 2000; Lachica-Alino et al., 2006; Libralato and Solidoro, 2009; Lin et al., 1999; Lin et al., 2006; Link et al., 2006; 2009; Megrey and Aydin, 2009; Moloney et al., 2005; Monaco and Ulanowicz, 1997; Morales-Zarate et al., 2004; Moreau et al., 2001; Morissette et al., 2009; Osterblom et al., 2007; Patricio and Marques, 2006; Pinnegar et al., 2005; Rochette et al., 2009; Rosado-Solorzano and Guzman del Proo, 1998; Rybarczyk and Elkaim, 2003; Rybarczyk et al., 2003; Sandberg, 2007; Sandberg et al., 2000; Skurzak, 2009; Tomczak et al., 2009; Vega-Candejas and Arreguin-Sanchez, 2001; Vidal and Pauly, 2004; Villanueva et al., 2006; Villanueva et al., 2008; Zetina-Rejon et al., 2003 and to determine food demand of the successive levels of the system in the descending order. This is the way the popular ECOPATH model is constructed (Ecopath with Ecosim (EwE) software, Christensen and Pauly, 1992; Christensen et al., 2008), applied primarily to show the position of fish and fisheries in the ecosystem context (Christensen and Pauly, 1993, 2004; Vasconcellos et al., 1997 and many others; Fig. 2.14).

The two model types produce an identical final effect, i.e., a balanced set of equations describing energy (matter) flow in an ecosystem. However, each model type has its specific properties. The bottom-up models require the operator to enter information on the fate of production of each biotic component at the ascending order of trophic levels. The drawback is that the fate of production of individual food web components, and particularly the quantification of production utilisation by other components, are seldom targeted by direct studies, therefore the relevant values needed by the model are usually calculated indirectly or selected by the trial and error method.

The situation is somewhat easier in the top-down models, because – rather than utilisation of production – the operator has to enter the components' food composition, the food being usually studied directly, at least at the upper trophic levels (fish, birds). However, proceeding from selected top trophic components down the trophic pyramid to its bottom, in a direction opposite to that of energy flow, creates a risk of ignoring important trophic web links which may

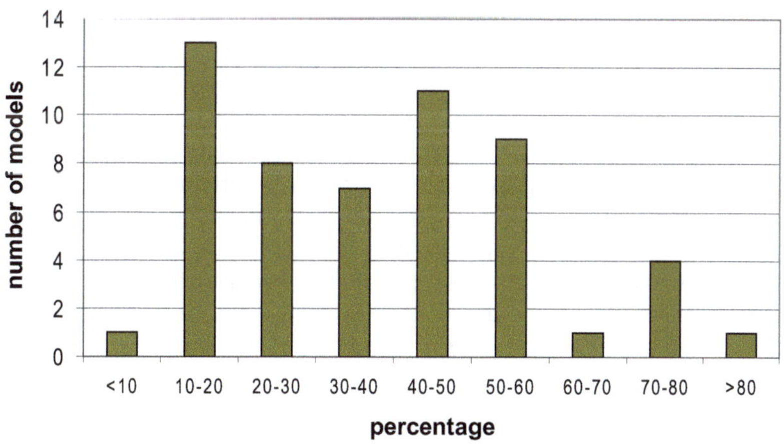

Figure 2.14 Percent contribution of fish components to the total number of food web components in 55 ecosystem models constructed with the ECOPATH software. In the Great Lagoon food web model, fish account for 13% of all the components. Data sources as in Fig. 2.13.

be bypassed by trophic chains leading to top predators. Most frequently such overlooked (or ignored) food web components are microorganisms (bacteria, protists) and the meiobenthos.

Although a competent researcher is capable of including all the important categories in either model type, a surprisingly high number of studies involving the ECOPATH programme ignored the components mentioned above. Only 12 out of the 41 inspected models, published in the recent two decades, included bacteria and protists, 14 including the meiobenthos. And yet, as shown by this study as well as by publications considering those components in the trophic web structure, microbes and meiobenthos may be responsible for 40-90% of respiration of the entire biota. Ignoring them in models seriously distorts different indicators developed to characterise the ecosystem, such as the total respiration, total system throughput, net system production, primary production to respiration ratio, total biomass of the biota, etc. Unfortunately, analyses and comparisons of ecosystems, based on such distorted indicators, have no merit whatsoever.

An additional shortcoming of ECOPATH, when it is used to modelling the entire ecosystem, is the fact that – in the case of primary producers – it uses the net production and consequently ignores the respiration of phytoplankton, phytobenthos, and phytoperiphyton. As a result, respiration of the pelagic or benthic (littoral) community cannot be directly compared with field data, as these include also respiration of plant components.

Although less experienced authors frequently do not pay enough attention to such details, a seasoned researcher is capable of dealing with those ECOPATH shortcomings and supplement or adjust the model accordingly. Due to its clarity, a user-friendly interface, a rich armoury of system analysis tools (e.g., the Network Analysis Plugin), and the general availability of the EwE software (**www.ecopath.org**), ECOPATH has become a standard in aquatic ecosystem modelling, particularly in fisheries-oriented models. Its great advantage is a possibility of switching from a mass-balanced static model (ECOPATH) to time dynamic (ECOSIM) and spatial (ECOSPACE) modelling. For this potential to be utilised to its full capacity, it is important that a starting, correctly constructed, mass-balance model, is available. Towards this end, a food web model developed with the use of the ECOPATH software is confronted with a bottom-up model, for example such as the model of the Great Lagoon food web described in this work.

2.4.2 The Quality and Availability of Empirical Data

To assess the quality of data sources used to develop the model described in this work, a system proposed by the User Guide of the ECOPATH with ECOSIM modelling software (Christensen et al., 2005) was used. Four data categories were assessed: biomass, P/B (Q/B), food composition, and catches. Individual parameters were assigned an index (from 0.0 to 1.0) depending on the origin and accuracy of the data used: from inferred to compilation to original (measurements and analyses conducted by the authors). The assessment criteria are shown in Table 2.20.

In the Great Lagoon, the best quality data were those on biomasses and fish catches (Table 2.21). Most of the data were derived from actual studies conducted in the Lagoon, although the scope of those studies not always allowed the assessment to be highly accurate. Unfortunately, there were no local biomass data on certain components (primarily microorganisms). The quality of P/B and food composition data scored in the assessment lower than the biomass did. Almost all the P/B values used were derived from literature compilations. The provenance of data on food composition of various food web components was very diverse. In the case of fish food, the data were collected in the *in situ* quantitative studies; the *in situ* qualitative studies supplied data on food composition of larval *Chironomus* sp., while the literature was the only source of information on feeding of the remaining components. With respect to food composition of invertebrates, particularly useful was the study of Monakov (1998). In all the data categories, the lack of information was most acutely felt with respect to microorganisms.

The review above allows to conclude that, to increase the accuracy and reliability of future estimates of matter/energy flows in the food web of the Great (Szczecin) Lagoon, particularly urgent are studies on biomass, production, and feeding of various groups of microorganisms inhabiting the area, and studies on feeding of planktonic and benthic invertebrates. It would be very useful to have access to current data on primary production of the phytoplankton, phytoperiphyton, and microphytobenthos in the Lagoon, as those groups – together constituting the base of the trophic pyramid – are the most important suppliers of easily assimilable organic matter in the ecosystem. Despite the high scores assigned to the phytoplankton, the data on its production used in the model were not collected during the period of study and were not entirely satisfactory in terms of methodology of their acquisition. Particularly useful would also be results of more comprehensive studies on the periphytic community and the littoral fauna. The relevant studies have been so far constrained by a low number of sites and a very short periods of study. Trophic links within those communities are far from understood as well.

The average collective score of all the four data categories was 0.421. It corresponds almost ideally to the mid-point (0.420) of the score range of 0.164-0.676 found by Coll et al. (2006) in their review of 50 applications of the ECOPATH model.

Table 2.20 A scale for classification of the quality of data sources used in the ECOPATH with ECOSIM model (Christensen et al., 2005).

Parameter	Score
Biomass	
Missing parameter (estimated by model)	0
From other models	0
Guesstimates	0
Approximate or indirect method	0.4
Sampling-based, low precision	0.7
Sampling-based, high precision	1
P/B and Q/B	
Missing parameter (estimated by model)	0
Guesstimates	0.1
From other models	0.2
Empirical relationship	0.5
Similar group/species, similar system	0.6
Similar group/species, same system	0.7
Same group/species, similar system	0.8
Same group/species, same system	1
Diet composition	
General knowledge of related group/species	0
From other models	0
General knowledge of the same group/species	0.2
Qualitative diet composition study	0.5
Quantitative, but limited diet composition study	0.7
Quantitative, detailed diet composition study	1
Catches	
Guesstimates	0
From other models	0
FAO statistics	0.2
National statistics	0.5
Local study, low precision/incomplete	0.7
Local study, high precision/complete	1

Table 2.21 Quality of data sources used in the food web model of the Great Lagoon.

Component	**Score**			
	(Bio)mass	**P/B; Q/B**	**diet**	**catches**
Pelagic:				
Allochthonous POM input	0.4			
Allochthonous DOM input	0.4			
Phytoplankton	1	0.7		
Bacterioplankton	0	0	0	
Protozooplankton	0	0	0.2	
Rotifers	0.7	0.5	0.2	
Non-predatory cladocerans	0.7	0.5	0.2	
Copepods	0.7	0.5	0.2	
Leptodora kindti	0.4	0.5	0.2	
Mysids	0	0.5	0.2	
Epiphytic:				
Epiphytic algae	0.7	0.5		
Epiphytic bacteria and fungi	0	0	0.2	
Protozooperiphyton	0	0	0.2	
Mesozooperiphyton	0.7	0.5	0.2	
Sessile macrozooperiphyton	0.7	0.5	0.2	
Motile macrozooperiphyton	0.7	0.5	0.2	
Benthic:				
Emergent plants	0.4	0.5		
Submerged and floating-leaved plants	0.4	0.5		
Microphytobenthos	0	0.1		
Benthic bacteria and fungi	0	0	0.2	
Benthic protozoans	0	0	0.2	
Meiobenthos	0.7	0.5	0.2	
Chironomus sp. larvae	1	0.5	0.5	
Macrobenthic ostracods	0.7	0.5	0.2	
Oligochaetes	1	0.5	0.2	
Dreissena polymorpha	1	1	0.2	
Other molluscs	1	0.5	0.2	
Remaining zoobenthos	0.7	0.5	0.2	

cont. Table 2.21 Quality of data sources used in the food web model of the Great Lagoon.

Component	Score			
	(Bio)mass	P/B; Q/B	diet	catches
Ichthyofauna:				
Larvae and fry of planktivorous fish	0.4	0.5	1	
Juvenile and adult planktivorous fish	0.7	0.5	1	0
Larvae and fry of benthivorous fish	0.4	0.5	1	
Juvenile and adult benthivorous fish	0.7	0.5	1	0.7
Larvae and fry of piscivorous fish	0.4	0.5	1	
Juvenile and adult piscivorous fish	0.7	0.5	1	0.7
Birds:				
Herbivorous birds	0.4	0.2	0.2	
Molluscivorous birds	0.4	0.2	0.2	
Omnivorous birds	0.4	0.2	0.2	
Piscivorous birds	0.4	0.2	0.2	
n	38	36	31	3
Average score	**0.497**	**0.392**	**0.358**	**0.467**

2.4.3 Comparison of the Great (Szczecin) Lagoon with Other Ecosystems

The comparison, conducted in Part 1 of this work (Section 1.14), of the trophic status and biological characteristics of the Great (Szczecin) Lagoon and other areas was limited mainly to the southern Baltic's coastal lagoons and lakes, dominated by the riverine water regime, where influxes of the brackish Baltic water are of secondary importance only. The freshwater nature of the Lagoon's flora and fauna allowed also to carry out biocoenotic comparisons with inland water bodies, particularly with dam reservoirs and larger lakes. The comparisons showed the biomass and production of the Lagoon biota components analysed to be high or very high. During the period of study (1998-2002), the area experienced no serious environmental stress, whereby the ecosystem was capable of functioning efficiently. This was evidenced by the high ecological efficiency of the great majority of the communities studied, corresponding to the efficiency observed in other highly productive temperate water bodies. However, the fully informative comparison of production potential of the Lagoon with that of other water bodies is possible only with the modelling approach in which units of measurements, system structure, assumptions adopted, etc. are formalised, and the matter/energy flow in various systems balanced out. There are a substantial number of publications on food web models of aquatic ecosystems in various climatic zones, which increases the representativeness of comparative analyses. This section reviews those aquatic ecosystems for which food web models have been developed.

To narrow down the analysis to ecosystems with physical conditions similar to those in the Great Lagoon, comparisons were performed on lagoon- or estuary-type water bodies, usually shallow and land-bound. Another prerequisite was that a model would adopt one year as unit time. Of the 68 aquatic ecosystem models described in the literature reviewed, the criteria were met by 20 ecosystems (one of which, the Pearl River Estuary, was analysed in two different periods of study) described in 16 publications (Table 2.22). The different ways of expressing weight of functional groups (wet weight, organic carbon, dry weight, ahs-free dry weight, organic nitrogen) used in the models were converted to organic carbon units using, whenever possible, conversion factors provided in the original papers; the remaining case, conversion factors published by Atkinson and Smith (1983) Vinogradov and Shushkina (1987), and Opitz (1996).

Although the ecosystems compared are situated in different regions of the world, all are located in the northern hemisphere. Eleven of them are in the temperate zone, the remaining 9 being confined to the tropical zone. Their size ranges extensively, from 3 to 72,600 km^2 (Table 2.22), most covering from several tens to few hundred km^2.

The ecosystems compared differed widely in their productivity. The primary production in the most oligotrophic of them amounted to as little as a several tens of gC m^{-2} yr^{-1} (the Pearl River estuary and some Gulf of Mexico lagoons), 3 thou. gC m^{-2} yr^{-1} being the primary production in the most eutrophic one (the Chiku Lagoon, Taiwan). Three quarters of the ecosystems showed net primary production (NPP) ranging within 100-1000 gC m^{-2} yr^{-1}. NPP of the Great Lagoon was in this range, closer to its upper boundary (538 gC m^{-2} yr^{-1}) (Fig. 2.15). Like in the Great Lagoon, the primary production in many systems was dominated by that of the phytoplankton, but some ecosystems (e.g., the Venice Lagoon) featured the phytobenthos as the major organic matter producer.

The ecosystems compared differed widely in terms of the zooplankton and zoobenthos production as well, the differences being 200-fold in the extreme cases. The Pearl River Estuary and the Tampamachoco Lagoon in the Gulf of Mexico supported the total production of zooplankton and macrozoobenthos of a few gC m^{-2} yr^{-1}, whereas the Chiku Lagoon and Lake Nokouke in the Gulf of Guinea showed several hundred gC m^{-2} yr^{-1} (Fig. 2.16). In a vast majority of the ecosystems compared, the combined

Table 2.22 Lagoons and estuaries productivities of which were compared in Figs. 2.15 – 2.18.

Ecosystem	Region	Latitude °N	Surface area (km^2)	Reference
Mondego estuary (south arm)	Portugal coast, E Atlantic	40	3	Patricio and Marques (2006)
Tapong Lagoon	SW Taiwanese coast, W Pacific	22	4.44	Lin et al. (2006)
Chiku Lagoon	W Taiwanese coast, W Pacific	23	9.6	Lin et al. (1999)
Tampamachoco Lagoon	Gulf of Mexico	21	15	Rosado-Solorzano and Guzman del Proo (1998)
Lake Gardno	Polish coast, S Baltic Sea	54.5	22	Skurzak (2009)
Celestun Lagoon	Yukatan coast, Gulf of Mexico	21	30	Vega-Candejas and Arreguin-Sanchez (2001)
Seine Estuary	The English Channel, NE Atlantic	50	34.6	Rybarczyk and Elkaim (2003)
Alvarado Lagoon	Gulf of Mexico	19	62	Cruz-Escalona et al. (2007)
Bay of Somme	English Channel, NE Atlantic	49.5	70	Rybarczyk et al. (2003)
Lake Nokoue	N Gulf of Guinea, E Atlantic	6	150	Villanueva et al. (2006)
Huizache-Caimanero Lagoon complex	Mexican coast, E Pacific	23	175	Zetina-Rejon et al. (2003)
Narragansett Bay	NW Atlantic	41	380	Monaco and Ulanowicz, 1997
Great Lagoon (Szczecin Lagoon)	S Baltic Sea	54	410	this study
Venice Lagoon	N Adriatic Sea	45.5	415	Libralato and Solidoro (2009)
Ebrie Lagoon	N Gulf of Guinea, E Atlantic	5	566	Villanueva et al. (2006)
Pärnu Bay	N Gulf of Riga, E Baltic Sea	58	700	Tomczak et al. (2009)
Vistula Lagoon	SE Baltic Sea	54.5	838	Kissler (2009)
Curonian Lagoon	SE Baltic Sea	55	1584	Tomczak et al. (2009)
Delaware Bay	NW Atlantic	39	2030	Monaco and Ulanowicz (1997)
Chesapeake Bay	NW Atlantic	38	11600	Monaco and Ulanowicz (1997)
Pearl River Estuary	NW South China Sea	22	72600	Duan et al. (2009)

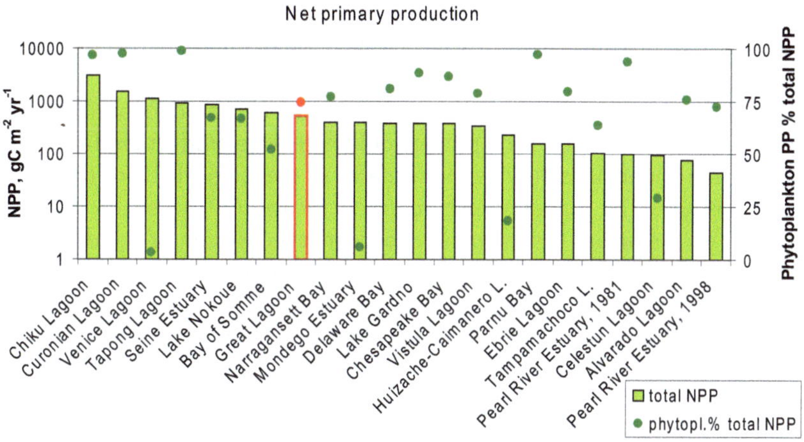

Figure 2.15 Primary production (NPP) in various lagoon- and estuary-type ecosystems, and contribution of phytoplankton production to total pelagic and benthic primary production.

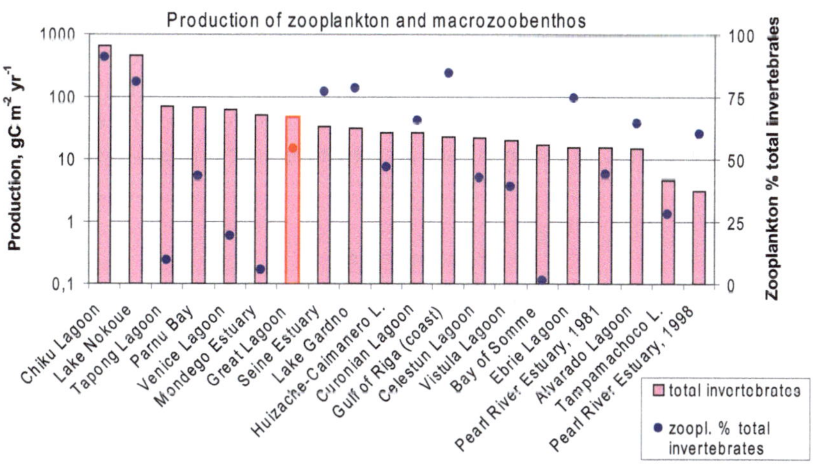

Figure 2.16 Combined production of zooplankton and macrozoobenthos in various lagoon- and estuary-type ecosystems, and contribution of zooplankton production to the combined zooplankton and macrozoobenthos production.

zooplankton and macrozoobenthos production was in the range of 10-100 gC m^{-2} yr^{-1}. The invertebrate production in the Great Lagoon (47 gC m^{-2} yr^{-1}) was in this range, too, the Lagoon being regarded as one of the more productive areas. The Lagoon zooplankton production was similar to that of the macrozoobenthos (25.5 and 21.5 gC m^{-2} yr^{-1}, respectively). Other ecosystems frequently show a distinct domination of production of one or the other ecological category.

The difference between the extreme values of fish production (800-fold) was still higher than that of the primary or invertebrate production. The lowest values were found in estuaries, particularly in the Seine Estuary (less than 0.1 gC m^{-2} yr^{-1}), the highest being reported for the Chiku Lagoon (close to 75 gC m^{-2} yr^{-1}). However, the fish production in most ecosystems ranged between 1 and 10 gC m^{-2} yr^{-1} (Fig.2.17). In this case, too, the Great Lagoon was in the mid-range (4.4 gC m^{-2} yr^{-1}).

As already discussed in Part 1, the Great Lagoon is characterised by particularly good conditions for a high fish production: a high production of the zooplankton and zoobenthos, i.e., the pool of natural food resources of fish, as well as a full availability of dietary items in the entire water column and throughout the surface of the bottom (no stratification, lack of anoxic zones). In addition, should edaphic and habitat conditions deteriorate locally, fish have a possibility to migrate to other areas of the Odra river mouth system.

Of all the parameters compared, the widest differences were observed in fish catches (Fig. 2.18). Those in the Celestun Lagoon in the Gulf of Mexico (0.026 gC m^{-2} yr^{-1}) were almost 3 000 times lower than in the Chiku Lagoon (76.4 gC m^{-2} yr^{-1}). Most ecosystems (including the Great Lagoon) showed, however, a much narrower range of catch magnitude, between 0.1 and 1 gC m^{-2} yr^{-1}. Apart from the Baltic coastal water bodies, in few estuaries only the catches were limited to fish, molluscs and crustaceans being usually exploited as well.

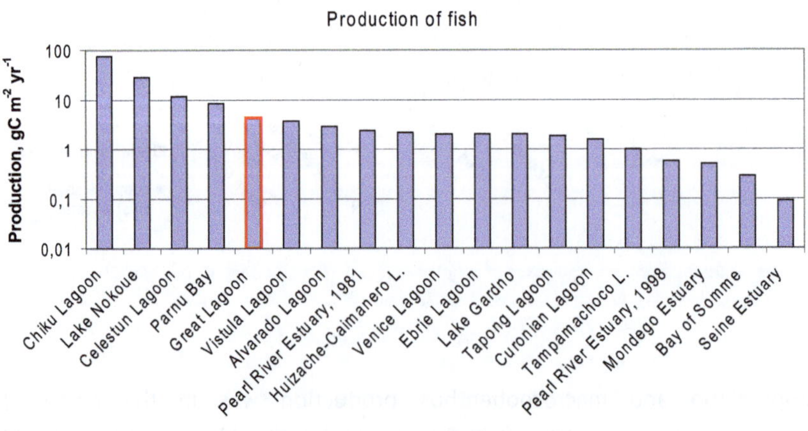

Figure 2.17 Fish production in various lagoon- and estuary-type ecosystems.

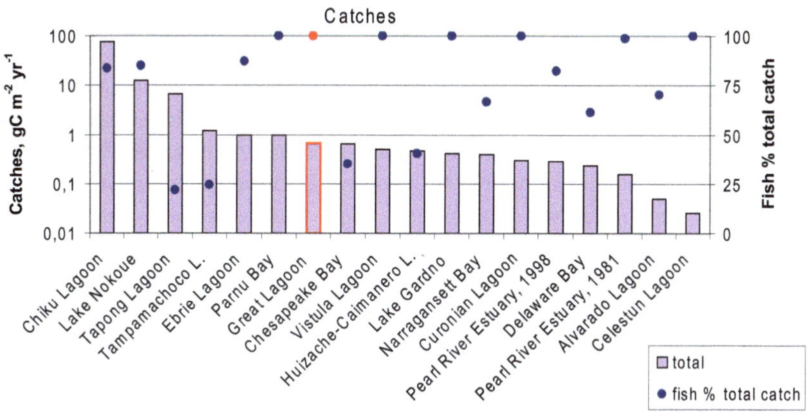

Figure 2.18 Total catches (landings) in various lagoon- and estuary-type ecosystems and contribution of fish to combined fish and shellfish catches.

In all the components compared, the highest production was typical of eutrophic water bodies from the tropical zone. The Chiku Lagoon inTaiwan was particularly productive. This is an ecosystem intensively fed with nutrients from the surrounding mangroves and culture ponds, in which the high primary production of the phytoplankton is very efficiently used in short trophic chains (the major fishery target are fish feeding on detritus and periphyton as well as oysters). In terms of the primary production, that of the Great Lagoon ecosystem was lower than the primary production in some most eutrophic tropical lagoons and some temperate lagoon-estuarine systems. However, the production of higher trophic levels, particularly of fish, and the fish catches in the Great Lagoon were only lower than those in warmer climates and in the Pärnu Bay in Estonia. As opposed to the latter, where a high proportion of catches is contributed by fish migrating from other Baltic areas to the Bay to spawn (herring), the Great Lagoon fisheries are based on the local fish fauna. Noteworthy is also the fact that the Great Lagoon supports no fish cultures, the total catches (landings) being the effects of exploitation of the wild stocks. The true magnitude of catches is doubtless higher than the official records, considering the by-catch of undersized fish and poaching (cf. Part 1, Sections 1.12.3 and 1.14.11). This evidences a high ecological efficiency of the Great Lagoon ecosystem which allows a relatively large portion of energy assimilated by producers to be transferred to the highest trophic levels, even when the trophic chains are not extremely short, and climatic conditions are less favourable, compared to other ecosystems.

2.5 Summary

The food web of the Great Lagoon (major basin of the Szczecin Lagoon) was modelled using a static model based on energy budget equations arranged from the bottom up to the top of a trophic pyramid. 45 components were differentiated in the food web. Five components involved autotrophs: the phytoplankton, emergent macrophytes, submerged and floating-leaved macrophytes, phytoperiphyton, and microphytobenthos. Nine components were associated with various forms of non-living organic matter (3 and 6 components consisted of allochthonous and autochthonous forms, respectively), the remaining 31 components being different functional/trophic groups of heterotrophs: 7 planktonic, 5 periphytic, 9 benthic, 6 formed by fish, and 4 by birds.

The autotrophic components and allochthonous sources of organic matter formed the base of the trophic pyramid. When constructing a network of energy flow between the components identified, biomass values of individual elements of the Great Lagoon biota were used in combination with the published values of their production rates (P/B), food assimilation efficiency (*AE*), and efficiency of utilisation of assimilated food for growth (K_2). The expert knowledge on feeding of various animal groups was applied as well, derived either from research conducted in the Lagoon (the major fish species, larval *Chironomus* sp.) or from literature reviews. The model was parameterized using data on 26 heterotrophic components out of the 45 components of the entire food web. Parameterization was carried out so that the difference between the animal production values calculated by the model and determined from empirical biomass data and the P/B values would not exceed 1%.

It should be mentioned that empirical biomasses of numerous components were only approximations and the bioenergetic indicators used were derived from studies conducted in other ecosystems, not infrequently on different species. For some components (bacteria, protists), no empirical data exist. It has then to be realised that the picture of the Great Lagoon food web produced by the model, although consistent with the measurement data and meeting the mass conservation requirement, is only an approximation of the true set of interactions; should other assumptions be adopted, a picture emerging would be more or less different that that painted in this work.

The net primary production of all the autotrophic Great Lagoon ecosystem components combined was estimated, for 1998-2002, at 538 gC m^{-2} yr^{-1}, the supply of allochthonous organic matter being estimated at 470 gC m^{-2} yr^{-1}. Thus, the total input of organic matter to the ecosystem amounted to about 1000 gC m^{-2} yr^{-1}, almost half of which being obtained from the outside, from the Szczecin Lagoon catchment area. As the Great Lagoon is a flow-through water body, the supply of organic matter from the catchment is accompanied by constant export to the Pomeranian Bay in the Baltic Sea. According to the model estimates, more than 60 % (about 300 gC m^{-2} yr^{-1}) of the allochthonous organic matter entering the Lagoon is exported unused. In addition, there was an export of about 50 gC m^{-2} yr^{-1} of the organic matter produced in the Lagoon, hence the amount of the organic matter exported from the Lagoon was by about ¼ lower than the organic matter supply to the area.

The non-living organic matter was assigned to trophic level 1, similarly to all the primary producers. The highest number of food web components (almost all the invertebrates) were placed between trophic levels 2 and 3. A higher trophic position, between trophic levels 3 and 4, was occupied by *Leptodora*, sessile macrozooperiphyton, planktivorous fish and benthivorous fish and birds. The top predators (predatory fish as well as omni- and piscivorous birds) were found at trophic level 4-4.5.

In terms of biomass, vascular plants, primarily the emergent species (mainly reed), were the dominant group in the Great Lagoon. The macrophyte biomass was several times higher than the phytoplankton biomass and contributed almost 80% to the total biomass of the flora and fauna of the Lagoon (not including the biomass of bacteria and protists). The highest contribution to the animal biomass (about ¼) was that of the zebra mussel. The zoobenthos accounted for almost 60% of the faunal biomass, fish, zooplankton, and zooperiphyton contributing 27, 8, and 5%, respectively; the waterfowl accounted for as little as about 0.1%. The biomass of fish larvae and fry contributed as little as 7% to the fish biomass in the ecosystem. Compared to the zoobenthos and the ichthyofauna, the zooplankton biomass was low, 7 times lower that the benthic fauna biomass and 3 times lower than the fish biomass. Thus, a large part of the flora and fauna biomass in the Great Lagoon was accumulated on the bottom, primarily in the littoral.

The dominant role in the energy flow in the Great Lagoon was played by unicellular organisms. Among the autotrophs, the most important was phytoplankton (74% of the net primary production), the heterotrophic part being dominated by bacteria. The consumption, secondary production, and respiration in both the water column and the bottom sediments were dominated by the bacterial components. Of about 1000 gC m^{-2} yr^{-1} of organic matter supplied to the system and available to the consumers, heterotrophs respired about 580 gC m^{-2} yr^{-1}. Of that heterotrophs' respiration, bacteria used more than 60%, further 10% being utilised by protists. Thus, the contribution of metazoan animals to the heterotrophic utilisation of organic matter supply amounted to less than 30%.

The benthic and the planktonic communities played comparable roles in the overall budget of the matter and energy flow in the ecosystem. The zoobenthos and the zooplankton utilised 46 and 47% of the organic matter used by all heterotrophs in the system, respectively, the zooperiphyton and fish using 4 and 3%, respectively. The higher biomass of benthic organisms was balanced by a higher metabolic rate of the planktonic organisms.

From the standpoint of fisheries management, it may be interesting to compare the production of macrozoobenthos and the crustacean zooplankton, as the two groups are the basic food of non-predatory fish. The macrozoobenthos production in the Great Lagoon was estimated at about 22 gC m^{-2} yr^{-1}, the crustacean zooplankton production amounting to about 20 gC m^{-2} yr^{-1}. The two values are similar; in this context it may be surprising to find a substantial (several-fold) domination of benthivorous over planktivorous fish in the catches. An explanation of this apparent contradiction is the fact that a high proportion of the crustacean zooplankton production is consumed by larvae and fry of all the fish species, including benthivores and predators. The amount of food consumed by larvae and fry of all fish categories was almost 10 times that consumed by juvenile and adult planktivorous fish. An additional factor is the zooplankton export from the Lagoon to the Pomeranian Bay, the export accounting for 10-30% of the crustacean zooplankton production.

The Great Lagoon ecosystem produced about 420 gC m^{-2} yr^{-1} of non-living organic matter in the form of faeces and remains of plants and animals. Together with the allochthonous organic matter, this pool formed a substantial food resource for decomposers. The detritus consumption in the Lagoon was estimated at about 590 gC m^{-2} yr^{-1} and was similar to the biomass consumption estimated at about 570 gC m^{-2} yr^{-1}. The detritus consumption was dominated (about 75%) by bacteria.

The model of the Great Lagoon food web developed allows to analyse the utilisation of production of various components in the ecosystem (the so-called ecotrophic efficiency). Production of most heterotrophic components was utilised to a high extent (80-95%). Low ecotrophic efficiency (30-70%) was typical of molluscs (including the zebra mussel) and sessile macrozooperiphyton, which can be explained by efficient protection against predation afforded by shells and exoskeletons. On the other hand, difficult to explain is the relatively low degree (55%) of utilisation of the oligochaete production. Perhaps those animals, while living among sediment particles, are better protected from predation and are prone to natural mortality much more than other organisms. However, a too high oligochaete production assumed in the model cannot be ruled out. The relatively high proportion of unused production of predatory fish (more than 30%), called the "additional mortality" in the model, may be an effect of incomplete catch reports, and may point out to the real catches being twice those officially reported. A particularly low ecotrophic efficiency was shown by the vascular plants, especially the emergent macrophytes, an otherwise well-known effect.

The ecological indicators that can be calculated for the entire Great Lagoon ecosystem using the model (e.g., the gross primary production to total community respiration ratio close to unity) point out to a considerable maturity of the system. Compared to 20 other ecosystems, the primary production of the Great Lagoon was lower than that in some most fertile tropical lagoons and some lagoon-estuarine temperate systems. However, in terms of higher trophic level production, particularly that of fish and fish catches, the Great Lagoon ecosystem scored less only than systems from warmer climatic regimes and the Pärnu Bay in Estonia. This comparison provides evidence of a high ecological efficiency of the Great Lagoon and Pärnu Bay ecosystems, which allows to transfer a relatively high portion of energy assimilated by the producers to the highest trophic levels, even at climatic conditions that are less favourable than those in other ecosystems.

References

Adamczak B., M. Luścińska, A.S. Oleksowicz, M. Rejewski, 1979. The primary production of the Lake Tynwałd. AUNC 47, Prace Limnol., 11: 3-9.

Aleksandrov S.V., 2003. Pervichnaya produktsiya planktona v Vislinskom i Kurskom zalivakh Baltiyskogo Mora i ee svyazi s ryboproduktivnostyu. *Primary production of plankton in the Baltic Sea's Vistula and Curonian Lagoons and its relationships to fish yields.* Ph.D. Thesis, Inst. Zool. Ros. Akad. Nauk, Sankt-Peterburg. (in Russian)

Aleksandrov S.V., 2005a. Pervichnaya produktsiya planktona v Kursskom i Vislinskom zalivakh Baltijskogo mora. *Primary production of plankton in the Baltic Sea's Curonian and Vistula Lagoons.* In: Gidrobiologicheskiye issledovaniya v Baltiyskom more, Atlanticheskom i Tikhom okeanakh na rubezhe tysachleti. *Hydrobiological research in the Baltic Sea and Atlantic and Pacific Oceans at the turn of millennia.* Part I. ATLANTNIRO, Kalinіningrad, 8-17. (in Russian)

Aleksandrov S.V., 2005b. Primary production of phytoplankton and chlorophyll concentration in the Vistula and Curonian Lagoons of the Baltic Sea. In: The Baltic Sea a changing ecosystem. 5th BSSC 2005, Sopot, Poland; Abstracts, 192-193.

Aleksandrov S., 2009. Long-term variability of biological production and the ecological state of the Curonian Lagoon and Vistula Lagoon under the impact of climate change and anthropogenic pollution. In: 7th BSSC 2009 Tallinn, Estonia; Abstract Book, 63.

Aleksandrov S.V., O.A. Dmitrieva, 2006. Primary production and phytopankton characteristics as eutrophication criteria of Kursiu Marios Lagoon, the Baltic Sea. Water Res., 33: 97-103.

Aleksandrov S.V., Yu. M. Senin, V.A. Smyslov, 2006. Pervichnaya produktsiya planktona, soderzhanye chlorofila i byogennych elementov kak pokazateli ekologicheskogo sostoyaniya Kurskogo i Vislinskogo zalivov Baltiyskogo mora. *Primary production of plankton and contents of chlorophyll and nutrients as indicators of the state of the Baltic Sea's Curonian and Vistula Lagoons.* Biol. Vnutr. Vod, 1: 41-47. (in Russian)

Alekseev V.R., 1987. Photoperiodism and seasonal development of the population of *Mesocyclops leuckarti* Claus, Crustacea Copepoda. Gidrobiol. Zh., 23: 22-28.

Alimov A.F., 1981. Funktsionalnaya ekologiya presnovodnykh molluskov. *The functional ecology of freshwater molluscs.* Izd. Nauka, Leningrad. (in Russian)

Aller R.C., 1998. Mobile deltaic and continental shelf muds as suboxic, fluidized bed reactors. Mar. Chem., 61: 143-155.

Ameryk A., B. Podgórska, Z. Witek, 2005. The dependence between bacterial production and environmental conditions in the Gulf of Gdańsk. Oceanologia, 47: 27-45.

Angelini R., A.A. Agostinho, 2005. Food web model of the Upper Parana River Floodplain: description and aggregation effects. Ecol. Model., 181: 109-121.

Andronikova I.N., V.G. Drapkova, K.N. Kuzmenko, N.F. Mikhailova, E.A. Stravinskaya, 1972. Biological productivity of the main communities of the Red Lake. In: Kajak Z., A. Hilbricht-Ilkowska (eds), Productivity problems of freshwaters. Proc. IBP-UNESCO Symp., PWN Polish Sci. Publ., Warszawa-Kraków, 57-71.

Andrulewicz E., 1997. An overview on lagoons in the Polish coastal area of the Baltic Sea. Int. J. Salt Lake Res., 5: 1-14.

Anon., 1991-2003. Zasoby Wodne Kraju. *Water Resources of Poland.* Inst. Meteorol. Gosp. Wod., Warszawa. (monthly bulletin). (in Polish)

Anon., 1992. Program racjonalnego wykorzystania i plan zagospodarowania ujścia Odry. *A programme of rational utilisation of the River Odra mouth area and its management plan.* Wyd. Inst. Mor., Gdańsk-Słupsk-Szczecin. (in Polish)

Anon., 1998-2002. Warunki środowiskowe polskiej strefy południowego Bałtyku w 1998, 1999, 2000 i 2001 r. *Environmental conditions in the Polish waters of the southern Baltic in 1998, 1999, 2000, and 2001.* Mat. Oddz. Mor. IMGW. Reports Maritime Branch, Inst. Meteorol. Water Manag., Gdynia. (in Polish)

Anon., 2003-2008. Bulletin of the State's Hydrological-Meteorogical Service 2003-2008. Inst. Meteorol. Gosp. Wod., Warszawa.

Anon., 2009. Stan środowiska w województwie zachodniopomorskim w 2008 roku. *The environmental status of the Distric of Western Pomerania in 2008.* http://www.wios.szczecin.pl/bip/chapter_16003.asp?soid=FAEB5BD9C3FA4DF9B674415D40778F46 (in Polish).

Aristova G.J., 1965. Bentos Kursskogo Zaliva. The b*enthos of the Curonian Lagoon.* Issl. Kurskom Visl. Zal., Tr. AtlantNIRO, 14: 19-39. (in Russian)

Arlt G., 1973. Zur produktionsbiologischen Bedeutung der Meiofauna in Küstengewässern. Wiss. Zeitschr. Univ. Rostock, Math.-Naturwiss. Reihe, 22: 1141-1145.

Arreguin-Sanchez F., E. Arcos, E.A. Chavez, 2002. Flows of biomass and structure in an exploited benthic ecosystem in the Gulf of California, Mexico. Ecol. Model., 156: 167-183.

Atkinson M.J., S.V. Smith, 1983. C:N:P ratios of benthic marine plants. Limnol. Oceanog., 28: 568-574.

Azam F., T. Fenchel, J.G. Field, J.S. Gray, L.A. Meyer-Reil, F. Thingstad, 1993. The ecological role of water-column microbes in the sea. Mar. Ecol. Prog. Ser., 10: 257-263.

Backiel T., 1971. Production and food consumption of predatory fish in the Vistula River. J. Fish. Biol., 3: 369-405.

Backiel T., J.F. Kitchel, J.E Thorpe, 1980. Fish. In: Le Cren E.D., R.H. Lowe-McConell (eds): The functioning of freshwater ecosystems. Int. Biol. Progr. 22. Cambridge Univ. Press, Cambridge.

Baird D.R., E. Ulanowicz, 1993. Comparative study on the trophic structure, cycling and ecosystem properties of four tidal estuaries. Mar. Ecol. Prog. Ser., 99:221-237.

Bangel H., G. Schernewski, A. Bachor, M. Landsberg-Uczciwek, 2004. Spatial pattern and long-term development of water quality in the Oder Estuary. In: Schernewski G., T. Dolch (eds): The Oder Estuary – against the background of the European Water Framework Directive. Mar. Sci. Rep., 57: 17-65.

Banse K., S. Mosher, 1980. Adult body mass and annual production/biomass relationships of field populations. Ecolog. Monogr., 50: 355-379.

Barica J., 1990. Factors controlling the evolution of salinity, productivity and ecological stability of eutrophic lakes in central-north America. In: Burkhardt L. (ed.): Int. Symp. "Evolution of Freshwater Lakes", Poznań, 8-13.

Barausse A., A. Duci, C. Mazzoldi, Y. Artioli, L. Palmeri, 2009. Trophic network model of the Northern Adriatic Sea: Analysis of an exploited and eutrophic ecosystem. Est. Coast. Shelf Sci., 83: 577-590.

Bartel R., A. Garbacik-Wesołowska, J. Waluga, H. Wilkońska, A. Wysokiński, 1998. Changes in catches of fishes in the Szczecin Lagoon in 1889-1994. In: The problems of management of fishery resources in the Polish and German waters of the Szczecin Lagoon and Pomeranian Bay. Proc. Polish-German Symp., Świnoujście, 19-20 Nov. 1966. Sea Fish. Inst., Gdynia, 84-104.

Beattie M, H.J. Bromley, M. Chambers, R. Goldspink, J. Vijverberg, N.P. van Zalinge, H.L. Golterman, 1972. Limnological studies on Tjeukemeer – a typical Dutch "polder resevoir". In: Kajak Z., A Hilbricht-Ilkowska (eds): Productivity problems of freshwaters. Proc. IBP-UNESCO Symp., PWN Polish Sci. Publ., Warszawa-Kraków, 421-446.

Berg C., F.A. Sibbing, J.W.M. Osse, W. Hogenboezem, 1992. Structure, development and function of the branchial sieve of the common bream, *Abramis brama*, white bream, *Blicca bjoerkna* and roach, *Rutilus rutilus*. Environ. Biol. Fish, 33: 105-124.

Berg C., J.G.M. Boogaart, F.A. Sibbing, E.H.R.R. Lammens, J.W.M. Osse, 1993. Shape of zooplankton and retention in filter-feeding: A quantitative comparison between industrial sieves and the branchial sieves of common bream (*Abramis brama*) and white bream (*Blicca bjoerkna*). Can. J. Fish. Aquat. Sci., 50: 716-724.

Bernatowicz S., E. Pieczyńska, 1965. Organic matter production of macrophytes in the Lake Tałtowisko (Mazurian Lakeland). Ekol. Pol. A, 13: 113-124.

Bernatowicz S., P. Wolny, 1974. Botanika dla limnologów i rybaków. *Botany for limnologists and fishermen.* PWRL, Warszawa. (in Polish)

Bloesch J., 1995. Mechanisms, measurement and importance of sediment resuspension in lakes. Mar. Freshw. Res., 46: 295-304.

Błaszczyk A., A. Toruńska, J. Kobos, B. Browarczyk-Matusiak, H. Mazur-Marzec, 2010. Ekologia toksycznych sinic. *Ecology of toxic cyanobacteria.* Kosmos, 59: 173-198. (in Polish)

Bogdanowicz R., 2004. Czynniki hydrologiczne wpływające na transport wybranych związków azotowych i fosforowych z Odry, Wisły i polskich rzek przymorskich do Bałtyku. *Hydrological factors influencing transport of selected nitrogen and phosphorus compounds from the Odra River, the Vistula River and Polish coastal rivers to the Baltic Sea.* Wyd. Uniw. Gdańskiego, Gdańsk. (in Polish)

Bohr R., 1962. Badania fitosocjologiczne perifitonu w Jeziorze Mamry. *Phytosociological studies on the periphyton in Lake Mamry.* Stud. Soc. Scient. Toruń. D (Botanica), 6: 1-44. (in Polish)

Bohr R., B. Adamczak, M. Luścińska, 1975. Primary production of algae in the lakes Jeziorak and Tynwałdzkie. Pol. Arch. Hydrobiol., 22: 107-111.

Borowski W., 2000. Zasoby ryb w Zalewie Wiślanym i warunki ich eksploatacji. *Fish resources in the Vistula Lagoon and conditions for their exploitation.* Stud. Mat. Mor. Inst. Ryb., Ser. B, 72: 9-33. (in Polish)

Borowski W., H. Dąbrowski, 1996. Przyłów leszcza i sandacza w połowach żakami węgorzowymi w Zalewie Wiślanym podczas zamkniętego sezonu wiosennego 1996 r. *By-catch of bream and pikeperch in the eel-directed fyke-net fishery in the Vistula Lagoon during the spring closed season in 1996.* Rap. Mor. Inst. Ryb., Gdynia, 50-54. (in Polish)

Borowski W., H. Dąbrowski, R. Zaporowski, 1998. The occurrence of young fish in fyke-net catches in the Vistula Lagoon in 1995. In: The problems of management of fishery resources in the Polish and German waters of the Szczecin Lagoon and Pomeranian Bay. Proc. Polish-German Symp., Świnoujście, 19-20 Nov. 1966. Sea Fish. Inst., Gdynia, 151-168.

Borówka R.K., A. Osadczuk, M. Latałowa, T. Duda, R. Woziński, A. Witkowski, J. Święta-Musznicka, B. Kosińska, 2005. Pokrywa osadowa Zalewu Szczecińskiego i jej wymowa paleograficzna. *Sedimentary cover of the Szczecin Lagoon and its palaeogeographic significance.* In: Borówka R.K., S. Musielak (eds), Środowisko przyrodnicze wybrzeży Zatoki Pomorskiej i Zalewu Szczecińskiego. *Natural environment of the Pomeranian Bay and Szczecin Lagoon coastal areas.* Szczecin, 93-102. (in Polish)

Bosselmann S., 1975. Population dynamics of *Eudiaptomus graciloides* in Lake Esrom. Arch. Hydrobiol., 75: 329-346.

Bosy S., 1990. Źródła zanieczyszczenia wód w rejonie planowanych wolnych obszarów celnych i specjalnych stref ekonomicznych. *Sources of water pollution in the area of planned duty-free and special economic zones.* Mat. I Sejmiku Ekol. PTNoZ, Szczecin, 69-82. (in Polish)

Boulion V.V., 2004a. Contribution of major groups of autotrophic organisms to primary production of water bodies. Water Res., 31: 92-102.

Boulion V.V., 2004b. Pervichnaya produktsiya v litoralnoy i otkrytoy chasti ozernykh ekosistem. *Primary production in the littoral and open parts of lacustrine ecosystems.* In: Alimov A.F., M.B. Ivanova (eds): Zakonomernosti gidrobiologicheskogo rezhima vodoyemov raznogo tipa. *Patterns of the hydrobiological regime in water bodies of different type.* Nauchny Mir, Moskva, 208-226. (in Russian)

Brabrand A., 1984. Microhabitat segregation between bream (*Abramis brama* (L.)) and white bream (*Blicca bjoerkna* (L.)) in a mesotrophic lake, SE Norway. Pol. Arch. Hydrobiol., 31: 99-108.

Bradford-Grieve J.M., P.K. Probert, S.D. Nodder, D. Thompson, J. Hall, S. Hanchet, P. Boyd, J. Zeldis, A.N. Baker, H.A. Best, N. Broekhuizen, S. Childerhouse, M. Clark, M. Hadfield, K. Safi, I. Wilkinson, 2003. Pilot trophic model for subantarctic water over the Southern Plateau, New Zealand: a low biomass, high transfer efficiency system. J. Exp. Mar. Biol. Ecol., 289: 223-262.

Brey T., H. Rumhor, S. Ankar, 1988. Energy content of macrobenthic invertebrates: general conversion factors from weight to energy. J. Exp. Mar. Biol. Ecol., 117: 271-278.

Brylinsky M., 1980. Estimating the productivity of lakes and reservoirs. In: Le Cren E.D., R.H. Lowe-Mc Connell (eds), The functioning of freshwater ecosystems. Int. Biol. Prog. 22. Cambrige Univ. Press., Cambridge, 411-454.

Buchholz W., 1990. Zmienność stanów i przepływów w ujściach rzek na przykładzie ujścia Odry. *Variability of water level and flow rate in river mouth as exemplified by the River Odra mouth.* Probl. Bad. Morza Reg. Szczecińskim, Mat. Kom. Bad. Morza PAN, Szczecin, 49-70. (in Polish)

Buchholz W. (ed.), 1991. Monografia Dolnej Odry. Hydrologia i hydrodynamika. *A monograph of the Lower Odra. Hydrology and hydrodynamics.* Pr. IBW PAN, Gdańsk, 25: 1-103 (in Polish)

Bucior A.M., 2009. Produkcja pierwotna toni wodnej Zalewu Szczecińskiego (Wielkiego Zalewu) w latach 2004-2006. *Water column primary production in the Szczecin Lagoon (Great Lagoon).* Ph.D. Thesis, Uniw. Szczeciński. (in Polish)

Bzoma S., 2008a. Śmiertelność węgorza wywoływana przez ptaki. *Bird-induced eel mortality.* In: Zebranie i opracowanie danych naukowych w formie założeń do planu gospodarki zasobami węgorza europejskiego w Polsce. *Collection and assessment of data to be used as groundwork for the management plan of the European eel resources in Poland.* Sea Fish. Inst. Techn. Rep., Gdynia. (in Polish)

Bzoma S., 2008b. Wpływ kormorana *Phalacrocorax carbo* na ichtiofaunę wybranych akwenów strefy przybrzeżnej Bałtyku. *Effects of the black cormorant* Phalacrocorax carbo *on the fish fauna in selected coastal areas of the Baltic Sea.* Sea Fish. Inst. Techn. Rep., Gdynia. (in Polish)

Calow P., 1977. Conversion efficiencies in heterotrophic organisms. Biol. Rev., 52: 385-409.

Carlson R.E., 1977. A trophic state index for lakes. Limnol. Oceanog., 22: 361-369.

Carrick H.J, G.L. Fahnenstiel, E.F. Stoermer, R.G. Wetzel, 1992. The importance of zooplankton-protozoan trophic couplings in Lake Michigan. Limnol. Oceanog., 36: 1335-1345.

Cerco C.F., D. Tillman, J.D. Hagy, 2010. Coupling and comparing a spatially- and temporally-detailed eutrophication model with an ecosystem network model: An initial application to Chesapeake Bay. Environ. Model. Softw., 25: 562-572.

Chen Z., Y. Qiu, X. Jia, S. Xu, 2008. Using an Ecosystem Modeling Approach to Explore Possible Ecosystem Impacts of Fishing in the Beibu Gulf, Northern South China Sea. Ecosystems, 11: 1318-1334.

Chlubek N., 1975. Hydrochemia a dynamika wód Zalewu Szczecińskiego. *Water chemistry versus dynamics of the Szczecin Lagoon.* Pr. Inst. Techn. Chem. 1, Wyd. Nauk. Politechn. Szczec. (in Polish)

Chojnacki J., 1989. Hydrobionty Zatoki Pomorskiej i estuarium Odry. *Hydrobionts of the Pomeranian Bay and the River Odra estuary.* In: Ochrona wód Bałtyku. *Protection of Baltic waters.* Polish-Swedish Seminar, Szczecin, 21-27. (in Polish)

Chojnacki J.C., 1991. Badania populacyjne *Neomysis integer* (Leach, 1815) morskiego wybrzeża wyspy Wolin. *Population studies on Neomysis integer (Leach, 1815) off the Wolin Island sea coast.* Zesz. Nauk. AR Szczecinie, 143: 49-58. (in Polish)

Christensen V., D. Pauly, 1992. Ecopath II – a software for balancing steady-state ecosystem models and calculating network characteristics. Ecol. Model., 61: 169–185.

Christensen V., D. Pauly (eds), 1993. Trophic models of aquatic ecosystems. ICLARM Conference Proceedings, Manila, 26.

Christensen V., D. Pauly, 2004. Placing fisheries in their ecosystem context, an introduction. Ecol. Model., 172: 103–107.

Christensen V., C.J. Walters, D. Pauly, 2005. Ecopath with Ecosim: A user's guide. Fisheries Centre, Univ. British Columbia, Vancouver, Canada.

Christensen V., C.J. Walters, D. Pauly, R. Forrest, 2008. Ecopath with Ecosim version 6. User guide. Lenfest Ocean Future Project.

Christensen V., 1995. Ecosystem maturity – towards quantification. Ecol. Model., 77: 3-32.

Chrobak M., 2004. Warunki środowiskowe polskiej strefy południowego Bałtyku w 2001 r. *Environmental conditions in the Polish waters of the southern Baltic in 2001.* Mat. Oddz. Mor. IMGW, Gdynia. (in Polish)

Chubarenko B.V., P. Margoński, 2008. The Vistula Lagoon. In: Schiewer U. (ed.): Ecology of Baltic coastal waters. Ecol. Studies 197. Springer Verlag, Berlin-Heidelberg, 167-196.

Chudecki Z., E. Niedźwiecki, 1987. Akumulacja związków chemicznych w osadach Zalewu Szczecińskiego. *Accumulation of chemical compounds in the sediments of Szczecin Bay.* Roczn. Glebozn., 38: 85-90. (in Polish)

Clark C. W., 2006. The worldwide crisis in fisheries. Economic models and human behaviour. Cambridge Univ. Press, Cambridge.

Cloern J.E., C. Grenz, L. Vidergar-Lucas, 1995. An empiricical model of the phytoplankton chlorophyll:carbon ratio – the conversion factor between productivity and growth rate. Limnol. Oceanog., 40: 1313-1321.

Cobb S.E., M.C. Watzin, 1998. Trophic interactions between yellow perch (*Perca flavescens*) and their benthic prey in a littoral zone community. Can. J. Fish. Aquat. Sci., 55: 28-36.

Coffin R.B., J.P. Connolly, P.S. Harris, 1993. Availability of dissolved organic carbon to bacterioplankton examined by oxygen utilization. Mar. Ecol. Prog. Ser., 101: 9-22.

Cole J.J., S. Findlay, M.L. Pace, 1988. Bacterial production in fresh and saltwater ecosystems: a cross-system overview. Mar. Ecol. Prog. Ser., 43: 1-10.

Coll M., I. Palomera, S. Tudela, F. Sarda, 2006. Trophic flows, ecosystem structure and fishing impacts in the South Catalan Sea, Northwestern Mediterranean. J. Mar. Sys., 59: 63-96.

Cornejo-Donoso J., T. Antezana, 2008. Preliminary trophic model of the Antarctic Peninsula Ecosystem (Sub-area CCAMLR 48.1). Ecol. Model., 218: 1-17.

Cramp S., K.E.L. Simmons, 1977. Birds of the western Palearctic. Vol. 1. Oxford Univ. Press, Oxford.

Cruz-Escalona V.H., F. Arreguin-Sanchez, M. Zetina-Rejon, 2007. Analysis of the ecosystem structure of Laguna Alvarado, western Gulf of Mexico, by means of a mass balance model. Est. Coast. Shelf Sci., 72: 155-167.

Czarnecka M., 2005. Ekologiczny status epifauny zasiedlającej sztuczne podłoża podwodne. *Ecological status of the epifauna inhabiting artificial submerged substrata.* Ph.D. Thesis, Nicolas Copernicus Univ., Toruń. (in Polish)

Czarnecka M., M. Tymolewski, 2003. Makrofauna trzciny pospolitej (*Phragmites communis* L.) północno-wschodniej części Zalewu Szczecińskiego (Estuarium Odry) w aspekcie zasoleniowym. *The macrofauna of the common reed* (Phragmites communis L.) *in the north-eastern part of the Szczecin Lagoon (River Odra estuary): effects of salinity.* Idee Ekolog., 15, Szkice, 8: 43-45. (in Polish)

Czarnecki M., 1993. Wpływ eutrofizacji na bentos jeziora Nowowarpieńskiego. *Effects of eutrophication on the Lake Nowowarpieńskie benthos.* Stud. Mat. Mor. Inst. Ryb., Świnoujście, Ser. S: 72-80. (in Polish)

Czeraszkiewicz R., A. Oleksiak (eds), 2003. Ptaki wodno-błotne na Pomorzu Zachodnim. Wyniki liczeń w sezonie 2002/2003, ekologia i ochrona. *The waterfowl in the Western Pomerania. The 2002/2003 census, ecology and protection.* Proj. Przyr. Łow., ZTO-PZŁ, Szczecin, 14-25. (in Polish)

Czeraszkiewicz R., A. Oleksiak (eds), 2004. Ptaki wodno-błotne na Pomorzu Zachodnim. Wyniki liczeń w sezonie 2003/2004, ekologia i ochrona. *The waterfowl in the Western Pomerania. The 2003/2004 census, ecology and protection.* Proj. Przyr. Łow., ZTO-PZŁ, Szczecin, 5-16. (in Polish)

Darwall W.R.T., E.H. Allison, G.F. Turner, K. Irvine, 2010. Lake of flies, or lake of fish? A trophic model of Lake Malawi. Ecol. Model., 221: 713-727.

Dermott R., J. Mitchell, I. Murray, E. Fear, 1993. Biomass and production of zebra mussels (*Dreissena polymorpha*) in shallow waters of Northeastern Lake Erie. In: Nalepa T.F., D.W. Schloesser (eds): Zebra Mussels. Lewis Publ., 399-413.

Dmitrieva O.A., 2005. Phytoplankton as indicator a condition of ecosystems of the Vistula anad Curonian Lagoons.). In: The Baltic Sea as a changing ecosystem. 5th BSSC 2005, Sopot, Poland; Abstracts, 217-218.

Dmitrieva O, A. Semionova, 2009. Impact of Cyanobacteria blooms on zooplankton community in offshore zone of Curonian Lagoon (Baltic Sea). In: 7th BSSC 2009, Tallinn, Estonia; Abstract Book, 221.

Dobrowolski Z., 2001. Density, biomass and distribution of benthic invertebrates in the mid-lake zone of coastal Lake Gardno. Oceanol. Stud., 20: 39-58.

Dombrovskiy K.O., 2009. Znachenia dreiseny u formovanni indyvidualnykh konsortsiy. *Importance of zebra mussel for the formation of individual consortia.* Visn. Zaporic. Nacional. Univ., Ukraina, 2: 30-38. (in Ukrainian)

Draganik B., 2000. Określenie wartości parametrów eksploatowanych populacji ryb, determinujących wydajność populacji sandacza, okonia, leszcza i płoci w Zalewie Szczecińskim; sandacza i leszcza w Zalewie Wiślanym. *Estimation of parameters describing exploited fish populations, which determine the yield of zander, perch, breach, and roach populations in the Szczecin Lagoon and the zander and bream populations in the Vistula Lagoon.* Sea Fish. Inst. Techn. Rep., Gdynia. (in Polish).

Drgas A., 2000. Rola meiofauny w biocenozach dennych Zatoki Gdańskiej ze szczególnym uwzględnieniem wolnożyjących nicieni (Nematoda). *The role of the meiofauna in benthic communities of the Gulf of Gdańsk, with a particular reference to free-living nematodes (Nematoda).* Ph.D. Thesis, Uniw. Gdański. (in Polish)

Drzycimski I., 1986. Niekorzystne zmiany w biocenozie Zalewu. *Unfavourable changes in the Szczecin Lagoon biota.* Aura, 7: 10-11. (in Polish)

Drzycimski I., 1989. Biocenoza estuarium Odry w latach 1984-1988. The River Odra estuary biota in 1984-1988. In: Ochrona wód Bałtyku. *Protection of Baltic waters.* Polish-Swedish Seminar, Szczecin, 15-20. (in Polish)

Duan L.J., S.Y. Liu, J. Moreau, V. Christensen, 2009. Modeling changes in the coastal ecosystem of the Pearl River Estuary from 1981 to 1998. Ecol. Model., 220: 2802-2818.

Dukowska M., 2007. Bentos i fauna naroślinna w rzece jako baza pokarmowa ryb. *Benthos and epiphytic fauna as a food resource for fish.* In: Hydromorfologiczna ocena ekosystemów wodnych. *Hydromorphological assessment of aquatic ecosystems.* XIV Ogólnop. Warsz. Bentol. PTH., Opole-Turawa, 16-17. (in Polish)

Ecopath with Ecosim, www.ecopath.org, 2012-08-08

Ejsmont-Karabin J., 1998. Empirical equations for biomass calculation of planktonic rotifers. Pol. Arch. Hydrobiol., 45: 513-522.

Elmgren R., 1984. Trophics dynamics in the enclosed, brackish Baltic Sea. Rapp. P.-v. Reun. Cons. Int. Explor. Mer, 183: 152-169.

Ezhova E., L. Żmudziński, K. Maciejewska, 2005. Long-term trends in the macrozoobenthos of the Vistula Lagoon, southeastern Baltic Sea. Species composition and biomass distribution. Bull. Sea Fish. Inst., 1 (164): 55-70.

Fal B., E. Bogdanowicz, W. Czernuszenko, I. Dobrzyńska, A. Koczyńska, 1997. Przepływy charakterystyczne głównych rzek polskich w latach 1951-1990. *Characteristic flows in major Polish rivers in 1951-1990.* Mat. Bad. IMGW, Ser. Hydrol. Oceanol., 21. (in Polish)

Fal B., E. Bogdanowicz, W. Czernuszenko, I. Dobrzyńska, A. Koczyńska, 2000. Przepływy charakterystyczne głównych rzek polskich w latach 1951-1995. *Characteristic flows in major Polish rivers in 1951-1995.* Mat. Bad. IMGW, Ser. Hydrol. Oceanol., 26. (in Polish)

Falandysz J., B. Wyrzykowska, J. Warzocha, I. Barska, A. Garbacik-Wesołowska, P. Szefer, 2004. Organochlorine pesticides and PCBs in perch *Perca fluviatilis* from the Odra/Oder river estuary, Baltic Sea. Food Chem., 87: 17-23.

Falkowska L., 1985. Dopływ związków azotu i fosoru z atmosfery do Bałtyku na podstawie badań w 1981 i 1983 r. *Atmospheric supply of nitrogen and phosphorus to the Baltic Sea, as elucidated by studies carried out in 1981 and 1983.* Stud. Mat. Oceanol., 48, Chemia Morza, 6: 5-28. (in Polish)

Fenchel T., 1987. Ecology of Protozoa. Sci. Tech. Publ., Madison/Springer Verlag, Berlin.

Fenchel T., B.J. Finlay, 1983. Respiration rates in heterotrophic, free-living Protozoa. Microbiol. Ecol., 9: 99-122.

Fenske Ch., 2002. The ecological importance of mussels, their effect on water quality and their possible use for coastal zone management. In: Schernewski G., U. Schiewer (eds): Baltic coastals ecosystems. Springer Verlag, Berlin-Heidelberg-New York, 53-64.

Fenske Ch., S. Dahlke, Ph. Riel, A. Woźniczka, 2010. Dynamics of mussel beds in the Szczecin Lagoon. Coast. Rep., 16: 49-54.

Fetahi T., S. Mengistou, 2007. Trophic analysis of Lake Awassa (Ethiopia) using mass-balance Ecopath model. Ecol. Model., 201: 398-408.

FishBase, www.fishbase.org, 2012-08-08

Fogarty M.J., S.A. Murawski, 1998. Large-scale disturbance and the structure of marine systems: Fishery impacts on Georges Bank. Ecol. Applic. Suppl., 8: 6-22.

Fry B., 2006. Stable isotope ecology. Springer, New York.

Gak D.Z., V.V. Gurvich, I.L. Korelyakova, L.E. Kostikova, N.A. Konstantinova, G.A. Olivari, A.D. Primachenko, Y.Ya. Tseeb, K..S. Vladimirova, L.L. Zimbalevskaya, 1972. Productivity of aquatic organism communities of different trophic levels in Kiev Reservoir. In: Kajak, Z., A Hilbricht-Ilkowska (eds), Productivity problems of freshwaters. Proc. IBP-UNESCO Symp., PWN Polish Scientific Publ., Warszawa Kraków, 445-455.

Galtsova V.V., 1991. Meiobentos v morskikh ekosistemakh. *The meiobenthos in marine ecosystems.* Tr. Zool. Inst. AN SSSR, 224, Leningrad. (in Russian)

Garbacik-Wesołowska A., 1969. Roślinność Zalewu Szczecińskiego. *Vegetation of the Firth of Szczecin.* Pr. Mor. Inst. Ryb., Gdynia, 15, Ser. A: 103-119. (in Polish)

Garbacik-Wesołowska A., 1973. Biomasa makrofitów Zalewu Szczecińskiego. *The macrophyte biomass in the Szczecin Lagoon.* In: Symp. Naukowe w Świnoujściu. Wyd. Mor. Inst. Ryb., Gdynia, 81-86. (in Polish)

Garbacik-Wesołowska A., N. Wolnomiejski, J. Wengrzyn, 1984. Badania skutków refulacji brzegów na faunę i florę Zalewu Szczecińskiego na przestrzeni Warnołęka-Brzózki. *Studies on effects of on-shore dredge spoil dumping on the Szczecin Lagoon fauna and flora from Warnołęka to Brzózki.* Sea Fish. Inst. Techn. Rep., Świnoujście. (in Polish)

Garbacik-Wesołowska A., J. Wengrzyn, C. Kaczewiak, 1985. Charakterystyka węgorza Zalewu Szczecińskiego. *Characteristics of the eel in the Szczecin Lagoon.* In: Możliwości eksploatacyjne zasobów rybnych Zalewu Szczecińskiego i Zatoki Pomorskiej z uwzględnieniem charakterystyki hydrologicznej. *Expoitation potential of fish resources of the Szczecin Lagoon and Pomeranian Bay against the background of hydrological characteristics.* Sea Fish. Inst. Techn. Rep., Gdynia. (in Polish)

Garbacik-Wesołowska A., C. Kaczewiak, J. Strąk, K. Sołtysik, 1995. Eksploatacja rybacka wód Zalewu Szczecińskiego w roku 1994. *The fishing exploitation of the waters of the Szczecin Lagoon in 1994.* Rap. Mor. Inst. Ryb. 1993-1994,. Gdynia, 109-124. (in Polish)

Garbacik-Wesołowska A., E. Boberski, A. Wysokiński, 1998. Bioróżnorodność i czasowo-przestrzenne zróżnicowanie zespołów ichtiofauny w Estuarium Odry. *Biodiversity and spatio-temporal variability of fish assemblages in the River Odra estuary.* Sea Fish. Inst. Techn. Rep., Świnoujście. (in Polish)

Garbacik-Wesołowska A., E. Boberski, 2000. Zasoby rybne w Zalewie Szczecińskim i w strefie przybrzeżnej Zatoki Pomorskiej oraz warunki ich eksploatacji. *Fish resources in the Szczecin Lagoon and the coastal zone of the Pomeranian Bay and conditions for their exploitation.* Stud. Mat. Mor. Inst. Ryb., Ser. B, 72: 77-104. (in Polish)

Garcia S.M., K.L. Cochrane, 2005. Fisheries and Aquaculture topics. The ecosystem approach to fisheries management. Topics Fact Sheets. FAO Fisheries and Aquaculture Department, 2012-01-02

Gasiûnaité Z.R., D. Daunys, S. Olenin, A. Razinkovas, 2008. The Curonian Lagoon. In: U. Schiewer (ed.): Ecology of Baltic coastal waters. Ecol. Studies 197, Springer Verlag, Berlin-Heidelberg, 197–215.

Gaten E., 1986. Life cycle of *Lymnaea peregra* (Gastropoda, Pulmonata) in the Leicester canal, U.K., with an estimate of annual production. Hydrobiologia, 195: 45-54.

Gayevskaya I.S., 1966. Rol vysshikh vodnykh rasteniy v pitanii zhivotnykh presnykh vodoyemov. *The role of higher aquatic plants in feeding of freshwater animals.* Izd. Nauka, Moskva. (in Russian)

Gerlach S.A., 1971. On the importance of marine meiofauna for benthos communities. Oecologia, 6: 176-190.

Giles N., M. Street, R.M. Wright, 1990. Diet composition and prey preference of tench, *Tinca tinca* (L.), common bream, *Abramis brama* (L.), perch, *Perca fluviatilis* L. and roach, *Rutilus rutilus* (L.), in two contrasting gravel pit lakes: potential trophic overlap with wildfowl. J. Fish Biol., 37: 945-957.

Girjatowicz J. P., 2005. Zlodzenie Zatoki Pomorskiej i Zalewu Szczecińskiego. *Ice cover of the Pomeranian Bay and the Szczecin Lagoon.* In: Borówka, R.K., S. Musielak (eds): Środowisko przyrodnicze wybrzeży Zatoki Pomorskiej i Zalewu Szczecińskiego. *Natural environment of the Pomeranian Bay and Szczecin Lagoon coastal areas.* Szczecin, 126-138. (in Polish)

Giziński A., R. Wiśniewski, 1971. An attempt to determine the dynamics of number, biomass and production of main components of the profundal fauna in the southern part of the Lake Jeziorak. AUNC 27, Limnol. Pap., 6: 117-132.

Giziński A., N. Wolnomiejski, L. Mandziara, 1976. The production of the zoobenthos in the "pseudolittoral" of Lake Jeziorak. AUNC 38, Limnol. Pap., 9: 3-15.

Giziński A., M. Jagodzińska, N. Wolnomiejski, I. Grygiel, 1980. Zoobenthos of the muddy areas of Szczecin Lagoon. AUNC, Toruń, Pr. Limnol., 12: 41-59.

Giziński A., L.A. Błędzki, A. Kentzer, R, Wiśniewski, W. Zawiślak, J. Żbikowski, R. Żytkowicz, 1992. Abiotic and zoocenotic conditions for the functioning of selected lake ecosystems with various trophic levels. In: Some Ecological Processes of the Biological Systems in North Poland. Nicolas Copernicis Univ. Press, Toruń, 437-479.

Gmitrzuk K., 2004. Wpływ kormorana *Phalacrocorax carbo* na ekosystemy wodne i leśne Wigierskiego Parku Narodowego. *Effects of the black cormorant,* Phalacrocorax carbo, *on aquatic and forest ecosystems of the Wigry National Park.* Parki Nar. Rez. Przyr., 23: 129-146. (in Polish)

Grabowski A., 1973. The biomass, organic matter contents and calorific values of macrophytes in the lakes of the Szeszupa Drainage Area. Pol. Arch. Hydrobiol., 20: 269-282.

Graca B., Z. Witek, D. Burska, I. Białkowska, A. Pawelec, J. Bolałek, 2006. Porewater nutrients (phosphate, ammonia and silicate) in the eastern part of the southern Baltic Sea. Oceanol. Hydrobiol. Stud., 35: 237-256.

Graneli W., K. Sundbäck, 1986. Can microbenthic photosynthesis influence below-halocline oxygen conditions in the Kattegat. Ophelia, 26: 195-206.

Guénette S., V. Christensen, D. Pauly, 2008. Trophic modelling of the Peruvian upwelling ecosystem: Towards reconciliation of multiple datasets. Prog. Oceanog., 79: 326-335.

Grelowski A., M. Pastuszak, 1996. Odpływ wody oraz zrzuty zanieczyszczeń z polskich rzek do Bałtyku w latach 1988-1994. *Water discharge by and pollutant runoff from Polish rivers to the Baltic Sea in 1988-1994.* Stud. Mat. Mor. Inst. Ryb., Gdynia, Ser. A, 34: 5-67. (in Polish)

Gröger J.P., H. Winkler, R.A. Rountree, 2007. Population dynamics of pikeperch (*Sander lucioperca*) and its linkage to fishery driven and climatic influences in a Southern Baltic lagoon of the Darss-Zingst Bodden Chain. Fish. Res., 84: 189-201.

Gruszka P., 1999. The River Odra Estuary as a gateway for alien species immigration to the Baltic Sea basin. Acta Hydroch. Hydrobiol., 27: 374-382.

Grzybkowska M., A. Temech, M. Dukowska, 1996. Chironomid production in four lowland rivers (Central Poland). Pol. Arch. Hydrobiol., 43: 245-255.

Grzybkowska M., 2002. Paradoksy w hydrobiologii. *A few remarks about paradoxes in hydrobiology.* Kosmos, 51 (254): 115-120. (in Polish)

Grzybkowska M., 2008. Larwy Chironomidae (Diptera): adaptacje morfologiczne, behawioralne i fizjologiczne. *Chironomidae (Diptera) larvae. Morphological, behavioral and physiological adaptations.* Kosmos, 57 (278-279): 135-141. (in Polish)

Haputhantri S.S.K., M.C.S. Villanueva, J. Moreau, 2008. Trophic interactions in the coastal ecosystem of Sri Lanka: An ECOPATH preliminary approach. Est. Coast. Shelf Sci., 76: 304-318.

Harvey C.J., S.P. Cox, T.E. Essington, S. Hansson, J.F. Kitchell, 2003. An ecosystem model of food web and fisheries interactions in the Baltic Sea. ICES J. Mar. Sci., 60: 939-950.

Heip C., P.M.J. Herman, 1988. The production of meiofauna. ICES, C.M../L: 26.

Herman P.M.J., C. Heip, 1985. Secondary production of the harpacticoid copepod *Paronynchocamptus nanus* in brackish-water habitat. Limnol. Oceanog., 30: 1060-1066.

Heymans J.J., D. Baird, 2000. Network analysis of the northern Benguela ecosystem by means of NETWRK and ECOPATH. Ecol. Model., 131: 97-119.

Heymans J.J., S. Guénette, V. Christensen, 2007. Evaluating network analysis indicators of ecosystem status in the Gulf of Alaska. Ecosystems, 10: 488-502.

Hilbricht-Ilkowska A., E. Pieczyńska, E. Pieczyński, 1971. Z badań nad producktywnością kilku jezior mazurskich oraz wpływem ryb na biocenozę jeziora. *Studies on the productivity of several Masurian lakes and on the effect of fish on lake biocenosis.* Wiad. Ekol., 17, 2: 127-146. (in Polish).

Hilbricht-Ilkowska A., Z. Kajak, J. Ejsmont-Karabin, A. Karabin, J. Rybak, 1975. Ecosystem of Mikołajskie Lake. The utilization of the consumer production by invertebrate predators in pelagic and profundal zones. Pol. Arch. Hydrobiol., 22: 53-64.

Hily Ch., 1991. Is the activity of benthic suspension feeders a factor controlling water quality in the Bay of Brest. Mar. Ecol. Prog. Ser., 66: 179-188.

Hopkinson C.S., B. Sherr, W.J. Wiebe, 1989. Size fractionated metabolism of coastal microbial plankton. Mar. Ecol. Prog. Ser., 51: 155-166.

Humborg C., K. Fennel, M. Pastuszak, W. Fennel, 2000. A box model approach for a long-term assessment of estuarine eutrophication, Szczecin Lagoon, Southern Baltic. J. Mar. Sys., 25: 387-403.

Humphreys W.F., 1979. Production and respiration in animal populations. J. Anim. Ecol., 48: 427-453.

Humphreys W.F., 1981. Towards a simple index based on live-weight and biomass to predict assimilation in animal populations. J. Anim. Ecol., 50: 543-561.

Ismann S., T. Mehner, 1997. Predation by underyearyling perch (*Perca fluviatilis*) on a *Daphnia galeata* population in a short-term enclosure experiment. Freshw. Biol., 38: 209-219.

Jakuczun U., F. Nowacki, 1994. Program kompleksowych przedsięwzięć proekologicznych na wyspie Wolin. Program zagospodarowania i rozwoju wyspy Wolin. Warsztaty Ekologiczne. *A programme of complex environment-friendly actions on the Wolin Island. A management and development programme of the Wolin Island. Ecological Workshop.* Międzyzdroje, 49-101. (in Polish)

Janukowicz H., 2006. Zawartość chlorofilu a i feopigmentów w fitoplanktonie Zalewu Szczeińskiego jako element monitoringu wód. *Content of chlorophyll α and pheopigments in phytoplankton of Szczecin Lagoon, as an element of the water monitoring.* Woda-Środowisko-Obszary Wiejskie, 6: 133-138. (in Polish)

Jarosiewicz A., D. Ficek, 2008. Seasonal nutrient dynamics in Łebsko Lake. In: Bajkiewicz-Grabowska E., D. Borowiak (eds): Anthropogenic and natural transformations of lakes. 2. Wyd. KLUG-PTLim., Gdańsk, 67-70.

Jażdżewski K., A. Konopacka, M. Grabowski, 2005. Native and alien malacostracan Crustacea along the Baltic Sea coast in the twentieth century. Oceanol. Hydrobiol. Stud., 36: 175-193.

Job S.V., V. Kannan, 1980. The detritus limnology of Sathiar Reservoir. Hydrobiologia, 72: 81-84.

Johnson R.K., B. Bostroem, W. Van de Bund, 1989. Interactions between *Chironomus plumosus* (L.) and the microbial community in surficial sediments of a shallow, eutrophic lake. Limnol. Oceanog., 34: 992-1003.

Jurkowski J., I. Złoczowska, E. Wierzchowska, K. Jurkowska, J. Żurawska, 2003. Ocena jakości wód. *Water quality assessment.* In: Landsberg-Uczciwek M. (ed.): Raport o stanie środowiska w województwie zachodniopomorskim w roku 2001. *Report on the environmental quality of the Western Pomeranian District in 2001.* (in Polish)

Kaczewiak C., N. Wolnomiejski, 1977. Fosforany wód Zalewu Szczecińskiego w aspekcie eutrofizacyjnym. *Phosphates in the Szczecin Lagoon water from the eutrophication standpoint.* Stud. Mat. Mor. Inst. Ryb., Gdynia, Ser. A, 17: 33-49. (in Polish)

Kaczewiak C., 1990. Ocena stanu stada leszcza Zalewu Szczecińskiego. *Assessment of the Szczecin Lagoon bream stock.* In: Biologiczne podstawy racjonalnej gospodarki zasobami rybnymi Zalewu Szczecińskiego i Zatoki Pomorskiej. *Biological groundwork for rational management of fish resources of the Szczecin Lagoon and the Pomeranian Bay.* Sea Fish. Inst. Techn. Rep., Gdynia. (in Polish)

Kajak Z., 1988. Bentos. *Benthos.* In: Tarwid K. (ed.), Ekologia wód śródlądowych. *Ecology of inland waters.* PWN Warszawa, 235-313. (in Polish)

Kajak Z., 2001. Hydrobiologia – limnologia. Ekosystemy śródlądowe. *Hydrobiology – limnology. Inland water ecosystems.* Wyd. Nauk. PWN, Warszawa. (in Polish)

Kajak Z., J.I. Rybak, 1966. Production and some trophic dependences in benthos against primary production and zooplankton production of several Masurian lakes. Verh. Int. Ver. Limnol., 16: 441-451.

Kajak Z., A. Hilbricht-Ilkowska, E. Pieczyńska, 1972. The production processes in several Polish lakes. In: Kajak Z., A. Hilbricht-Ilkowska (eds): Productivity problems of freshwaters. Proc. IBP-UNESCO Symp., PWN Polish Scientific Publ., Warszawa-Kraków, 129-147.

Kajak Z., G. Bretschko, F. Schiemer, Ch. Leveque, 1980. Zoobenthos. In: Le Cren E.D., R.H. Lowe-McConnell (eds): The functioning of freshwaters ecosystems. Cambridge Univ. Press, Cambridge, 285-301.

Kakareko T., 2000. Ekologia leszcza *Abramis brama* (L.) w Zbiorniku Włocławskim. *Ecology of common bream Abramis brama (L.) in Włocławek Reservoir.* Ph.D. Thesis. Nicolaus Copernicus Univ., Toruń. (in Polish).

Kakareko T., 2001. The diet, growth and condition of common bream *Abramis brama* (L.) in Włocławek Reservoir. Acta. Ichthyol. Pisc., 21: 37-53.

Kaliciuk J., 2004. Obszary specjalnej ochrony ptaków sieci Natura 2000 w województwie zachodniopomorskim. *NATURA 2000 Special conservation areas in the Province of Western Pomerania.* In: Czeraszkiewicz R., A. Oleksiak (eds): Ptaki wodno-błotne na Pomorzu Zachodnim. Wyniki liczeń w sezonie 2003/2004, ekologia i ochrona. *Waterfowl in Western Pomerania. Results of censuses in 2003/2004, ecology and conservation..* Proj. Przyr.-Łow., ZTO-PZŁ, Szczecin, 34-37. (in Polish)

Kalisiński M., R. Czeraszkiewicz, J. Kaliciuk, D. Marchowski, A. Oleksiak, A. Staszewski, D. Wysocki, 2004. Zalew Szczeciński. *The Szczecin Lagoon.* In: Sidło P., B. Blaszkowska, P. Chylarecki (eds): Ostoje Ptaków o znaczeniu europejskim w Polsce. *IBAs of EU importance in Poland.* Ogólnopolskie Tow. Ochr. Ptaków, Warszawa, 52-56. (in Polish)

Kamiński M., 1992. Life cycle of the freshwater bryozoan *Plumatella fungosa* (Pall.). 2. Reproduction and general conclusions. Acta Hydrobiol., 34: 239-252.

Kasza H., 1995. Symptomy eutrofizacji a sukcesja ekologiczna w Goczałkowickim Zbiorniku Zaporowym. *Symptoms of eutrophication versus ecological succession in the Goczałkowice Dam Reservoir.* In: Zalewski M. (ed.): Procesy biologiczne w ochronie i rekultywacji nizinnych zbiorników zaporowych. *Biological processes in conservation and revitalization of dam reservoirs.* PIOŚ, Bibl. Monit. Środow., 143-153. (in Polish)

Kazantseva T.I., 2004. Balansovyy podkhod izucheniya potokov energii v ekosisteme presnovodnykh vodoyemov. *A budget-based approach to studying energy flows in freshwater ecosystems.* In: Alimov A.F., M.B. Ivanova (eds): Zakonomernosti gidrobiologicheskogo rezhima vodoyemov raznogo tipa. *Patterns of the hydrobiological regime in water bodies of various type.* Nauchny Mir, Moskva, 190-208. (in Russian)

Khlebovich T.V., 1979. Kolichestvennye pokazateli pitanya infusoriy. *Quantitative indicators of ciliate feeding.* In: Winberg G.G. (ed.): Obshchiye osnovy izuchenyja ekosistem. *Fundamentals of ecosystem research.* Izdat. Nauka, Leningrad, 100-106. (in Russian)

Kissler K., 2009. Model sieci troficznej Zalewu Wiślanego. *A model of the Vistula Lagoon trophic web.* M.Sc. Thesis, Akademia Pomorska w Słupsku, Słupsk. (in Polish)

Kitchell J.F., S.P. Cox, C.J. Harvey, T.B. Johnson, D.M. Mason, K.K. Schoen, K. Aydin, C. Bronte, M. Ebener, M. Hansen, M. Hoff, S. Schram, D. Schreiner, C.J. Walters, 2000. Sustainability of the Lake Superior fish community: Interactions in a food web context. Ecosystems, 3: 545–560.

Kluchareva O.A., A.A. Svetovidova, 1968. Zavisimost rosta ryb ot osobennostey kormovoy bazy v ozerakh yuga Sakhalinskoj oblasti. *Relationship between fish growth and food resource characteristics in lakes of the southern part of the Sakhalin region.* Vopr. Ikhtiol., 8: 1022-1033. (in Russian)

Knasiak M., D. Rutkowski, A. Tadajewski, 1990. Wpływ warunków hydrochemicznych Zalewu Szczecińskiego na chemizm wod Zatoki Pomorskiej z uwzględnieniem układów hydrologicznych estuarium Odry. *Effects of the Szczecin Lagoon water chemistry on the Pomeranian Bay chemical regime, with a consideration of hydrography of the Odra estuary.* In: Problemy Badań Morza w Regionie Szczecińskim. *Problems of Marine Research in the Szczecin Region.* Mat. Kom. Bad. Morza PAN, Szczecin, 71-79. (in Polish)

Kolasa J.M., 1972. Fauna denna strefy przybrzeżnej Roztoki Odrzańskiej. *Bottom fauna of the near-shore zone of Roztoka Odrzanska.* Zesz. Nauk. Wydz. BiNoZ Uniw. Gdańskiego, 1: 73-96. (in Polish)

Kompowski A., Z. Neja, 1990. Wstępna analiza możliwości eksploatacji ryb mniej wartościowych w Zalewie Szczecińskim. *A preliminary analysis of possibilities to exploit less valuable fish species in the Szczecin Lagoon.* Wyd. AR, Szczecin, Oprac. Tematyczne, 52: 3-20. (in Polish)

Kornijów R., 1996. Znaczenie bezkręgowców jako konsumentów zielenic nitkowatych i moczarki w litoralu jeziornym. *The importance of invertebrates as consumers of filamentous chlorophytes and* Elodea *in the lacustrine littoral.* Wyd. AR w Lublinie, Rozpr. Nauk., 188. (in Polish)

Koronkiewicz A., A. Garbacik-Wesołowska, E. Boberski, C. Kaczewiak, H. Czykieta, A. Wysokiński, 1997. Biologiczno-rybacka charakterystyka głównych gatunków ryb w Zalewie Szczecińskim i w Zatoce Pomorskiej w sezonie wegetacyjnym 1996 r. *Biological and fisheries characteristics of the principal fish species of the Szczecin Lagoon and Pomeranian Bay in the growing season of 1996.* Rap. Mor. Inst. Ryb. 1996, Gdynia, 183-212. (in Polish)

Kotta I., J. Kotta, 2001. Vertical migrations of mysids in the Gulf of Riga. Proc. Estonian Acad. Sci., Biol. Ecol., 50: 248-255.

Kotta I., J. Kotta, K. Herkül, 2007. Seasonal changes in the population structure and life cycle of *Neomysis integer* in Pärnu Bay, northeastern Baltic Sea. Proc. Estonian Acad. Sci., Biol. Ecol., 56: 312-325.

Kowalczewski A., 1975. Algal primary production in the zone of submerged vegetation of a eutrophic lake. Verh. Int. Verein Limnol., 19: 1305-1308.

Kowalewska-Kalkowska H., B. Wiśniewski, M. Kowalewski, 2005. Meteorologiczne uwarunkowania zmienności warunków hydrologicznych w rejonie ujściowym Odry – badania modelowe. *Meteorological effects in hydrogaphic variability in the River Odra mouth: model studies.* In: Borówka R.K., S. Musielak (eds): Środowisko przyrodnicze wybrzeży Zatoki Pomorskiej i Zalewu Szczecińskiego. *Natural environment of the Pomeranian Bay and Szczecin Lagoon coastal areas.* Szczecin, 103-112. (in Polish)

Kozlova M.F., V.A. Panasenko, 1977. Godovyye ratsiony i velichina izyatiya kormovykh organizmov populatsyami leshcha *Abramis brama* (L.) i yersha *Acerina cernua* (L.) v Kurskom Zalive. *Annual rations and the magnitude of removal of food organisms by populations of bream* Abramis brama *(L.) and ruffe* Acerina cernua *(L.) in the Curonian Lagoon.* Vopr. Ikhtiol., 7: 437-444. (in Russian)

Kraska M., 1990. Makrofity Jeziora Budzyńskiego. *Macrophytes in Lake Budzyńskie.* In: Kajak Z. (ed.): Funkcjonowanie ekosystemów wodnych, ich ochrona i rekultywacja. *Functioning of aquatic ecosystems, their conservation and revitalization.* Wyd. SGGW-AR, Warszawa, Part II, 23-35. (in Polish)

Krylova O.I., V.V. Ten, 1992. Mnogoletnaya dinamika i sovremennoye sostoyaniye zoobentosa Vislinskogo Zaliva. *Long-term dynamics and current status of the Vistula Lagoon zoobenthos.* In: Naumenko E.N. (ed.): Ekologicheskiye rybokhozaystvennyye issledovaniya v Vislinskom Zalive Baltiyskogo Mora. *Ecological and fisheries-related studies in the Vistula Lagoon of the Baltic Sea.* Sb. Nauchn. Tr. AtlantNIRO, 52-64. (in Russian)

Krzywosz T., 2008. Wpływ kormorana na zasoby naszych jezior. *Effects of the black cormorant on fish resources in our lakes.* Użytk. Ryb.-Nowa Rzeczyw., PZW, 90-96. (in Polish)

Krzyżanek E., H. Kasza, G. Pająk, 1993. The effect of water blooms caused by blue-green algae on the bottom macrofauna in the Goczałkowice Reservoir (southern Poland) in 1992. Acta Hydrobiol., 35:221-230.

Kuczyńska-Kippen, N., B. Mesyasz, B. Nagengast, 2004. Struktura zespołów peryfitonu w Jeziorze Wielkowiejskim. *The structure of the periphytic communities of the Wielkowiejskie Lake.* Roczn. Akad. Roln. Poznaniu, 363: 175-191. (in Polish)

Kuparinen J., A. Heinanen, 1993. Inorganic nutrient and carbon controlled bacterioplankton growth in the Baltic Sea. Est. Coast. Shelf Sci., 37: 271-285.

Lachica-Alino L., M. Wolff, L.T. David, 2006. Past and future fisheries modeling approaches in the Philippines. Rev. Fish. Biol. Fisheries, 16: 201–212.

Lacroix G., F. Lescher-Moutoué, A. Bertolo, 1999. Biomass and production of plankton in shallow and deep lakes: are there general patterns? Ann. Limnol., 35: 111-122.

Lajtha K., R.H. Michener, 1994. Stable isotopes in ecology and environmental science. Blackwell Sci. Publ., Oxford.

Lammens E.H.R.R., W. Hoogenboezem, 1991. Diets and feeding behaviour. In: Winfeld I.J., J.Nelson (eds): Cyprinid fishes: systematics, biology and exploitation. Chapman and Hall, London, 353-376.

Lampert W., U. Sommer, 2001. Ekologia wód śródlądowych. *Ecology of inland waters.* Wyd. Nauk. PWN, Warszawa. (Polish edition)

Latała A., 1993. Fotosynteza i oddychanie roślin. *Plant phytosynthesis and respiration*. In: Korzeniewski K. (ed.): Zatoka Pucka. *The Puck Bay*. Inst. Oceanogr. UG, Gdynia, 475-492. (in Polish)

Lazarenko N.N., A. Majewski (eds.), 1975. System hydrometeorologiczny Zalewu Wiślanego. *The hydrometeorological system of the Vistula Lagoon*.Wyd. Komun. Łączn., Warszawa. (in Polish)

Leipe T., J. Eidam, R. Lampe, H. Meyer, T. Neumann, A. Osadczuk, W. Janke, T. Puff, T. Blanz, F.X. Gingele, D. Dannenberger, G. Witt, 1998. Das Oderhaff-Beiträge zur Rekonstruktion der holozänen und geologischen Entwicklung und anthropogenen Beeinflussung des Oder Ästuars. Mar. Sci. Rep., 28: 1-83.

Lewandowski K., 1982. O zmiennych liczebnościach *Dreissena polymorpha* (Pall.). *On the variable numbers of* Dreissena polymorpha *(Pall.)*. Wiad. Ekol., 28: 141-154. (in Polish)

Lewandowski K., A. Stańczykowska, 1975. The occurrence and role of bivalves of the family Unionidae in Mikołajskie Lake. Ekol. Pol., 23: 317-334.

Lewis W.M., 2011. Global primary production of lakes: 19th Baldi Memorial Lecture. Inland Wat., 1: 1-28.

Libralato S., C. Solidoro, 2009. Bridging biogeochemical and food web models for and end-to-end representation of marine ecosystem dynamics: the Venice Lagoon case study. Ecol. Model., 220: 2960-2971.

Lin H.-J., K.-T. Shao, S.-R. Kuo, H.-L. Hsieh, S.-L. Wong, I.-M. Chen, W.-T. Lo, J.-J. Hung, 1999. A trophic model of a sandy barrier lagoon at Chiku in Southwestern Taiwan. Est. Coast. Shelf Sci., 48: 575–588.

Lin H.-J., X.-X. Dai, K.-T. Shao, H.-M. Su, W.-T. Lo, H.-L. Hsieh, L.-S. Fang, J.-J. Hung, 2006. Trophic structure and functioning in a eutrophic and poorly flushed lagoon in southwestern Taiwan. Mar. Envir. Res., 62: 61–82.

Link J.S., Griswold, C.A., Methratta, E.M., Gunnard, J. (eds), 2006. Documentation for the energy modeling and analysis eXercise (EMAX). Northeast Fish. Sci. Cent. Ref. Document., 06–15.

Link J., L. Col, V. Guida, D. Dow, J. O'Reilly, J. Green, W. Overholtz, D. Palka, C. Legault, J. Vitaliano, C. Griswold, M. Fogarty, K. Friedland, 2009. Response of balanced network models to large-scale perturbation: Implications for evaluating the role of small pelagics in the Gulf of Maine. Ecol. Model., 220: 351–369.

Ložys L., 2003. Seasonal migrations of pikeperch (*Sander lucioperca* L.) from the Curonian Lagoon to the Baltic Sea and advantages of the phenomenon. Acta Zool. Lituan., 13: 188-198.

Luścińska M., 2005. Phytoplankton of the Szczecin Lagoon (The Great Lagoon). Oceanol. Hydrobiol. Stud., 34. 1: 3-29.

Maciejewska K., K. Opaliński, 2002. Daily energy requirement in the mysid shrimp *Neomysis integer* in the Vistula Lagoon. Bull. Sea Fish. Inst., 3 (157): 19-31.

Maciejewska K., K.W. Opaliński, 2004. Bioenergetics of early life stages of fish species from Vistula Lagoon. In: Aquaterra. Mat. Conf. Russian Ministry Natural Res., St. Petersburg, 132-134.

Maciejewska K., K.W. Opaliński, 2010. Does ecological energetics lead to trophic competition in juvenile fish? A case study from the Vistula Lagoon (southern Baltic Sea). Mar. Poll. Bull., 61: 178-182.

Maciejewska K., K.W. Opaliński, P. Margoński, 2001. Energy budget in early stages of smelt *Osmerus eperlanus* (L.). Bull. Sea Fish. Inst., 2 (153): 53-61.

Mackie G.L., W.N. Gibbons, B.W. Muncaster, I.M. Gray, 1989. The zebra mussel, *Dreissena polymorpha*: a synthesis of European experiences and a preview for North America. Rep. Water Resources Branch, Great Lakes Section. Queen's Printers for Ontario.

Madsen N.N., P. Nilsson, K. Sundbäck, 1993. The influence of benthic microalgae on the stability of subtidal sediment. J. Exp. Mar. Biol. Ecol., 170: 159-177.

Maitland P.S., N.W. Charles, N.C. Morgan, K. East, M.C. Gray, 1972. Preliminary research on the production of Chironomidae in Loch Leven. In: Kajak Z., A Hilbricht-Ilkowska (eds): Productivity problems of freshwaters. Proc. IBP-UNESCO Symp., PWN Polish Sci. Publ., Warszawa-Kraków, 795-812.

Majewski A., 1972. Charakterystyka hydrologiczna estuariowych wód u polskiego wybrzeża. *Hydrological characteristics of estuarine areas at the Polish coast.* Pr. PIHM, 105: 3-40. (in Polish)

Majewski A. 1974. Charakterystyka hydrologiczna Zatoki Pomorskiej. *A hydrological characteristics of the Pomeranian Bay.* Wyd. Komun. Łączn., Warszawa. (in Polish)

Majewski A. (ed.), 1980. Zalew Szczeciński. *The Szczecin Lagoon.* Wyd. Komun. Łączn., Warszawa. (in Polish)

Malej J., 1974. Fauna denna w zanieczyszczonym estuarium. *The bottom fauna in a polluted estuary.* Stud. Mat. Mor. Inst. Ryb. Gdyni, Ser. A 13: 7-83. (in Polish)

Mann K.H., 1971. Use of the Allen curve method for calculating benthic production. In: Edmondson W.T., G.G. Winberg (eds): A manual on methods for the assessment of secondary productivity in freshwaters. IBP Handbook, 17: 160-165.

Mann K.H., 1988. Production and use of detritus in various freshwater, estuarine and coastal marine ecosystems. Limnol. Oceanog., 33: 910-930.

Mann K.H., 2000. Ecology of coastal waters, with implications for management. Blackwell Sci. Inc.

Mann K.H., R.H. Britton, A. Kowalczewski, T.J. Lack, Ch.P. Mathews, I. McDonald, 1972. Productivity and energy flow at all trophic levels in the River Thames, England. In: Kajak Z., A Hilbricht Ilkowska (eds): Productivity problems of freshwaters. Proc. IBP-UNESCO Symp., PWN Polish Sci. Publ., Warszawa-Kraków, 579-596.

Marten G.G., 1979. Predator removal: Effect on fisheries yields in Lake Victoria (East Africa). Science, 203: 646–648.

Martyniak A., J. Terlecki, U. Szymańska, B. Wziątek, P. Hliwa, J. Kozłowski, S. Boroń, H. Kamińska, 1998. Odżywianie się kormorana czarnego *Phalacrocorax carbo sinensis* z kolonii legowej usytuowanej w Kątach Rybackich. *Feeding of the black cormorant* Phalacrocorax carbo sinensis *from a breeding colony at Kąty Rybackie.* In: Symp. Ptaki wodne a rybactwo. *Symp. Waterfowl versus Fisheries.* Gdynia, 2-3. (in Polish)

Margoński P., K. Horbowa, S. Gromisz, Z. Witek, M. Zalewski, J. Warzocha, 2003. Report on trophic and ecological status of the Vistula Lagoon. MANTRA-East Deliverable D3eVL, 47.

Masłowski J., 1992. Bottom macrofauna of the Szczecin Lagoon (north-western Poland). Acta Hydrobiol., 34: 253-274.

Masłowski J., 1993. Long-term changes in the bottom macrofauna of the Szczecin Lagoon (north-western Poland). Acta Hydrobiol., 35: 341-355.

Masłowski J., 2010. Makrofauna denna Zatoki Pomorskiej w latach 1976-2005. *Macrobenthos of the Pomeranian Bay in 1976-2005.* Wyd. Adam Marszałek, Toruń. (in Polish).

Mastitsky S.E., V.M. Samoilenko, 2005. Larvae of chironomids (Insecta, Diptera) encountered in the mantle cavity of zebra mussels *Dreissena polymorpha* (Bivalvia, Dreissenidae). Int. Rev. Hydrobiol., 90: 42-50.

Mattice J.S., 1972, Production of a natural population of *Bithynia tentaculata* L. (Gastropoda, Mollusca). Ekol. Pol., 29: 525-539.

McCauley E., 1984. The estimation of the abundance and biomass of zooplankton in samples. In:. Downing J.A., F.H. Rigler (Eds). A manual on methods for the assesment of secondary productivity in fresh waters. 2nd Edition. Blackwell Sci. Publ. Oxford, England: 228-265

McIntyre A.D., 1969. Ecology of marine meiobenthos. Biol. Rev., 44: 245-288.

McLusky D.S., M. Elliott, 2004. The estuarine ecosystem: ecology, threats, and management. Oxford Univ. Press, Oxford.

Megrey B.A., K.Y. Aydin, 2009. A macrodescriptor perspective of ecological attributes for the Bering and Barents Seas. Deep-Sea Res. II, 56: 2132–2140.

Mehner T., R. Thiel, 1999. A review of predation impact by 0+ fish on zooplankton in fresh and brackish waters of the temperate northern hemisphere. Envir. Biol. Fish., 56: 169-181.

Meire P.M., 1993. The impact of bird predation on marine and estuarine bivalve populations: a selective review of patterns and underlying causes. NATO ASI Series, G33: 197-243.

Meire P.M., J. Seys, J. Buijs, J. Coosen, 1994. Spatial and temporal patterns of intertidal macrobenthic populations in the Oosterschelde: Are they influenced by the construction of the storm-surge barrier? Hydrobiologia, 282-283: 157-182.

Merilainen J.J., 1988. Meiobenthos in relation to macrobenthic communities in a low saline, partly acidified estuary, Bothnian Bay, Finland. Ann. Zool. Fenn., 25: 277-292.

Meyer H., R. Lampe, 1999. The restricted buffer capacity of a south Baltic estuary – the Oder estuary. Limnologica, 29: 242-248.

Miączyńska M., 1997. Peryfiton rozwijający się na pasmach folii polietylenowej zatopionych w Jeziorze Wicko Małe w pobliżu dopływu wód pościekowych z Międzyzdrojów. *Periphyton developing on polyethylene foil bands deployed in Lake Wicko Małe in the vicinity of the Międzyzdroje sewage treatment plant discharge.* M.Sc. Thesis, Uniw. Szczeciński. (in Polish)

Michelsen K., J. Pedersen, K. Christoffersen, F. Jensen, 1994. Ecological consequences of food partitioning for the fish population structure in a eutrophic lake. Hydrobiologia, 291: 35-45.

Middelboe M., B. Nielsen, M. Sondergaard, 1992. Bacterial utilization of dissolved organic carbon (DOC) in coastal waters – determination of growth yield. In: Bjørnsen P.K., Riemann B. (eds): Microbial ecology of pelagic environments. Ergeb. Limnol. Adv. Limnol., 37: 51-61.

Mikulski J.St., B. Adamczak, I. Bittel, R. Bohr, D. Bronisz, W. Donderski, A. Giziński, M. Luścinska, M. Rejewski, E. Strzelczyk, N. Wolnomiejski, W. Zawiślak, R. Żytkowicz, 1975. Basic regularities of productive processes in the Iława Lakes and in the Gopło Lake from the point of view of utility values of the water. Pol. Arch. Hydrobiol., 22: 101-122.

Minning M., 2004. Der Einfluss des Schiffahrtskanals auf den Stoffhaushalt des Oderhafts. In: Schernewski G., T. Dolch (eds): The Oder Estuary – against the background of the European Water Framework Directive. Mar. Sci. Rep., 57: 179-222.

Młodzińska Z., 1980. Hydrochemia. *Hydrochemistry.* In: Majewski A. (ed.): Zalew Szczeciński. *The Szczecin Lagoon.* Wyd. Komun. Łączn., Warszawa, 248-276. (in Polish)

Moloney C.L., A. Jarre, H. Arancibia, Y.-M. Bozec, S. Neira, J.-P. Roux, L.J. Shannon, 2005. Comparing the Benguela and Humboldt marine upwelling ecosystems with indicators derived from inter-calibrated models. ICES J. Mar. Sci., 62: 493-502

Monaco M.E., R.E. Ulanowicz, 1997. Comparative ecosystem trophic structure of three US. mid-Atlantic estuaries. Mar. Ecol. Prog. Ser., 161: 239-254.

Monakov A.V., 1974. Osnovnyye rezultaty issledowaniy IBVV AN SSSR po pitaniyu vodnykh bezpozvonochnykh. *Principal results of studies of IBVV AN SSSR on feeding of aquatic invertebrates.* In: Butorin I.V. (ed.): Biologia i produktivnost presnovodnykh bezpozvonochnykh. *Biology and productivity of freshwater invertebrates.* Tr. Inst. Biol. Vnutr. Vod, 25: 3-36. (in Russian)

Monakov A.V., 1998. Pitaniye presnovodnykh bezpozvonochnykh. *Feeding of the freshwater invertebrates.* Ros. Akad. Nauk, Moskva. (in Russian)

Mooij W.M., W.L.T. Van Densen, E.H.R.R. Lammens, 1996. Formation of year-class strength in the bream population in the shallow eutrophic Lake Tjeukemeer. J. Fish. Biol., 48: 30-39.

Morales-Zárate M.V., F. Arreguin-Sánchez, J. López-Martinez, S.E. Lluch-Cota, 2004. Ecosystem trophic structure and energy flux in the Northern Gulf of California, México. Ecol. Model., 174: 331–345.

Moreau J., K. Mavuti, T. Daufresne, 2001. A synoptic Ecopath model of biomass flows during two different static ecological situations in Lake Nakuru (Kenya). Hydrobiologia, 458: 63–74.

Morgan N.C., T. Backiel, G. Bretschko, A. Duncan, A. Hilbricht-Ilkowska, Z. Kajak, J.F. Kitchell, P. Larson, C. Lévêque, A. Nauwerck, F. Schiemer, J.E. Thorpe, 1980. Secondary production. In: Le Cren E.D., R.H. Lowe-McConnell (eds): The functioning of freshwater ecosystems. Int. Biol. Prog. 22, Cambrige Univ. Press, Cambridge, 247- 339.

Mohrholz V., M. Pastuszak, S. Sitek, K. Nagel, H.U. Lass, 1998. The exceptional Oder flood in summer 1997 – riverine mass nutrient transport into the Pomeranian Bight. Ger. J. Hydrogr., 50: 129-144.

Morissette L., T. Pedersen, M. Nilsen, 2009. Comparing pristine and depleted ecosystems: The Sørfjord, Norway versus the Gulf of St. Lawrence, Canada. Effects of intense fisheries on marine ecosystems. Prog. Oceanog., 81: 174–187.

Moss B., M. Tims. 1989. Predation, sediment stability and food availability as determinants of the benthic invertebrate fauna in two shallow lakes. Hydrobiologia. 185: 249-257.

Mudryk Z (ed.), 2003. Jezioro Gardno *Lake Gardno.* Wydawn. PAP. Słupsk. (in Polish).

Musielak S., S. Wochna, 2005. Typy i rodzaje brzegów Zalewu Szczecińskiego. *Shore types and classes in the Szczecin Lagoon.* In: Borówka R.K., S. Musielak (eds):

Środowisko przyrodnicze wybrzeży Zatoki Pomorskiej i Zalewu Szczecińskiego. Natural environment of the Pomeranian Bay and Szczecin Lagoon coastal areas. Szczecin, 79-92. (in Polish).

Mutko T. (ed.), 1986. Raport o stanie zanieczyszczenia Zalewu Szczecińskiego. *Report on pollution of the Szczecin Lagoon.* PIOŚ, Szczecin. (in Polish)

Mutko T., J. Stechman, M. Landsberg-Uczciwek, A. Garbacik-Wesołowska, M. Protasowicki, E. Niedźwiecki, W. Ciereszko, E. Meller, 1994. Zalew Szczeciński, Wielki Zalew. Zmiany jakościowe w wieloleciu. *The Szczecin Lagoon. The Great Lagoon. Long-term water quality changes.* Bibl. Monit. Środ., PIOŚ, Warszawa. (in Polish)

Nagelkerke L.A.J., F.A. Sibbing, 1996. Efficiency of feeding on zebra mussel (*Dreissena polymorpha*) by common bream (*Abramis brama*), white bream (*Blicca bjoerkna*), and roach (*Rutilus rutilus*): The effects of morphology and behaviour. Can. J. Fish. Aquat. Sci., 53: 2847-2861.

Nagięć M., 1964. Wzrost i próba oceny śmiertelności sandacza (*Lucioperca lucioperca* (L) z Wisły. *Growth and an attempt to estimate the mortality rate of zander (*Lucioperca lucioperca *(L.) in River Vistula.* Roczn. Nauk. Roln., 84-B-2: 329-344. (in Polish)

Niedźwiecka J., 2000. Fitoplankton północno-wschodniej części Zalewu Szczecińskiego (Zalewu Wielkiego). *Phytoplankton in the north-eastern part of the Szczecin Lagoon (the Great Lagoon).* M.Sc. Thesis, Uniw. im. M. Kopernika, Toruń. (in Polish)

NMFS, 1999. Ecosystem-based fishery management: A Report to Congress by the Ecosystem Principles Advisory Panel. U.S. Dep. Commer., NOAA, National Marine Fisheries Service, Silver Spring, MD.

Nowak B., 1980. Zawiesiny. *Suspended particulates.* In: Majewski A. (ed.): Zalew Szczeciński. *The Szczecin Lagoon.* Wyd. Komun. Łączn., Warszawa, 239-247. (in Polish)

Obolewski K., 2002. Organizmy epifityczne (peryfiton) zamieszkujące trzcinę *Phragmites australis* i pałkę *Typha latifolia* oraz sztuczne podłoże w Jeziorze Lubowidzkim – badania wstępne *Epiphytic organisms (periphyton) inhabiting reed* Phragmites australis *and* Typha latifolia *and artificial substrates in Lake Lubowidzkie – initial research.* Słupsk. Pr. Przyr., Ser. Limnol., 1: 71-82. (in Polish)

Odum E.P., 1969. The strategy of ecosystem development. Science, 104: 262–270.

Odum E.P., 1971. Fundamentals of ecology. W.B. Saunders Co., Philadelphia.

Oleksowicz A.S., 1986. Planktonic and epiphytic algal communities in three, limnology different lakes of Tuchola Forests area. II. Characteristics of components, seasonal quantitative changes and species diversity. AUNC, Prace Limnol., 15: 49-73.

Olsza E., 1973. Analiza połowów w Zalewie Szczecińskim w latach 1948-1971. *Analysis of fish catches in the Szczecin Lagoon in 1948-1971.* In: Symp. Nauk. Świnoujściu. Wyd. Mor. Inst. Ryb., Gdynia, 48-56. (in Polish)

Olszewski, P 1971. Trofia i saprobia. *Trophic status and saproby.* Zesz. Nauk. WSR, Olsztyn, Ser. C, suppl. 3: 3-19. (in Polish)

Opaliński K.W., K. Maciejewska, A. Krajewska-Sołtys, D.P. Fey, 2004. Production and oxygen consumption in the early life stages of herring and smelt in the Vistula Lagoon (Baltic Sea). Bull. Sea Fish. Inst., 2 (162): 13-22.

Opaliński K.W., K. Maciejewska, P. Bieńkowski, 2008. Energetic value of the early developmental stages of smelt from Vistula Lagoon. Tr. VI Jubil. Mezhd. Nauch. Konf. "Innovatsii v nauke i obrazovanii-2008". Part 1. Izd. KGTU, Kaliningrad.

Opitz S., 1996. Trophic interactions in Caribbean coral reefs. ICLARM Tech. Rep., 43.

Orlova M., S. Golubkov, L. Kalinina, N. Ignatieva, 2004. *Dreissena polymorpha* (Bivalvia, Dreissenidae) in the Neva Estuary (eastern Gulf of Finland, Baltic Sea): Is it biofilter or source of pollution? Mar. Poll. Bull., 49: 196-205.

Osadczuk A., 2004. Zalew Szczeciński – środowiskowe warunki współczesnej sedymentacji lagunowej. *The Szczecin Lagoon: environmental conditions of recent lagoonal sedimentation.* Rozpr. Stud. Uniw. Szczec., 1123, 549: 5-156. (in Polish)

Osadczuk A., S. Musielak, R.K. Borówka, 2007 Why should the Odra River mouth area not be regarded as an estuary? A geologist's point of view. Oceanol. Hydrobiol. Stud., 26: 87-99.

Osadczuk A., R. Lampe, H. Meyer, D. Sobieraj, 1996. Characteristic of recent sediments from the Zalew Szczeciński. Rozpr. Hydrotechn., 60. IBW PAN Gdańsk.

Österblom H., S. Hansson, U. Larsson, O. Hjerne, F. Wulff, R. Elmgren, C. Folke, 2007. Human-induced trophic cascades and ecological regime shifts in the Baltic Sea. Ecosystems, 10: 877–889.

Ozimek T., 1983. Biotic structure and processes in the lake system of R. Jorka watershed (Masurian Lakeland, Poland). X. Biomass and distribution of submerged macrophytes. Ekol. Pol., 31: 781-792.

Parsons T.R., M Takahashi, B. Hargrave, 1977. Biological oceanographic processes. Pergamon Press, Oxford.

Pastuszak M., Z. Witek, K. Nagel, M. Wielgat, A. Grelowski, 2005. Role of the Oder estuary (southern Baltic) in transformation of the riverine nutrient loads. J. Mar. Sys., 57: 30-54.

Pastuszak M., S. Sitek, A. Grelowski, 1988. The exceptional Oder flood in summer 1997 – Nutrient concentrations in the Świna Strait during the years 1996 and 1997 – with emphasis on the flood event. Ger. J. Hydrogr., 50: 183-202.

Pastuszak M., D.J. Conley, Ch. Humborg, Z. Witek, S. Sitek, 2008. Silicon dynamics in the Oder estuary, Baltic Sea. J. Mar. Sys., 73: 250-262.

Pastuszak M., Z. Witek, 2009a. Odpływ wód i substancji biogenicznych Wisła i Odrą z obszaru Polski. *The discharge of water and nutrients withe the Vistula and Odra runoff from Poland.* In: Igras J., M. Pastuszak (eds): Udział polskiego rolnictwa w emisji związków azotu i fosforu do Bałtyku. *Contribution of Polish agriculture to emission of nitrogen and phosphorus compounds to the Baltic Sea.* IUNG-PIB, Puławy, 271-306. (in Polish)

Pastuszak M., Z. Witek, 2009b. Rola estuariów Odry i Wisły w retencji ładunków azotu i fosforu wprowadzanych do Bałtyku. *The role of Odra and Vistula estuaries in the retention of nitrogen and phosphorus loads discharged into the Baltic Sea.* In: In: Igras J., M..Pastuszak (eds): Udział polskiego rolnictwa w emisji związków azotu i fosforu do Bałtyku. *Contribution of Polish agriculture to emission of nitrogen and phosphorus compounds to the Baltic Sea.* IUNG-PIB, Puławy, 307-329. (in Polish).

Paszkiewicz C., W. Jaworski, 1980. Falowanie wiatrowe. *Wind-induced wave action.* In: Majewski A. (ed.): Zalew Szczeciński. *The Szczecin Lagoon.* Wyd. Komun. Łączn., Warszawa, 157-166. (in Polish)

Patricio J., J.C. Marques, 2006. Mass balanced models of the food web in three areas along a gradient of eutrophication symptoms in the south arm of the Mondego estuary (Portugal). Ecol. Model., 197: 21–34.

Paturej E., 2006. Estuaries of the Polish Baltic coastal zone. Baltic Coast. Zone, 10: 83-96.

Pauly D., V. Christensen, J. Dalsgaard, R. Froese, F. Torres Jr., 1998. Fishing down marine food webs. Science, 279: 860–863.

Pawlik A., 1976. Makroproblemy. Aura, 4: 2-4. (in Polish).

Pełszyńska A., 1973. Fitoplankton Zalewu Szczecińskiego. *The Szczecin Lagoon phytoplankton.* M.Sc. Thesis, Uniw. im. M. Kopernika, Toruń. (in Polish)

Pempkowiak J., G. Kupryszewski, 1980. The input of organic matter to the Baltic from the Vistula River. Oceanologia, 12: 79-98.

Penczak T., 1992. Fish production in the Warta River, Poland: a preimpoundment study. Hydrobiologia, 237: 117-129.

Penczak T., 2000. Wybrane zagadnienia z ekologii ryb słodkowodnych. *Selected problems of freshwater fish ecology.* In: Brylińska M. (ed.): Ryby słodkowodne Polski. *The freshwater fish of Poland.* Wyd. Nauk. PWN, Warszawa, 87-95. (in Polish)

Penczak T., M. Zalewski, M. Moliński, M. Gajos, 1977. The ecology of roach, *Rutilus rutilus* (L.) in the barbel region of the polluted Pilica River. IV. Elements of production and food consumption. Ekol. Pol., 25: 241-255.

Penczak T., E. Suszycka, M. Moliński, 1982. Production, consumption and energy transformation by fish populations in a small lowland river. Ekol. Pol., 30: 111-137.

Penczak T., K. Kusto, D. Krzyżanowska, M. Moliński, E. Suszycka, 1984. Food consumption and energy transformations by fish populations in two small lowland rivers in Poland. Hydrobiologia, 108: 135-144.

Penczak T., I. Tátrai, 1985. Contribution of bream *Abramis brama* (L.) to the nutrient dynamics of Lake Balaton. Hydrobiologia, 126: 59-64.

Pereyra-Ramos E., 1981. The ecological role of Characeae in the lake littoral. Ekol. Pol., 29: 167-209.

Pęcherzewski K., 1991. Substancje wnoszone z atmosfery do wód południowego Bałtyku. *Substances deposited from the atmosphere in the southern Baltic waters.* Rozprawy AR Szczecinie, 141. (in Polish)

Pęczalska A., 1973. Ichtiofauna Zalewu Szczecińskiego. Jej specyfika i zmiany zachodzące w składzie gatunkowym ryb w powojennym 25-leciu. *The Szczecin Lagoon fish fauna, its specificity and changes in species composition in the 25 post-war years.* In: Symp. Nauk. Świnoujściu. Ośr. Wyd. Mor. Inst. Ryb., Gdynia, 32-37. (in Polish)

Pęczalska A., 1974. Nie wykorzystane zasoby ryb w Zalewie Szczecińskim. *Unutilised fish resources in the Szczecin Lagoon.* Techn. Gosp. Mor., 24: 205-206. (in Polish)

Pęczalska A., W. Kraczkiewicz, 1972. Wyniki badań nad zasobami rybnymi i gospodarką rybacką na Zalewie Szczecińskim i Zatoce Pomorskiej. *Results of studies on fish resources and fisheries in the Szczecin Lagoon and Pomeranian Bay.* In: Ekosyst. Morskie. Badania zalewów, 4. Mor. Inst. Ryb., Gdynia, 133-196. (in Polish)

Piasecki W., 1993. Próba określenia stanu Jeziora Nowowarpieńskiego i południowej części Zalewu Szczecińskiego na podstawie badań zooplanktonu i testów biologicznych. *An attempt to assess the environmental status of Lake Nowowarpieńskie and the southern part of the Szczecin Lagoon from zooplankton studies and biotests.* Stud. Mat. Mor. Inst. Ryb., Świnoujście, Ser. S: 65-71. (in Polish)

Pieczyńska E., 1964. Investigations on colonization of new substrates by nematodes (Nematoda) and some other periphyton organisms. Ekol. Pol. A, 12: 185-234.

Piesik Z., 1992. Biologiczna i ekologiczna rola organizmów porośIowych (perifiton) zasiedlających sztuczne podłoża w różnych typach wód. *The biological and ecological role of the fouling organisms inhabiting artificial substratas in various types of water.* Rozpr. Stud. Uniw. Szczecińskiego, CXCVI. (in Polish)

Piesik Z., K. Obolewski, 2000. Epiphytic organisms (periphyton) inhabiting reed, *Phragmites australis* and artificial substrates in Lake Kopań. Baltic Coast. Zone, 4: 73-86.

Piesik Z., K. Obolewski, 2001. Characteristics on an epiphytic assemblage inhabiting reed, *Phragmites australis* (Cav.) Trin. ex. Steud. in Bukowo Lake. Baltic Coast. Zone, 5: 41-49.

Piesik Z., K. Obolewski, M. Wiśniewska, 2002/2003. Fouling organisms (periphyton) inhabiting common reed, *Phragmites australis* in Lake Jamno. Baltic Coast. Zone, 5: 91-100.

Piesik Z., B. Wawrzyniak-Wydrowska, 2003. Organizmy porośłowe (perifiton) zasiedlające trzcinę (*Phragmites australis*) w ujściowym odcinku rzek Odry i Gunicy. *Reed (Phragmites australis) periphyton in the mouth area of rivers Odra and Gunica.* In: Rogalska S.M., J. Domagała (eds): Człowiek i środowisko przyrodnicze Pomorza Zachodniego. *The man and the natural environment in Western Pomerania.* Uniw. Szczec., 161-172. (in Polish)

Pikitch E.K., C. Santora, E.A. Babcock, A. Bakun, R. Bonfil, D.O. Conover, P. Dayton, P. Doukakis, D. Fluharty, B. Heneman, E.D. Houde, J. Link, P.A. Livingston, M. Mangel, M.K. McAllister, J. Pope, K.J. Sainsbury, 2004. Ecosystem-Based Fishery Management. Science, 305: 346-347.

Pinnegar J.K., J.L. Blanchard, S. Mackinson, R.D. Scott, D.E. Duplisea, 2005. Aggregation and removal of weak-links in food-web models: system stability and recovery from disturbance. Ecol. Model., 184: 229–248.

Piotrowska H., 1966. Stosunki geobotaniczne wysp Wolina i południowo-wschodniego Uznamu. *The geobotanical regime of the Island of Wolin and the south-eastern Usedom.* Monogr. Botan., 22. PWN Warszawa. (in Polish)

Pivnička K., 1982. Long-termed study of fish populations in the Kličava Reservoir. Přirodov. Práce ustavu Českosl. Akad. Věd., Brno., Acta Sc. Nat. Brno, 10. Czech. Acad. Sci., Praha.

Pliński M., 1972a. Zmiany dominacji gatunków w fitoplanktonie Zalewu Szczecińskiego w latach 1968 i 1969. *Changes in the domination structure in the Szczecin Lagoon phytoplankton in 1968 and 1969.* In: Siudziński K. (ed.), Ekosyst. Morskie, 4. PAN-MIR, Gdynia, 37-62. (in Polish)

Pliński M., 1972b. Badania nad glonami południowego Bałtyku oraz Zalewu Szczecińskiego i Wiślanego. *Studies on algae of the southern Baltic and the Szczecin and Vistula Lagoons.* Zesz. Nauk. Wydz. BiNoZ Uniw. Gdańsk., 1: 59-72. (in Polish)

Pliński M. 1979. Kierunki zmian strukturanych fitoplanktonu estuariów Bałtyku Południowego. *Trends in changes of the phytoplankton structure in southern Baltic estuaries.* Uniw. Gdański. Rozpr. Monogr., 15, 1-136. (in Polish)

Pliński M., B. Kreńska, T. Wnorowski, 1978. Floristic relations and biomass of vascular plants in the Vistula Lagoon. Stud. Mater. Oceanol., 21, Biologia Morza, 4: 161-196.

Pliszka F., 1956. Znaczenie organizmów wodnych jako pokarmu ryb w świetle badań polskich. *The importance of aquatic organisms as fish food in the light of Polish research.* Pol. Arch. Hydrobiol., 3: 429-458. (in Polish)

Poddubnyy A.G., A.J. Bakanov, 1980. O kolichestvennoy otsenke wyedaniya bentosa rybami. *On the quantitative estimation of fish feeding on the benthos.* Vopr. Ikhtiol., 20: 888-896. (in Russian)

Podolska A.Z., 1977. Zooplankton Zalewu Szczecińskiego. *Zooplankton of the Szczecin Lagoon.* M.Sc. Thesis, Uniw. im. M. Kopernika, Toruń. (in Polish)

Podsetchine V., G. Schernewski, I. Tejakusma, 2004. Simulations of water flow, transport and temperature in the Oder (Szczecin) Lagoon. In: Schernewski G., T. Dolch (eds): The Oder Estuary – against the background of the European Water Framework Directive. Mar. Sci. Rep., 57: 85-97.

Poleszczuk G., 1994. Kryteria oceny stanu środowiska wodnego w otulinie Wolińskiego Parku Narodowego. *Criteria for the aquatic environment status assessment in the buffer zone of the Wolin National Park.* In: Program zagospodarowania i rozwoju wyspy Wolin. *A programme of management and development of the Wolin Island.* Warszt. Ekol. Międzyzdroje. Wyd. "Dokument", Szczecin, 131-135. (in Polish)

Poleszczuk G., 1995. Zarys hydrografii i wprowadzenie do charakterystyki hydrochemicznej warunków środowiska wodnego estuarium rzeki Odry. *The outline of hydrography and introduction to hydrochemical characterization of the condition of the aquatic environment in the River Odra estuary.* Stud. Mat.. Mor. Inst. Ryb., Gdynia, Ser. A, 33: 7-20. (in Polish)

Poleszczuk G., 1996. On variability of mineral composition of water in Roztoka Odrzańska (Odra River mouth, Poland). Oceanol. Stud., 3: 39-53.

Poleszczuk G., 1997. Chemiczna charakterystyka wód Zalewu Szczecińskiego (Zalewu Wielkiego) jako środowiska ichtiofauny. *Chemical characteristics of waters in the Szczecin Lagoon (The Great Lagoon) as an ichthyofauna settlement.* Akad. Roln., Szczecin. Rozprawy 179. (in Polish)

Poleszczuk G., 1998. Środowisko abiotyczne toni wodnej Zalewu Szczecińskiego. *Abiotic environment in water of the Szczecin Lagoon.* Uniw. Szczec. Rozprawy Studia, 292. (in Polish)

Poleszczuk G., 2007. Procesy kształtujące jakość biotopu toni wodnej w estuarium bez pływów na przykładzie Zalewu Szczecińskiego. *Processes controlling water column biotope quality in an atidal estuary, as exemplified by the Szczecin Lagoon.* Zesz. Nauk. Uniw. Szczec., 503, Acta Biol., 14: 185-198. (in Polish)

Poleszczuk G., A. Garbacik Wesołowska, T. Mutko, 1993. Wpływ cofki wiatrowej na stan natlenienia wód wybranych akwenów estuarium Odry. *Effects of wind-induced backflow on the water oxygen regime in selected areas of the River Odra estuary.* Stud. Mat. Mor. Inst. Ryb., Świnoujście, Ser. S: 102-110. (in Polish)

Poleszczuk G., Z. Piesik, 2000. On differences in chemical composition occurring between surface and near bottom water in the Szczecin Lagoon. Baltic Coast. Zone, 4: 27-43.

Poleszczuk G., S. Sitek, 1993. Próba zastosowania klasycznych mierników trofii do określania stanu wód Zalewu Szczecińskiego w latach 1990 i 1992. *An attempt to apply classic trophic status indicators to assess the state of the Szczecin Lagoon water in 1990 and 1992.* Stud. Mat. Mor. Inst. Ryb., Świnoujście, Ser. S: 111-114. (in Polish)

Porębski J., A. Szkudlarek, 1995. Larwy śledzia wiosennego *Clupea harengus* i babki Gobiidae w wodach Świny i w Zalewie Szczecińskim. *Larvae of spring herring* Clupea harengus *and goby Gobiidae in the waters of the Świna and the Szczecin Lagoon.* Rap. Mor. Inst. Ryb. 1993-1994, Gdynia, 342-353. (in Polish)

Porębski J., A. Szkudlarek-Pawełczyk, 1997. Ichtioplankton Zatoki Pomorskiej w latach 1996 i 1996, ze szczególnym uwzględnieniem śledzia wiosennego. *Ichthyoplankton of the Pomeranian Bay in 1995 i 1996 with a special regard to spring herring.* Rap. Mor. Inst. Ryb. 1996, Gdynia, 7-34. (in Polish)

POSTNOTE, 2010. EU Fisheries Management. The Parliamentary Office of Science and Technology. London, UK POSTNOTE 357 May 2010 [Cited 2 January 2012]. www.parliament.uk/post

Poznańska M., J. Kobak, N. Wolnomiejski, T. Kakareko, 2010. Macrozoobenthos communities from two types of land-water transition zones in a European lowland dam reservoir. Fundam. Appl. Limnology (Arch. Hydrobiol.), 176: 115-126.

Priymachenko A.D., M.A. Litvinova, 1968. Raspredeleniye i dinamika sinezelenykh vodorosley v Dneprovskikh vodokhranilishchakh. *Distribution and dynamics of blue-green larvae in Dneper reservoirs.* In: Cveteniye vody. *Water blooms.* Izd. Naukova Dumka, Kiev, 42-67. (in Russian)

Protasowicki M., 1987. Wybrane metale ciężkie w rybach Bałtyku Południowego. *Selected heavy metals in fishes of the southern Baltic.* Rozprawy, 110. AR Szczecin. (in Polish)

Protasowicki M., 1991. Long-term studies of heavy metals in aquatic organisms from the River Odra mouth area. Acta Ichthyol. Pisc., 21, suppl.: 301-309.

Protasowicki M., E. Niedźwiecki, 1991. Zagrożenia ekologiczne w strefie ujścia Odry ze szczególnym uwzględnieniem metali ciężkich. *Environmental hazards in the Odra mouth area, with a special reference to heavy metals.* In: Smoczyńska A. (ed.): Substancje toksyczne w środowisku. *Toxic substances in the environment.* ART, Olsztyn, 3-6. (in Polish)

Protasowicki M., E. Niedźwiecki, 1995. Metale ciężkie w zanieczyszczeniu osadów dennych Rzeki Odry (wyniki badań wieloletnich). *The heavy metals in pollution load of Odra River bottom sediment (the effects of multiyear studies).* In: Chojnacki J.C., E.J. Pałyga (eds): Europejski ład ekologiczny a problemy ochrony środowiska krajów nadbałtyckich. *The European ecological order versus environmental protection problems in the Baltic countries.* NFOŚGW, Warszawa, 122-127. (in Polish)

Prus T., 1993. Zawartość energii w materiałach biologicznych. *Energy content in biological materials.* In: Klekowski R.Z., Z. Fischer (eds): Bioenergetyka ekologiczna zwierząt zmiennocieplnych. *Ecological bioenergetics of poikilotherms.* Wydawn. PAN, Warszawa, 203-229. (in Polish)

Psuty-Lipska I., W. Blady, E. Boberski, H. Dąbrowski, B. Draganik, K. Trella, 2004. Stabilność wybranych populacji z ichtiocenoz Zalewu Wiślanego i Szczecińskiego w odniesieniu do czynników warunkujących eksploatacje rybacką. *Stability of selected populations in fish communities of the Vistula and Szczecin Lagoons in relation to factors controlling the fisheries.* Sea Fish. Inst. Techn. Rep., Gdynia. (in Polish)

Psuty-Lipska I., N. Madsen, B. Draganik, W. Blady, 2006. Gill net selectivity for perch (*Perca fluviatilis*) in the Szczecin Lagoon, Poland. Fish. Res., 80: 339-344.

Pustelnikovas O., 1998. Geochemistry of sediments of the Curonian Lagoon (Baltic Sea). Publ. Inst. Geography, Vilnius.

Radziejewska T., I. Drzycimski, 1988. Meiobenthic communities of the Szczecin Lagoon. Kieler Meeresf., Sonderh. 6: 162-172.

Radziejewska T., I. Drzycimski, 1990. Dynamics of meiobenthic communities in a eutrophic and polluted estuary. Limnologica, 2: 83-88.

Radziejewska T., M. Chabior, 2004. Climatic and hydrological controls over the zoobenthos in a southern Baltic coastal lagoon. Hydrobiologia, 514: 171-181.

Radziejewska T., G. Schernewski, 2008. The Szczecin (Oder-) Lagoon. In: Schiewer U. (ed.): Ecology of Baltic coastal waters. Ecol. Studies 197. Springer Verlag, Berlin-Heidelberg, 115-129.

Radziejewska T., Ch. Fenske, B. Wawrzyniak-Wydrowska, P. Riel, A. Woźniczka, P. Gruszka, 2009. The zebra mussel (*Dreissena polymorpha*) and the benthic community in a coastal Baltic lagoon: another example of enhancement? Mar. Ecol., 30 Suppl. 1: 138-150.

RIEP, Regional Inspectorate of Environmental Protection, 2007. Stan środowiska w województwie zachodniopomorskim w 2006 roku. (in Polish)

Rejewski M., 1979. The structure and function of macrophytic communities in two eutrophic lakes in the Iława Lake District. AUNC 47, Prace Limnol., 11: 19-34.

Rejewski M., 1988. Littoral trophy and bioenergetics of macrophytes. AUNC 65, Prace Limnol., 16: 27-37.

Renk H., S. Ochocki, M. Zalewski, 2001. Environmental factors controlling primary production in the Polish part of the Vistula Lagoon. Bull. Sea Fish. Inst., Gdynia, 1 (152): 77-95.

Repechka R., J. Maksimov, S. Toliushis, 1998. Fish and the fishery in the Lithuanian part of the Curonian Lagoon. In: The problems of management of fishery resources in the Polish and German waters of the Szczecin Lagoon and Pomeranian Bay, Proc. Polish-German Symp., Świnoujście, 19-20 Nov. 1966. Sea Fish. Inst., Gdynia, 169-178.

Ricardi N., M. Mangoni, 1999. Considerations on the biochemical composition of some freshwater zooplankton species. J. Limnol., 58: 58-65.

Ringer Z., 1959. Doświadczenia w celu klasyfikacji biomasy flory litoralu w Zalewie Wiślanym na podstawie badań prowadzonych w roku 1955. *Experiments aiming to classify the littoral flora biomass in the Vistula Firth on the basis of investigations carried on in 1955*. Pr. Mor. Inst. Ryb., Gdynia, 10/A: 193-214. (in Polish)

Robakiewicz W. (ed.), 1993. Warunki hydrodynamiczne Zalewu Szczecińskiego i cieśnin łączących Zalew z Zatoką Pomorską. *Hydrodynamic conditions of the Szczecin Lagoon and straits connecting the Lagoon with the Pomeranian Bay*. IBW PAN, Gdańsk, Hydrotechnika, 16: 1-287. (in Polish)

Rochette S., J. Lobry, M. Lepage, Ph. Boët, 2009. Dealing with uncertainty in qualitative models with a semi-quantitative approach based on simulations. Application to the Gironde estuarine food web (France). Ecol. Model., 220: 122–132.

Rodgers D.W., S.U. Qadri, 1977. Seasonal variation in calorific values of some littoral benthic invertebrates of the Ottawa River, Ontario. Can. J. Zool., 55: 881-884.

Rosado-Solorzano R., S.A. Guzman del Proo, 1998. Preliminary trophic structure model for Tampamachoco lagoon, Veracruz, Mexico. Ecol. Model., 109: 141–154.

Rödiger S., 2004. Die Makrofauna des Oderhaffts – Vorschläge für ein Monitoring nach EU-WRRL. In: Schernewski G., T. Dolch (eds): The Oder Estuary – against the background of the European Water Framework Directive. Mar. Sci. Rep., 57: 127-178.

Rumohr H., T. Brey, S Ankar, 1987. A compilation of biometric conversion factors for benthic invertebrates of the Baltic Sea. BMB Publ., 9.

Rutkowski D., 1993. Badania hydrochemiczne Roztoki Odrzańskiej w latach 1985-88 i 1990-92. *Hydrochemical research in the Roztoka Odrzańska in 1985-1988 and 1990-1992*. Stud. Mat. Mor. Inst. Ryb., Świnoujście, Ser. S: 97-101. (in Polish)

Rybarczyk H., B. Elkaïm, 2003. An analysis of the trophic network of a macrotidal estuary: the Seine Estuary (Easter Channel, Normandy, France). Est. Coast. Shelf Sci., 58: 775-791.

Rybarczyk H., B. Elkaïm, L. Ochs, N. Loquet, 2003. Analysis of the trophic network of a macrotidal ecosystem: the Bay of Somme (Easter Channel). Est. Coast. Shelf Sci., 58: 405-421.

Rychert K., M. Wielgat-Rychert, D. Szczurowska, M. Myszka, M. Bochyńska, K. Krawiec, 2012. The importance of ciliates as a trophic link in shallow, brackish and eutrophic lakes. Polish Journal of Ecology, 60(4): 767-776.

Sandberg J., 2007. Cross-ecosystem analyses of pelagic food web structure and processes in the Baltic Sea. Ecol. Model., 201: 243–261.

Sandberg J., R. Elmgren, F. Wulff, 2000. Carbon flows in Baltic Sea food webs – a re-evaluation using a mass balance approach. J. Mar. Sys., 25: 249–260.

Scheffer M., 2001. Ecology of shallow lakes. Kluwer Academic Publ., Dordrecht-Boston-London.

Schreckenbach K., R. Knösche, K. Ebert, 2001. Nutrient and energy content of freshwater fishes. J. Appl. Ichthyol., 17: 142-144.

Schindler D.W., 1972. Production of phytoplankton and zooplankton in Canadian Shield Lakes. In: Kajak Z., A Hilbricht-Ilkowska (eds): Productivity problems of freshwaters. Proc. IBP-UNESCO Symp., PWN Polish Scientific Publ., Warszawa-Kraków, 311-331.

Schiewer U. (ed.), 2008. Ecology of Baltic coastal waters. Ecological Studies, 197. Springer Verlag, Berlin-Heidelberg.

Schwaerter S., M. Sondergaard, B. Riemann, L. Möller-Jensen, 1988. Respiration in eutrophic lakes: the contribution of bacterioplankton and bacterial growth yield. J. Plankton Res., 10: 515-531.

Sissenwine M.P., A. Rosenberg, 1993. Marine fisheries at a critical juncture. Fisheries, 18: 6–14.

Skowronek A., 2011. Level of sediment contamination in relation to dredging and hydrotechnical works (an example of Szczecin Lagoon). Workshop: Facing The Future; Generation BALT Project, Gdynia.

Skóra K., 1996. A comparison of changes in the composition of fish catches in the Polish lagoons in 1960-1989. Proc. Polish-Swedish Symp. "Baltic coastal fisheries". Sea Fish. Inst., Gdynia, 225-241.

Skurzak K., 2009. Model sieci troficznej jeziora Gardno. *A trophic web model of Lake Gardno.* M.Sc. Thesis, Akademia Pomorska w Słupsku, Słupsk. (in Polish)

Smith V.H., 1979. Nutrient dependence of primary productivity in lakes. Limnol. Oceanog., 24: 1051 – 1064.

Sokolova N.Yu., 1983. Motyl *Chironomus plumosus* L. (Diptera, Chironomidae). Sistematika, morfologia, ekologia, produktsiya. *The insect* Chironomus plumosus L. *(Diptera, Chironomidae). Systematics, morphology, ecology, production.* Izd. Nauka, Moskva. (in Russian)

Sokołowski A., 2009. Tracing the flow of organic matter based upon dual stable isotope technique, and trophic transfer of trace metals in benthic food web of the Gulf of Gdańsk (southern Baltic Sea). Wyd. Uniw. Gdańskiego, Gdańsk.

Sołtysik K., 1989. Charakterystyka biologiczno-rybacka stada płoci Zalewu Szczecińskiego. *Biological and fisheries-related characteristics of the roach stock in the Szczecin Lagoon.* Sea Fish. Inst. Techn. Rep., Świnoujście. (in Polish).

Sołtysik K., 1995. Liczebność narybku sandacza *Stizostedion lucioperca* L. i okonia *Perca fluviatilis* L. z Zalewu Szczecińskiego w r. 1994. *The number of the fry of pikeperch* Stizostedion lucioperca *L. and perch* Perca fluviatilis *L. from the Szczecin Lagoon in 1994.* Rap. Mor. Inst. Ryb. 1993-1994, Gdynia, 379-390. (in Polish).

Sokolova N.Yu., 1971. Life cycles of chironomids in the Uchinskoe Reservoir. Limnologica (Berl.), 8: 151-155.

Sorokin Y.I., 1972. Biological productivity of the Rybinsk reservoir. In: Kajak Z., A. Hilbricht-Ilkowska (eds): Productivity problems of freshwaters. Proc. IBP-UNESCO Symp, PWN Polish Sci. Publ., Warszawa-Kraków, 493-503.

Soszka G.J., 1975. Ecological relations between invertebrates and submerged macrophytes in the lake littoral. Ekol. Pol., 23 393-415.

Soszka H., 2002. Ocena i klasyfikacja wód powierzchniowych – ewolucja podejścia do zagadnienia. *Assessment and classification of surface waters; evolution of approaches to the problem.* Ochr. Środ. Zas. Natur., 23/24: 33-46. (in Polish)

Stallone J.N., 1979. Seasonal changes in the water metabolism of woodrats. Oecologia (Berl.), 38: 203-216.

Stälnacke P., S.M. Vandsemb, A. Wassiliev, A. Grimwall, G. Jolankai, 2004. Changes in nutrient levels in some Eastern European rivers in response to large-scale changes in agriculture. Water Sci. Technol., 49: 29-36.

Stańczykowska A., 1961. Szybki spadek liczebności racicznicy *Dreissena polymorpha* (Pall.) w niektórych jeziorach mazurskich koło Mikołajek. *Rapid reduction of* Dreissena polymorpha (Pall.) *densities in some Masurian lakes near Mikołajki.* Ekol. Pol. B, 7: 151-153. (in Polish)

Stańczykowska A., 1976. Biomass and production of *Dreissena polymorpha* (Pall.) in some Masurian lakes. Ekol. Pol., 24: 103-112.

Stańczykowska A., 1977. Ecology of *Dreissena polymorpha* (Pall.) (Bivalvia) in lakes. Pol. Arch. Hydrobiol., 24: 461-530.

Stańczykowska A., W. Ławacz, 1976. Caloric value of the *Dreissena polymorpha* (Pall.); dry body weight in some Masurian Lakes. Pol. Arch. Hydrobiol., 23: 271-275.

Stańczykowska A., M. Planter, 1985. Factors affecting nutrient budget in lakes of the r. Jorka watershed (Mazurian Lakeland, Poland). X. Role of mussel *Dreissena polymorpha* (Pall.) in N and P cycles in a lake ecosystem. Ekol. Pol., 33: 345-354.

Stead T.K., J.M. Schmid-Araya, A.G. Hildrew, 2005. Secondary poduction of a stream metazoan community: Does the meiofauna make a difference? Limnol. Oceanog., 50: 398-403.

Stempniewicz L., W. Meissner, 1999. Assessment of the zoobenthos biomass consumed yearly by diving ducks wintering in the Gulf of Gdańsk (southern Baltic Sea). Ornis Svecica, 9: 143-154.

Szczepańska W., 1970. Periphyton of several lakes of the Mazurian Lakeland. Pol. Arch. Hydrobiol., 17: 397-418.

Szczerbowski J., 1985. Rybactwo jeziorowe i rzeczne. *Lacustrine and riverine fisheries.* PWRL, Warszawa. (in Polish)

Szczerbowski J., 1993. Rybactwo śródlądowe. *Inland-water fishery.* Wyd. IRS, Olsztyn (in Polish)

Szefer P., M. Domagała-Wieloszewska, J. Warzocha, A. Garbacik-Wesołowska, T. Ciesielski. 2003. Distribution and relationships of mercury, lead, cadmium, copper, and zinc in perch (*Perca fluviatilis*) from the Pomeranian Bay and Szczecin Lagoon, southern Baltic. Food Chem., 81: 73-83.

Szlinder-Richert J., Z. Usydus, W. Pelczarski, 2010a. Organochlorine pollutants in European eel (*Anguilla anguilla* L.) from Poland. Chemosphere, 80: 93-99.

Szlinder-Richert J., I. Barska, Z. Usydus, R. Grabic, 2010b. Polybrominated diphenyl ethers (PBDEs) in selected fish species from the southern Baltic Sea. Chemosphere, 78: 695-700.

Szkudlarek-Pawełczyk A., 2001. Wczesne stadia rozwojowe sandacza (*Stizostedion lucioperca* (L.) jako element oceny efektywności tarła. *Early development stages of pikeperch* (Stizostedion lucioperca (L.) *as an element of the spawning efficiency evaluation.* Ph.D. Thesis, Mor. Inst. Ryb., Gdynia. (in Polish)

Szkudlarek-Pawełczyk A., 2003. Early development stages of pikeperch in the Szczecin Lagoon: selected aspects of their biology and ecology from 1994 to 1998. Bull. Sea Fish. Inst., Gdynia, 1 (158): 3-21.

Szlauer L., 1996. Determination method of the volume of periphyton components. Pol. Arch. Hydrobiol., 43, 1: 3-8.

Szlauer B., L. Szlauer, 1994. Zooplankton rzeki Odry w Szczecinie. *Zooplankton of the river Odra in Szczecin.* Zesz. Nauk. AR Szczecinie, 164. Ryb. Mor. Technol. Żywn., 21: 23-39. (in Polish)

Świerzawska-Wiktorowa K., 1957. Zooplankton Zalewu Szczecińskiego. Część I. *Zooplankton of the Szczecin Lagoon. Part I.* Pr. Mor. Inst. Ryb., Gdynia, 9: 501-541. (in Polish)

Świerczyński M., I. Czerniawska, 1992. Impact of blue-greens from genera *Microcystis* on some aquatic animals. Acta Ichthyol. Piscat. 22: 163-180.

Tadajewski A., M. Knasiak, T. Mutko, 1989. Warunki hydrochemiczne ekosystemu estuarium Odry. *Hydrochemical conditions in the Odra estuarine system.* In: Ochrona Wód Bałtyku. *Protection of Baltic waters.* Polish-Swedish Seminar, Szczecin, 5-14. (in Polish)

Tejakusuma I., 2004. Investigations into the hydrography and dynamics of suspended particulate matter and sediments in the Oder Lagoon, Southern Baltic Sea. IKZM – Oder Berichte 1. Warnemünde.

Ten W.W., 2004. Mizidy. *Mysids.* In: Alimov A.F., M.B. Ivanova (eds): Zakonomernosti gidrobiologicheskogo rezhima vodoyemov raznogo tipa. *Patterns of the hydrobiological regime in water bodies of different type.* Nauchny Mir, Moskva, 164-168. (in Russian)

Tomaszewski J.B., 1990. Środowisko przyrodnicze odrzańskiego obszaru estuariowego. *The natural environment of the Odra estuary.* In: Bronk H. (ed.): Estuarium Odry i Zatoka Pomorska w rozwoju społeczno-gospodarczym Polski. *The Odra estuary and the Pomeranian Bay in social and economic development of Poland.* Praca zlec. MEN, Uniw. Szczeciński, 11-46. (in Polish)

Tomczak M.T., B. Müller-Karulis, L. Järv, J. Kotta, G. Martin, A. Minde, A. Põllumäe, A. Razinkovas, S. Straże, M. Bučas, T. Blenckner, 2009. Analysis of trophic networks and carbon flows in south-eastern Balic coastal ecosystems. Prog. Oceanog., 81: 111-131.

Treasurer J.W., 1997. A review of recent research on the percids, perch *Perca fluviatilis* L. and ruffe *Gymnocephalus cernuus* (L.), in Scotland. Glasgow Natur., 23: 41-48.

Trzosińska A., E. Andrulewicz, 1997. Doraźne skutki powodzi 1997 roku w środowisku wodnym Zatoki Gdańskiej i Pomorskiej. *Immediate effects of the 1997 flood in the aquatic environment of the Gulf of Gdańsk and the Pomeranian Bay.* Sea Fish. Inst. Techn. Rep., Gdynia. (in Polish)

Tyluś K., 2006. Struktura ekologiczna mezozooplanktonu w II-rzędowym estuarium Odry ze szczególnym uwzględnieniem jego roli jako bazy pokarmowej dla ryb planktonożernych (w latach 2003-2005). *Ecological structure of the mesozooplankton in the second order Odra estuary, with a particular reference to its role as a food resorce for planktivorous fish (in 2003-2005).* Ph.D. Thesis, Zachodniop. Uniw. Technol., Szczecin. (in Polish)

Ulanowicz R.E., 1986. Growth and development: ecosystem phenomenology. Springer, New York.

Valiela I., 1995. Marine ecological processes. Springer, New York.

Van de Bund W.J., D. Groenendijk, 1994. Seasonal dynamics and burrowing of littoral chironomid larvae in relation to competition and predation. Arch. Hydrobiol., 132: 213-225.

Vasconcellos M., S. Mackinson, K. Sloman, D. Pauly, 1997. The stability of trophic mass-balance models of marine ecosystems: a comparative analysis. Ecol. Model., 100: 125-134.

Vega-Cendejas M.E., F. Arreguin-Sanchez, 2001. Energy fluxes in a mangrove ecosystem from a coastal lagoon in Yucatan Peninsula, Mexico. Ecol. Model., 137: 119–133.

Vidal L., D. Pauly, 2004. Integration of subsystems models as a tool toward describing feeding interactions and fisheries impacts in a large marine ecosystem, the Gulf of Mexico. Ocean Coast. Manag., 47: 709–725.

Villanueva M.C., M. Isumbisho, B. Kaningini, J. Moreau, J.-C. Micha, 2008. Modeling trophic interactions in Lake Kivu: What roles do exotics play? Ecol. Model., 212: 422–438.

Villanueva M.C., P. Laleye, J.-J. Albaret, R. Lae, L. Tito de Morais, J. Moreau, 2006. Comparative analysis of trophic structure and interactions of two tropical lagoons. Ecol. Model., 197: 461–477.

Vinogradov M.E., E.A. Shushkina, 1987. Funktsyonirovaniye planktonnykh soobsshchestv epipelagiali okeana. *The functioninng of planktonic communities in the oceanic epipelagial.* Izd. Nauka, Moskva. (in Russian).

Vladimirova T.M., 1974. Produktsiya zooplanktona Rybinskogo vodokhranilishcha. *The zooplankton production of the Rybinsk reservoir.* In: Butorin I.V. (ed.): *Biologia i produktivnost presnovodnykh bezpozvonochnykh. Biology and productivity of freshwater invertebrates.* Tr. Inst. Biol. Vnutr. Vod, 25(28): 37-42. (in Russian)

Vollenweider R.A., 1970. Scientific fundamentals of the eutrophication of lakes and flowing waters, with particular reference to nitrogen and phosphorus as factors in eutrophication. Rep. OECD, Paris.

Wasmund N., A. Kowalczewski, 1982. Production and distribution of benthic microalgae in the littoral sediments of Mikołajskie Lake. Ekol. Pol., 30: 287-301.

Waters T.F., 1977. Secondary production in inland waters. Adv. Ecol. Res., 10: 91-164.

Wawrzyniak-Wydrowska B., 1998. Zmiany w faunie dennej Zalewu Szczecińskiego na podstawie badań przeprowadzonych w latach 1989 i 1991. *Changes in the Szczecin Lagoon benthos as seen in studies carried out in 1989 and 1991.* In: Konf. Nauk. "Stan Środowiska Przyrodniczego Pomorza w badaniach Wydziału Nauk Przyrodniczych". *Sci. Conf. "The status of the natural environment of Pomerania as reflected by research at the Faculty of Natural Sciences".* Uniw. Szczeciński, Mat. Konf., 33: 201-230. (in Polish)

Wawrzyniak-Wydrowska B., P. Gruszka, 2005. Population dynamics of alien gammarid species in the River Odra Estuary. Hydrobiologia, 539: 13-25.

Westlake D.F., H. Casey, H. Dawson, M. Ladle, R.H.K. Mann, A.F.H. Marker, 1972. The chalk-stream ecosystem. In: Kajak Z., A Hilbricht-Ilkowska. (eds): Productivity problems of freshwaters. Proc. IBP-UNESCO Symp., PWN Polish Sci. Publ., Warszawa-Kraków, 615-635.

Westphal H., B. Lenk, 1998. Die räumlich-zeitliche Verteilung von Primärproduktion und Bakterienkeimzahlen. Greifsw. Geogr. Arb., 16: 228-249.

Wetzel R.G., 2001. Limnology. Lake and river ecosystems. Elsevier-Academic Press, London.

Wetzel R.G., G.E. Likens, 1995. Limnological analyses. Second Edition. Springer-Verlag, New York-Berlin-Heidelberg.

Wengrzyn J., 1978. Charakterystyka biologiczno-rybacka stada sandacza Zalewu Szczecińskiego. *Biological and fisheries-related characteristics of the pikeperch stock in the Szczecin Lagoon.* Sea Fish. Inst. Techn. Rep., Świnoujście. (in Polish).

Węsławski J.M., 1981. Obserwacje nad tworzeniem się skupisk *Neomysis vulgaris* (Thompson), 1928 w warunkach naturalnych. *Observations of the formation of* Neomysis vulgaris *(Thompson), 1928 aggregations under natural conditions.* Zesz. Nauk. Wydz. BiNoZ UG, Oceanogr., 8: 109-126. (in Polish)

Wielgat M., 2003. Model obiegu substancji biogenicznych w Zalewie Szczecińskim. *A model of nutrient cycling in the Szczecin Lagoon.* Ph.D. Thesis, Uniw. Gdański, Gdynia. (in Polish)

Wielgat M., G. Schernewski, 2002. Models as tools in coastal water management: eutrophication of the large, shallow Szczecin Lagoon. In: IHP-OHP. Intern. Symp. "Low-lying coastal areas, hydrology and integrated coastal zone management". Bremerhaven, Germany, 147-154.

Wielgat M., Z. Witek, 2004. A dynamic box model of the Szczecin Lagoon nutrient cycling and its first application to the calculation of the nutrient budget. In: Schernewski G., T.Dolch (eds): The Oder Estuary – against the background of the European Water Framework Directive. Mar. Sci. Rep., 57: 99-125.

Wielgat-Rychert M., K. Rychert, 2008. Annual primary production of Lake Gardno and Lake Łebsko. In: Bajkiewicz-Grabowska E., D. Borowiak (eds): Anthropogenic and natural transformations of lakes. 2. Wyd. KLUG-PTLim., Gdańsk, 211-214.

Wielgat-Rychert M., K. Rychert, D. Ficek, 2010. Factors controlling pelagic production and respiration in a shallow polymictic lake. Pol. J. Ecol., 58: 369-375.

Wielgosz S., M. Tadajewska, 1988. Factors determining diet composition and food availability for bream *Abramis brama* (L.) and white bream *Blicca bjoerkna* (L.) in Włocławek Dam Reservoir. Acta Ichthyol. Piscat., 18, 1: 79-100.

Wieser E., 1979. Zooplankton Zalewu Szczecińskiego. *The zooplankton of the Szczecin Lagoon.* Sea Fish. Inst. Techn. Rep., Świnoujście. (in Polish)

Wiktor J., 1962. Ilościowe i jakościowe badania fauny dennej Zalewu Szczecińskiego, Część II. *Quantitative and qualitative investigations of the Szczecin Firth bottom fauna. Part II.* Pr. Mor. Inst. Ryb., Gdynia, 11/A: 81-112. (in Polish)

Wiktor J., 1969. Biologia racicznicy *Dreissena polymorpha* (Pall.) i jej znaczenie biologiczne w Zalewie Szczecińskim. *The biology of* Dreissena polymorpha *(Pall.) and its biological importance in the Firth of Szczecin.* Stud. Mat. Mor. Inst. Ryb., Gdynia, Ser. A, 5: 1-88. (in Polish)

Wiktor J., A. Garbacik-Wesołowska, 1993. Gospodarka zasobami rybnymi Zalewu Szczecińskiego. *Management of fish resources of the Szczecin Lagoon.* Stud. Mat. Mor. Inst. Ryb., Świnoujscie, Ser. S: 7-12. (in Polish)

Wiktorowie J.K., 1954. Jakościowe i ilościowe badania fauny dennej Zalewu Szczecińskiego. Część 1. *Qualitative and quantitative investigations of the Szczecin Bay bottom fauna. Part 1.* Pr. Mor. Inst. Ryb., Gdynia, 7: 127-152. (in Polish)

Wiktor K., 1959. Zooplankton Zalewu Szczecińskiego. Cz. II. *The Szczecin Firth zooplankton. Part II.* Pr. Mor. Inst. Ryb., Gdynia, 10/A: 229-258. (in Polish)

Wiktor K., 1961. Obserwacje nad biologią *Neomysis vulgaris* (Thompson) w Zalewie Szczecińskim i Zatoce Pomorskiej. *Observations on the biology of* Neomysis vulgaris *(Thompson) in Zalew Szczeciński (Stettiner Haff) and Zatoka Pomorska (Pomeranian Bay).* Przegl. Zool., 5: 36-42. (in Polish).

Wiktor K., 1964. Związki miedzy produkcją zooplanktonu a odżywianiem ryb planktonożernych. *Relationships between zooplankton production and feeding of planktonic fish.* Stud. Mat. Mor. Inst. Ryb., Gdynia, Ser. A, No. 2. (in Polish)

Wiktor K., 1967. Poziomy troficzne Bałtyku południowego i polskich zalewów przybałtyckich – Zalewu Szczecińskiego i Zalewu Wiślanego. *Trophic levels in the southern Baltic and Polish Baltic Lagoons: the Szczecin and Vistula Lagoons.* Stud. Mat. MIR, 4 A: 117-132. (in Polish)

Wiktor K., 1971. The primary production of Szczecin Bay. Pol. Arch. Hydrobiol., 18: 1-15.

Wiktor K., 1972. Zmiany w rozwoju planktonu Zalewu Szczecińskiego jako wynik zmian w warunkach fizyko-chemicznych środowiska, spowodowanych gospodarka człowieka. *Changes in the development of the Szczecin Lagoon plankton as a result of anthropogenically induced changes in habitat physico-chemical regime.* In: Ekosyst. Morskie. Badania zalewów, 4. Mor. Inst. Ryb., Gdynia, 63-132. (in Polish)

Wiktor K., 1976. Zmiany w biocenozach wód przybrzeżnych i przyujściowych Bałtyku jako wynik wzrostu zanieczyszczeń. *Changes in communities of Baltic coastal and estuarine waters as a result of increasing pollution.* Stud. Mat. Oceanol., 15, Biol. Morza, 3: 143-168. (in Polish)

Wiktor K., 1980. Ogólna charakterystyka flory i fauny. Zmiany w biocenozie pod wpływem gospodarki człowieka. *A general characteristics of flora and fauna. Changes in the biota resulting from anthropogenic effects.* In: Majewski A. (ed.): Zalew Szczeciński. *The Szczecin Lagoon.* Wyd. Komun. Łaczn., Warszawa, 286-303. (in Polish)

Wiktor K., D. Zembrzuska, 1959. Materiały do hydrografii Zalewu Szczecińskiego. *Materials for hydrography of Firth of Szczecin.* Pr. Mor. Inst. Ryb., Gdynia, 10/A: 259-269. (in Polish)

Wilkońska H., 1996. Ocena składu gatunkowego, struktury wieku, potencjału rozrodczego ryb słodkowodnych – aktualnie poławianych w Zalewie Wiślanym i Zatoce Gdańskiej celem opracowania kryteriów optymalnego wykorzystania produktywności tych gatunków. *Assessment of species composition, age structure, and reproductive potential of freshwater fishes currently exploited in the Vistula Lagoon and the Gulf of Gdańsk to develop criteria of optimal utilization of the species' productivity.* Sea Fish. Inst. Techn. Rep., Gdynia-Olsztyn. (in Polish)

Winberg G.G., 1971. Symbols, units and conversion factors in studies of freshwater productivity. IBP Centr. Office, London.

Winberg G.G., 1972. Some interim results of Soviet IBP investigations on lakes. In: Kajak Z., A Hilbricht-Ilkowska (eds): Productivity problems of freshwaters. Proc. IBP-UNESCO Symp., PWN Plish Sci. Publ., Warszawa-Kraków, 363-381.

Winberg G.G. (ed.), 1979. Obshchye osnovy izucheniya vodnykh ekosistem. *Fundamentals of research on aquatic ecosystems.* Izd. Nauka, Leningrad. (in Russian)

Winberg G.G., V.A. Babitsky, S.I. Gavrilov, G.V. Gladky, I.S. Zakharenkov, R.Z. Kovalevskaia, T.M. Mikheieva, P.S. Nievyadomskaia, A.P. Ostapenia, P.G. Petrovich, J.S. Potaenko, O.F. Yakushko, 1972. Biological productivity of different types of lakes. In: Kajak Z., A. Hilbricht-Ilkowska (eds): Productivity problems of freshwaters. Proc. IBP-UNESCO Symp., PWN Polish Sci. Publ., Warszawa-Kraków, 383-404.

Winberg G.G., G.A. Pechen, G.A. Shushkina, 1965. Produktsya planktonnykh rakoobraznykh w trekh ozerakh raznogo tipa. *Production of planktonic crustaceans in three lakes of different type.* Zool. Zh., 44: 676-687. (in Russian)

Winkler H.M., 1996. Peculiarities of fish communities in estuaries of the Baltic Sea: the Darss-Zingst-Bodden chain as an example. Limnologica, 26: 199-206.

Wiśniewski R., 1976. The production of *Chironomus f.l. plumosus* larvae in the bay part of Gopło Lake. AUNC 38, Pr. Limnol., 9: 99-111.

Wiśniewski R., 1995. Rola resuspensji osadów dennych w funkcjonowaniu ekosystemów wodnych. *The role of bottom sediment resuspension in the functioning of aquatic ecosystems.* Wyd. Uniw. M. Kopernika, Toruń, Rozprawy. (in Polish)

Wiśniewski R., 1980. Pseudolittoral of Gopło lake. Part II. Biological characteristics. AUNC,. Pr. Limnol., 12: 83-116.

Witek M., 1998. Annual changes of abundance and biomass of planktonic ciliates in the Gdańsk Basin, southern Baltic. Int. Rev. Hydrobiol., 83: 163-182.

Witek Z. (ed.), 1993. Structure and functioning of the marine ecosystem in Gdańsk Basin region in 1987. Stud. Mat. Oceanol., 63, Biologia Morza, 8.

Witek Z., 1995. Produkcja biologiczna i jej wykorzystanie w morskim ekosystemie zachodniej części Głębi Gdańskiej. *Biological production and its utilization within a marine ecosystem in the western Gdańsk Basin.* Wyd. Mor. Inst. Ryb., Gdynia. (in Polish)

Witek Z., M. Zalewski, M. Wilegat-Rychert, 2010. Nutrient stocks and fluxes in the Vistula Lagoon at the end of the twentieth century. Sea Fish. Inst., Gdynia, Pomer. Univ., Słupsk.

Wolańczyk S., 1997. Peryfiton rozwijający się na pasmach folii polietylenowej zatopionych w miejscu przejścia jeziora Wicko Małe w Wicko Wielkie. *Periphyton developing on polyethylene foil bands deployed at the junction of lakes Wicko Małe and Wicko Wielkie.* M.Sc. Thesis, Uniw. Szczeciński. (in Polish)

Wolnomiejski N., 1970. The effects of *Dreissena polymorpha* Pall. aggregations on the differentiation of the benthonic macrofauna. Zesz. Nauk. UMK, 25: 31-39.

Wolnomiejski N., 1994. Ekologiczne studium makrofauny dna mulistego Zalewu Szczecińskiego (1982-1992). *Ecological study on muddy bottom macrofauna of the Szczecin Lagoon (1982-1992).* Stud. Mat. Mor. Inst. Ryb., Gdynia, Ser. A, 37: 1-126. (in Polish)

Wolnomiejski N., 1997. Hydrobiologiczna charakterystyka otwartych wód Wolińskiego Parku Narodowego (cz.1. Zalew Szczeciński). *The hydrobiological characteristics of open waters of the Wolin National Park (Part 1. The Szczecin Lagoon).* Sea Fish. Inst. Techn. Rep., Świnoujście. (in Polish)

Wolnomiejski N., A. Giziński, M. Jermołowicz, 1976. The production of the macrobenthos in the psammolittoral of Lake Jeziorak. AUNC 38, Prace Limnol., 9: 17-26.

Wolnomiejski N., I. Grygiel, 1989. Makrofauna dna mulistego Zalewu Szczecińskiego. Macrofauna of the muddy bottom of the Szczecin Lagoon. Stud. Mat. Mor. Inst. Ryb., Gdynia, Ser. A 29: 5-42. (in Polish)

Wolnomiejski N., I. Grygiel, 1992. Makrofauna pelofilna Zalewu Szczecińskiego w latach 1987-1990. *Pelophilous macrofauna in the Szczecin Lagoon in 1987-1990.* Stud. Mat. Mor. Inst. Ryb., Ser.B, 60: 5-23. (in Polish)

Wolnomiejski N., I. Grygiel, 1994a. Podstawowe siedliska makrofauny dennej w Zalewie Szczecińskim. Main habitats of bottom macrofauna in the Szczecin Lagoon. In: Badania bentosu i pokarmu ryb Zalewu Szczecińskiego. *Studies on benthos and fish food in the Szczecin Lagoon.* Stud. Mat. Mor. Inst. Ryb., Gdynia, Ser. B, 67: 7-26. (in Polish)

Wolnomiejski N., I. Grygiel, 1994b. Pokarm okonia (*Perca fluviatilis*) w Zalewie Szczecińskim. *Food of the perch* (*Perca fluviatilis*) *in the Szczecin Lagoon.* In: Badania bentosu i pokarmu ryb Zalewu Szczecińskiego. *Studies on benthos and fish food in the Szczecin Lagoon.* Stud. Mat. Mor. Inst. Ryb., Gdynia, Ser. B, 67: 27-44. (in Polish)

Wolnomiejski N. I. Grygiel, 1994c. Pokarm jazgarza (*Gymnocephalus cernuus*) w Zalewie Szczecińskim. *Food of ruffe* (*Gymnocephalus cernuus*) *in the Szczecin*

Lagoon. In: Badania bentosu i pokarmu ryb Zalewu Szczecińskiego. *Studies on benthos and fish food in the Szczecin Lagoon.* Stud. Mat. Mor. Inst. Ryb., Gdynia, Ser. B, 67: 45-65. (in Polish)

Wolnomiejski N., I. Grygiel, 1998a. *Chironomus* f.l. *plumosus* as the benthos and food of fish component in the Szczecin Lagoon (Great Lagoon). In: The problems of management of fishery resources in the Polish and German waters of the Szczecin Lagoon and Pomeranian Bay. Proc. Polish-German Symp., Świnoujście, 19-20 Nov. 1966. Sea Fish. Inst., Gdynia, 52-57.

Wolnomiejski N., I. Grygiel, 1998b. Muddy bottom macrofauna of the Szczecin Lagoon over the period which followed the peak of the flood. Deutsch. Hydrogr. Zeitsch., 50: 287.

Wolnomiejski N., M. Luścińska, I. Grygiel, 2000. Algae in plankton and muddy sediments as well as in food content of the larvae *Chironomus f.l. plumosus* (L.) of the Szczecin Lagoon (The Great Lagoon). Oceanol. Stud., 29: 43-56.

Wolnomiejski N., I. Grygiel, 2002. Food of common bream (*Abramis brama* L) in the Szczecin Lagoon (Great Lagoon). Bull. Sea Fish. Inst., 1 (155): 61-68.

Wolnomiejski N., G. Poleszczuk, 2003. Specyfika wieloletniej dynamiki biomasy pelofilnego makrobentosu Zalewu Szczecińskiego (Estuarium Odry): efekt zjawiska fluidyzacji i wynoszenia osadów? *Specificity of long-term dynamics of pelophilous macrobenthos biomass in the Szczecin Lagoon (River Odra estuary): an effect of sediment fluidization and removal?* 19 Zjazd Hydrobiol. Polskich. Warszawa, 208. (in Polish)

Wolnomiejski N., A. Woźniczka, 2003. Effect of zebra mussel (*Dreissena polymorpha* Pall.) aggregations on hydrobiology of selected parts of the Szczecin Lagoon (a southern Baltic estuary). Baltic Sea Sci. Congr., Helsinki, 268.

Wolnomiejski N., A. Woźniczka, 2008. A drastic reduction in abundance of *Dreissena polymorpha* Pall. in the Skoszewska Cove (Szczecin Lagoon, River Odra estuary): effects in the population and habitat. Ecolog. Quest., 9: 103-111.

Woroń J., 1995. Dopływ zanieczyszczeń z atmosfery. *Atmospheric input of pollutants.* In: Warunki środowiskowe polskiej strefy południowego Bałtyku w 1994 r. *Environmental conditions in the Polish zone of the southern Baltic in 1994.* IMGW Gdynia, 94-103. (in Polish)

Woźniczka A., N. Wolnomiejski, 2005. Zebra mussel (*Dreissena polymorpha* Pall.) in the River Odra Estuary: the current status. 5th Baltic Sea Sci. Congr., Sopot, Poland, 373-374.

Woźniczka A., N. Wolnomiejski, S. Sitek., 2006. How a large zebra mussel population can control eutrophication effects, an example from the River Odra Estuary. Intern. Symp. "Research and management of eutrophication in coastal ecosystems". Nyborg, Denmark, 46-48.

Wróbel S., 1975. Some limnological aspects of the dam reservoir at Goczałkowice. Pol. Arch. Hydrobiol., 22: 271-283.

Wulff F., J.G. Field, K.H. Mann, 1989. Network analysis in marine ecology: methods and applications. Springer, New York.

Wypych K., 1980. Osady denne Zalewu Szczecińskiego. *Bottom sediments of the Szczecin Lagoon.* In: Majewski A. (ed.): Zalew Szczeciński. *The Szczecin Lagoon.* Wyd. Komun. Łączn., Warszawa, 54-72. (in Polish)

Wysokiński A., 1998. Some factors affecting the interaction between young pikeperch and perch in the Szczecin Lagoon. In: The problems of management of fishery resources in the Polish and German waters of the Szczecin Lagoon and Pomeranian Bay. Proc. Polish-German Symp., Świnoujście, 19-20 Nov. 1966. Sea Fish. Inst., Gdynia, 105-119.

Wysokiński A., 2000. Ichtiofauna i jej ochrona w wodach Wolińskiego Parku Narodowego w obszarze estuarium Odry. *The ichthyofauna and its protection in the Odra estuary waters of the Wolin National Park.* Klify, 4: 185-198. (in Polish)

Wysokiński A., A.Garbacik-Wesołowska, 1995a. Ocena pławnic w odniesieniu do ochrony młodzieży sandacza i okonia w wodach estuarium Odry. *The evaluation of gill-nets with regard to the protection of juvenile pikeperch and perch in the waters of the Odra estuary.* Rap. Mor. Inst. Ryb. 1993-1994, Gdynia, 434-456. (in Polish)

Wysokiński A., A. Garbacik-Wesołowska, 1995b. Wpływ głębokowodnych żaków węgorzonych na stada sandacza i okonia w Zalewie Szczecińskim. *The influence of deep-water eel trap-nets on the pikeperch and perch stock in the Szczecin Lagoon.* Rap. Mor. Inst. Ryb. 1993-1994, Gdynia, 457-477. (in Polish)

Wysokiński A., A. Garbacik-Wesołowska, E. Boberski, A. Koronkiewicz, 1999. Dynamics of the numbers and distribution of juvenile pikeperch in the Szczecin Lagoon and Pomeranian Bay in 1995-1996. Arch. Ryb. Pol., 7: 169-186.

Wziątek B., A. Martyniak, 2009. Wpływ kormorana czarnego na ichtiofaunę Zbiornika Koronowskiego. *Effects of the black cormorant on the fish fauna in the Koronowo Dam Reservoir.* Techn. Rep. Woj. Fund. Ochr. Środ. Gosp. Wodnej, Toruń, Olsztyn. (in Polish)

Zaiko A., D. Daunys, S. Olenin, 2009. Habitat engineering by the invasive zebra mussel *Dreissena polymorpha* (Pallas) in a boreal coastal lagoon: impact on biodiversity. Helgol. Mar. Res., 63: 85-94.

Zaiko A., R. Paškauskas, A. Krevš, 2010. Biogeochemical alteration of the benthic environment by the zebra mussel *Dreissena polymorpha* (Pallas). Oceanologia, 52: 649-667.

Zawiślak W., 1972. Production of crustacean zooplankton in Moty Bay, Lake Jeziorak. Part II. Estimation of production of the predominating species. Pol. Arch. Hydrobiol., 19: 193-202.

Zawiślak W., 1977. The modified Winberg et al. (1965) method as applied to the calculation of the pelagial crustacean zooplankton production in the Jeziorak Lake. Pol. Arch. Hydrobiol., 24: 177-189.

Zembrzuska D., 1962. Fitoplankton Zalewu Szczecińskiego. *Szczecin Firth phytoplankton.* Pr. Mor. Inst. Ryb. Gdynia, 11A: 137-158. (in Polish)

Zetina-Rejon M.J., F. Arreguin-Sanchez, E.A. Chavez, 2003. Trophic structure and flows of energy in the Huizache-Caimanero lagoon complex on the Pacific coast of Mexico. Est. Coast. Shelf Sci., 57: 803–815.

Zettler M.L., D. Daunys, 2007. Long-term macrozoobenthos changes in a shallow boreal lagoon: Comparison of a recent biodiversity inventory with historical data. Limnologica, 27: 170-185.

Zhang Y., E.E. Prepas, 1995. Regulation of dominance of planktonic diatoms and cyanobacteria in four eutrophic hardwater lakes by nutrients, water column stability and temperature. Aquatic Sci., 53: 621-633.

Żmudziński L., 2000. Long-term changes in macrozoobentos of the Vistula Lagoon. Sea Envir. Jura ir Aplinka, Klaipeda University, 1(3): 46-50.

Żmudziński L. 1990. Świat zwierzęcy Bałtyku. *Animal world of the Balic Sea.* Wyd. Szkolne i Pedag., Warszawa. (in Polish)

Żmudziński L., S. Chubarova-Solovieva, Z. Dobrowolski, P. Gruszka, S. Olenin, N. Wolnomiejski, 1997. Expansion of the spionid polychaete *Marenzelleria viridis* in the southern part of the Baltic Sea. In: Andrušaitis A. (ed.): Proc. 13th Baltic Mar. Biol. Symp., Univ. Latvia, 127-129.

Żytkowicz R., 1976. Production of macrobenthos in LakeTynwałd. AUNC 38, Prace Limnol., 9: 75-97.

Żurek R., 2007. The basic paths of energy flow and matter transformations in a lowland dam reservoir ecosystem. Ocean. Hydrob. Stud., 26, suppl. 2: 5-147.

Index

(Bold page numbers refers to tables and figures)

N

O

P

Bei Fragen zur Produktsicherheit wenden Sie sich bitte an:
If you have any questions regarding product safety,
please contact:

Walter de Gruyter GmbH
Genthiner Straße 13
10785 Berlin
productsafety@degruyterbrill.com